Problems for
Concepts and Models of Inorganic Chemistry
Third Edition

Bodie E. Douglas
University of Pittsburgh

Darl H. McDaniel
University of Cincinnati

John J. Alexander
University of Cincinnati

John Wiley & Sons, Inc.
New York Chichester Brisbane Toronto Singapore

Copyright © 1994 by John Wiley & Sons, Inc.

All rights reserved.

Reproduction or translation of any part of this work beyond that permitted by Sections 107 and 108 of the 1976 United States Copyright Act without the permission of the copyright owner is unlawful. Requests for permission or further information should be addressed to the Permissions Department, John Wiley & Sons, Inc.

ISBN 0-471-63784-X

Printed in the United States of America

10 9 8 7 6 5 4 3 2 1

Printed and bound by Malloy Lithographing, Inc.

Preface

This is the second edition of the problems book designed as a self-study aid. Detailed solutions are presented for the problems in our text, *Concepts and Models of Inorganic Chemistry*, Third Edition, Wiley, 1994, with a few additional problems added to each problem set to provide more complete coverage of an area or to give a greater challenge. The topics covered are usually included in upper level undergraduate or graduate inorganic courses. References to the literature and to our text (cited as DMA, 3rd ed.) are given for elaboration on topics covered by the problems. Complete solutions are presented here. In cases where similar examples appear in DMA, 3rd ed., reference to that source is given for background and related information. In order to illustrate how chemists think about a problem, some solutions are more detailed than the reader might be expected to provide.

The necessary data and figures are included so that the problems book can be used independently or with another text.

Bodie E. Douglas
Darl H. McDaniel
John J. Alexander

Acknowledgement

We appreciate comments and advice of colleagues and users of the former editions of *Concepts and Models of Inorganic Chemistry* and *Problems For Inorganic Chemistry*. Your comments and advice are welcome.

We want to express our appreciation to Kevin Murphy for the design of the cover of this book and that of DMA, 3rd ed. and to AT&T and Arthur F. Hebard of AT&T Bell Laboratories for the drawing. The drawing, based on work at AT&T, was the cover of *Physics Today*, November, 1992. It is an experimental electron density plot of the 100 plane of solid C_{60}. The circular depressions at the center and corners are cross sections of the rotating "bucky balls" on a face of the face-centered cubic (*ccp*) unit cell.

Bodie E. Douglas
Darl H. McDaniel
John J. Alexander

CONTENTS

Part I. SOME BASIC CONCEPTS

1 Atomic Structure and
  The Periodic Table 1

2 Molecular Models 21

3 Symmetry 40

Part II. BONDING AND STRUCTURE

4 Discrete Molecules
  Molecular Orbitals 62

5 Inorganic Solids
  The Ionic Model 83

6 Solid State Chemistry 96

Part III. CHEMICAL REACTIONS

7 Acids and Bases 109

8 Oxidation–Reduction Reactions . .119

Part IV. COORDINATION CHEMISTRY

9 Models and Stereochemistry of
  Coordination Compounds 133

10 Spectra and Bonding of
  Coordination Compounds 148

11 Reaction Mechanisms of
  Coordination Compounds 169

Part V. ORGANOMETALLIC CHEMISTRY

12 General Principles of
  Organometallic Compounds 197

13 Survey of Organometallic
  Compounds 216

14 Organometallic Reactions,
  Mechanisms, and Catalysis 225

Part VI. SELECTED TOPICS

15 Metals
  Chemistry and Periodic Trends . . .245

16 Chemistry of Some Nonmetals . . . 257

17 Cluster and Cage Compounds 273

18 Bioinorganic Chemistry 299

I. SOME BASIC CONCEPTS
1
Atomic Structure and The Periodic Table

1.1 Problem: Use the *Handbook of Chemistry and Physics* as a data source to indicate the nature of the periodic variation of the following properties of the elements in Groups 1, 2, 11, 12, 15, 16, and 18: **(a)** melting points; **(b)** boiling points; **(c)** formulas of oxide; **(d)** colors of oxides.

1.1 Solution: This problem is designed to give a feel for pattern recognition. It is open-ended to allow individual experience with selecting data, criteria for comparison, conclusions to be drawn, or explanations which may be offered. We suggest here only that the periodic table provides a powerful framework for such an exercise—a cyclic exercise which reverses steps in the creation of the periodic table which was a far greater leap. From the foregoing we may conclude that the solutions to this problem can take many directions. The following is not "the answer", but only an approach.

With the advantage of hindsight, we often find that the first member of any periodic group is anomalous in its physical properties. (It is to the credit of Mendeleev and Mayer that they were able to see broad trends in properties which established periodic relationships without becoming bogged down by the anomalies overlaid on these trends.) Treating these properties as sorting characteristics, the Group 18 elements are distinct in that all are gases at room temperature. The melting (mp) and boiling points (bp) (no melting point for He) increase with increasing atomic weight for this family of nonmetals. The very low temperatures at which the noble gases melt and boil are indicative of extremely weak interatomic forces which increase with atomic weight.

Groups 1 and 2, in contrast, have sufficiently high melting points that all members are solids at room temperature even though Cs melts barely above. (In fact, the mp's of Na–Cs are quite low, being comparable to those of some of the nonmetals such as P. This fact led to some reluctance to classify the alkali metals as metals when they were first discovered in the early part of the nineteenth century.) The trend (which is more regular for mp's than for bp's) is that both melting and boiling temperatures *decrease* with increasing atomic weight. This behavior is generally observed for metals. Again, the first members of the families are somewhat anomalous with both Li and Be melting and boiling at considerably higher temperatures than Na and Mg, respectively. These facts indicate the the forces holding metal atoms together must be stronger than those inolved in the noble gases, but that they are still rather weak for the alkali metals and decrease as the atomic weights increase. These same trends are evident in the mp's of Group 11 and 12 metals, the decrease in mp with atomic weight increase being sufficient to make Hg a liquid at room temperature.

In Group 15, we observe the nonmetal pattern of increasing mp's and bp's with increasing atomic weight for N and P with the opposite trend appearing for more metallic As, Sb, and Bi. The chalcogens of Group 16 are nonmetals whose mp's and bp's increase with atomic weight. The first members of both groups, nitrogen and oxygen, melt and boil at temperatures lower than might be extrapolated from those of other family members.

Even though families show some degree of distinction on the basis of their physical properties, it is really in the formulas of oxides (their chemistry) that sorting of the elements becomes most apparent. A variety of oxides exists for several elements; for example, sodium forms Na_2O and Na_2O_2 while nitrogen forms N_2O, NO, N_2O_3, NO_2, N_2O_5. Thus, it was a problem of no small magnitude to decide which oxides to compare when sorting the elements on their chemical properties. Mendeleev recognized that the highest oxide was the appropriate choice, and his periodic table gave the formula of the highest oxide at the top of each family. This way of classifying the elements led to placement of A and B group elements in the same family. For example, the oxides of both Cu and Na have the formula M_2O. (However, formulas of some Group 11 oxides are anomalous; although Cu_2O and Ag_2O exist, so do CuO and Au_2O_3.) The formulas of the highest oxides of groups 2 and 12 are MO, of Group 15 E_2O_5, and of Group 16 EO_3. For the noble gases only Xe has so far proven sufficiently reactive to form a highest oxide with formula XeO_4 (not very stable).

The preceding properties of the elements are quantitative, but we find that even qualitative properties can often be correlated with periodicity. The colored oxides arise in one or two situations: odd-electron molecules (such as NO_2) and the oxides of elements in the lower rows of the table. If we consider the metal oxides, we note that metal cations in upper rows as well as oxide ions are hard. Hard ions result from high values of absolute electronegativity and electron affinity; they tend to bond their own electrons tightly. As we descend in any family, the metal ions become softer (polarizable); their large orbitals allow electron density to be shifted from perfect spherical distribution. One could envisage repulsion of the metal electron cloud by oxide negative charge creating a region of enhanced exposure of the metals's positive charge toward the anion. Absorption of light in the visible region could result in a short-lived excited state in which oxide electrons are transferred to the metal (so-called charge transfer). This effect is enhanced for cations of Groups 11 and 12 because of imperfect shielding by the d electrons and the resulting higher values of the effective nuclear charge. (See Slater's rules, DMA, 3rd ed., p. 40.) The more polarizable sulfide anion leads to more colored sulfides than oxides. For the nonmetals of Groups 15 and 16, the electron-deficiency of the central atom increases as indicated by calculation of oxidation state, making the energy necessary for charge transfer in the region of visible light.

1.2 Problem: Bohr postulated that lines in the emission spectrum of hydrogen in highly excited states, such as $n = 20$, would not be observed under ordinary laboratory conditions because of the large size of such atoms and the much greater probability of atom collision deactivation as compared with radiation emission. Using the Bohr model, calculate the ratio of the cross-sectional area of a hydrogen atom in the $n = 20$ state to that of one in the $n = 1$ state.

1.2 Solution: According to the Bohr model, the radius of a hydrogen-like atom, r, is proportional to the square of the quantum number, n. The cross-sectional area is given by πr^2; thus the cross-sectional area is proportional to n^4. Hence a hydrogen atom in the $n = 20$ state would have a cross-sectional area 20^4, or 160,000, times that of the ground state.

States of an atom or molecule which has a highly excited electron are called *Rydberg states*, and are spectroscopically similar to highly excited atomic hydrogen. (For a fascinating discussion of research on these "floppy, fragile, and huge" atoms see D. Kleppner, M.G. Littman, and M. S. Zimmerman, *Sci. Am.* **1981,** May, 130.)

1.3 Problem: According to the Bohr model of the atom, what would be the size of a Ne^{9+} ion? What would the ionization energy be for this ion? What would the excitation energy be for the first excited state?

1.3 Solution: The Ne^{9+} ion is a hydrogen-like atom; it has one electron and a nuclear charge, Z, of +10. According to the Bohr model (and without introducing the correction for the reduced mass) the radius of such atoms is proportional to n^2/Z and the energy to Z^2/n^2. Thus the ground state of Ne^{9+} would be predicted to have a radius of $(1^2/10)a_0$ or $0.1 a_0$ and an energy of $-[(10^2)/1^2]E_0$ where a_0 and E_0 are atomic unit equal to the ground state radius and energy of the hydrogen atom. The ionization energy is the energy required to remove the electron from its ground state, and is thus 1.36×10^3 eV. The energy required to excite the Ne^{9+} to the first excited state would be $1.36 \times 10^3 (1/1^2 - 1/2^2)$ eV or 1020 eV.

During the 1920's R. A. Millikan did extensive research on the spectra of "stripped atoms". Such studies not only confirmed Bohr's predictions for hydrogen-like atoms, but demonstrated that isoelectronic atomic species had similar spectra which were displaced to higher energies with an increase in Z in accordance with Moseley's law for characteristic x-rays. For an account conveying the excitement of these discoveries see Millikan's *Electrons (+ and −), Protons, Photons, Neutrons, and Cosmic Rays,* University of Chicago Press, 1935.

1.4 Problem: Calculate the energy (eV) released in the transition of a hydrogen atom from the state $n = 3$ to $n = 1$. The wavelength of the radiation emitted in this transition may be found from the relation λ (in Å) = 12,398/E (eV) (often remembered as the approximate λ = 12345/E (eV)). Calculate the wavelengths of all spectral lines that could be observed from a collection of hydrogen atoms excited by a potential of 12V.

1.4 Solution: $\Delta E = \left(\dfrac{1}{n_i^2} - \dfrac{1}{n_f^2}\right) E_0 = \left(\dfrac{1}{3^2} - \dfrac{1}{1^2}\right)(-13.60 \text{eV})$

ΔE = 12.08 eV released in going from the $n = 3$ state to the ground state; 12.08 eV is the excitation potential required to populate the $n = 3$ state. If exactly 12 eV were used to excite the collection of H atoms only the $n = 2$ state would be populated.

$$\Delta E = \tfrac{3}{4}(13.60) = 10.20 \text{ eV}$$

Wavelength of emitted radiation = $\dfrac{12398}{10.20}$ = 1215 Å

1.5 Problem: Assuming a screening of 0.5 for an *s* electron, calculate the wavelength expected for the K_α x-ray line of Tc.

1.5 Solution: For Tc, Z = 43; for σ = 0.5, Z_{eff} = 42.5. The K_α line arises from a transition of a 2*s* electron to the 1*s* level

$$\Delta E \cong 13.6(42.5)^2 \left(\frac{1}{1^2} - \frac{1}{2^2}\right) = 18424 \text{ eV}$$

$$\lambda \cong \frac{12398}{18424} \text{ Å} = 0.673 \text{ Å}$$

This calculation ignores the change in σ for a $2s$ compared to a $1s$ electron, which would give $\approx$ 0.65 Å.

1.6 Problem: The Balmer series in the hydrogen spectrum originates from transitions between $n = 2$ states and higher states. Compare the wavelengths for the first three lines in the Balmer series with those expected for similar transitions in Li^{2+}.

1.6 Solution: If the difference in reduced mass for H and Li is ignored, the increase in Z will cause the corresponding transitions in Li^{2+} to occur at an energy 9 times greater, or a wavelength of one ninth, the transitions for H.

A more refined value could be obtained using the Rydberg constant 109,677.581 cm^{-1} for ^{1}H and the value 109,737.31 cm^{-1} (asuming infinite nuclear mass) for Li. Still better, a reduced mass for an ^{7}Li nucleus and an electron gives a Rydberg constant of 109,729 cm^{-1}. Thus, the following wavelengths may be calculated for the indicated transitions

$$\omega = RZ^2 \left(\frac{1}{n_f^2} - \frac{1}{n_i^2}\right) \text{cm}^{-1} \qquad \lambda = \frac{1}{\omega}$$

$$\lambda(\text{Å}) = \frac{10^8}{\omega \text{ (cm}^{-1}\text{)}}$$

Transition ($n_i \to n_f$)	$3 \to 2$	$4 \to 2$	$5 \to 2$
H (Z = 1) R = R_H	6562 Å	4863	4342
Li (Z = 3) R = R_{Li}	729	540	482

1.7 Problem: Obtain expressions for $(\partial E/\partial Z)_n$ and $(\partial^2 E/\partial Z^2)_n$ using the relationship $E = -2\pi^2 m Z^2 e^4 / n^2 h^2$ obtained from the Bohr model. Using the ionization energies (IE's) for isoelectronic sequences of atoms and ions given below, and finite differences to obtain $(\Delta E/\Delta Z)_n$ and $(\Delta^2 E/\Delta Z^2)_n$, show that the $n = 1$ level is filled when the electron occupancy is 2. How would you estimate unknown IE's using a finite-difference approximation? Estimate the electron affinities (EA's) of F and O by this procedure.

Ionization energies of the elements
(in eV; 1 eV/atom = 96.4869 kJ/mol)

Z	Element	I	II	III	IV	V	VI
1	H	13.598					
2	He	24.587	54.416				
3	Li	5.392	75.638	122.451			
4	Be	9.322	18.211	153.893	217.713		
5	B	8.298	25.154	37.930	259.368	340.217	
6	C	11.260	24.383	47.887	64.492	392.077	489.981
7	N	14.534	29.601	47.448	77.472	97.888	552.057
8	O	13.618	35.116	54.934	77.412	113.896	138.116
9	F	17.422	34.970	62.707	87.138	114.240	157.161
10	Ne	21.564	40.962	63.45	97.11	126.21	157.93
11	Na	5.139	47.286	71.64	98.91	138.39	172.15
12	Mg	7.646	15.035	80.143	109.24	141.26	186.50
13	Al	5.986	18.828	28.447	119.99	153.71	190.47
14	Si	8.151	16.345	33.492	45.141	166.77	205.05
15	P	10.486	19.725	30.18	51.37	65.023	220.43
22	Ti	6.82	13.58	27.491	43.266	99.22	
35	Br	11.814	21.8	36	47.3	59.7	
55	Cs	3.894	25.1				

1.7 Solution: The constants in the Bohr equation may be grouped together to give:

$$E = \frac{-13.6\, Z^2}{n^2} \text{ eV}$$

for hydrogen-like atoms. For IE's from the nth level we have

$$\text{IE} = \frac{13.6\, Z^2}{n^2} \text{ eV}$$

$$\left[\frac{\partial (\text{IE})}{\partial Z}\right]_n = \frac{2(13.6)Z}{n^2} \qquad \left[\frac{\partial^2 (\text{IE})}{\partial Z^2}\right]_n = \frac{27.2}{n^2}$$

n = quantum level occupied; in an isoelectronic sequence, screening may be assumed constant, so $\Delta Z_{\text{eff}} = \Delta Z$.

Using IE data for 1-electron atoms (hydrogen-like atoms) we have the following

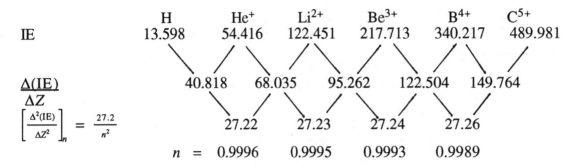

For 2-electron atoms we have

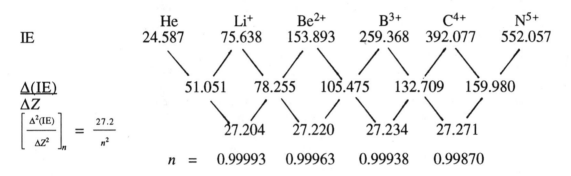

For 3-electron atoms we have

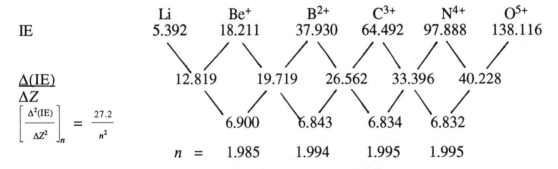

From the above it is clear that the electron undergoing ionization has an n quantum number of 1 for 1-electron and 2-electron atoms, but an n quantum number of 2 for 3-electron atoms. For further discussion of trends in IE's see G. P. Haight, *J. Chem. Educ.* **1967,** *44*, 468.

EA's may be estimated in similar fashion. In our illustration, a third difference has been used to compensate for slight shifts in shielding. The estimated values (in parentheses) are off by 0.2 to 0.3 eV for the EA's.

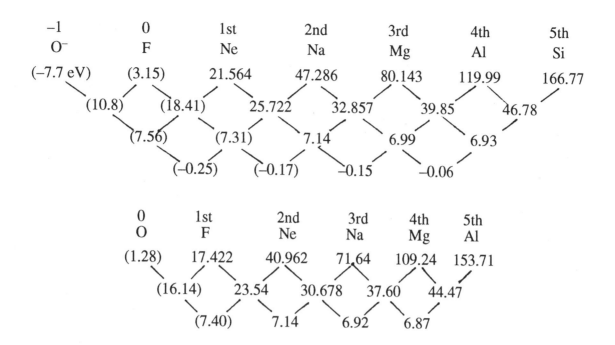

1.8 Problem: Calculate the values of r for which $2p$ and $4d$ hydrogenlike orbitals have radial nodes. The radial part of the wavefunction is

$$R_{n,l} = \text{constant} \times \left(\frac{2Zr}{na_0}\right)^l \times e^{-Zr/na_0} \times L_{n+1}^{2l+1}(x)$$

Associated Laguerre polynomials $\left(x = \dfrac{2Zr}{na_0}\right)$

	n	l	L_{n+1}^{2l+1}
$1s$	1	0	$-1!$
$2p$	2	1	$-3!$
$3d$	3	2	$-5!$
$4d$	4	2	$720x - 4320$

1.8 Solution: At a radial node the radial function vanishes. Hence for $2p$ orbitals $R_{2,1} = 0$ only when $r = 0$. For $4d$ orbitals $R_{4,2} = 0$ when $r = 0$ and also when the associated Laguerre polynomial vanishes—that is when $r = 6na_0/2Z = 12a_0/Z$. (It can be shown that the number of radial nodes is $(n - l - 1)$ in addition to the one at $r = 0$.

1.9 Problem: Using Φ_{ml} functions for H and the expressions for d orbitals of well-defined m_l values, calculate the values of θ and ϕ for which $2p_z$ and $3d_{x^2-y^2}$ have angular nodes.

Φ_{m_l} Functions for the H atom

$$\Phi_0(\phi) = \frac{1}{\sqrt{2\pi}} \quad \text{or} \quad \Phi_0(\phi) = \frac{1}{\sqrt{2\pi}}$$

$$\Phi_1(\phi) = \frac{1}{\sqrt{2\pi}} e^{i\phi} \quad \text{or} \quad \Phi_{1\,\cos}(\phi) = \frac{1}{\sqrt{\pi}} \cos\phi$$

$$\Phi_{-1}(\phi) = \frac{1}{\sqrt{2\pi}} e^{-i\phi} \quad \text{or} \quad \Phi_{1\,\sin}(\phi) = \frac{1}{\sqrt{\pi}} \sin\phi$$

$$\Phi_2(\phi) = \frac{1}{\sqrt{2\pi}} e^{i2\phi} \quad \text{or} \quad \Phi_{2\,\cos}(\phi) = \frac{1}{\sqrt{\pi}} \cos 2\phi$$

$$\Phi_{-2}(\phi) = \frac{1}{\sqrt{2\pi}} e^{-i2\phi} \quad \text{or} \quad \Phi_{2\,\sin}(\phi) = \frac{1}{\sqrt{\pi}} \sin 2\phi$$

Expressions for d orbitals of well-defined m_l values

$$d_{-2} = \sqrt{\frac{3}{8}} \frac{(x-iy)^2}{r^2} \qquad d_2 = \sqrt{\frac{3}{8}} \frac{(x+iy)^2}{r^2}$$

$$d_{-1} = \sqrt{\frac{3}{2}} \frac{z(x-iy)}{r^2} \qquad d_1 = -\sqrt{\frac{3}{2}} \frac{z(x+iy)}{r^2}$$

$$d_0 = \sqrt{\frac{1}{4}} \frac{(3z^2 - r^2)}{r^2}$$

1.9 Solution: Since $z = r\cos\theta$, the angular function for p_z vanishes when $\cos\theta = 0$, that is when $\theta = \pi/2$ (everywhere in the xy plane). $x^2 - y^2 = r^2\sin^2\theta[\cos^2\phi - \sin^2\phi]$. Thus, angular nodes will occur when $\sin^2\theta = 0$; this will happen when $\sin\theta = 0$ which occurs when $\theta = 0$ or π. The function also vanishes when the expression in square brackets vanishes; this happens for $\phi = \pi/4, 3\pi/4, 5\pi/4$ and $7\pi/4$. More simply, a node in p_z occurs for $z = 0$; nodes in $d_{x^2-y^2}$ occur when $x^2 - y^2 = 0$.

1.10 Problem: Explain briefly the observation that the energy difference between the $1s^2 2s^1$ $^2S_{1/2}$ state and the $1s^2 2p^1$ $^2P_{1/2}$ state for Li is 14,904 cm^{-1}, whereas for Li^{2+} the $2s^1$ $^2S_{1/2}$ and the $2p^1$ $^2P_{1/2}$ states differ by only 2.4 cm^{-1}.

1.10 Solution: The Li^{2+} ion is a 1-electron atom; Z_{eff} is equal to Z for all quantum states and the energy depends to a first approximation on n alone. The Li atom has a filled $1s^2$ core. The energy difference tells us that Z_{eff} is higher for the $2s^2 2s^1$ configuration than for the $2s^2 2p^1$ configuration. This may be rationalized on the basis of the greater degree of penetration of the $1s^2$ core by the $2s^1$ electron than by the $2p^1$ electron. (See page 38 DMA, 3rd ed.)

1.11 Problem: Using the p orbitals for an example, distinguish between the angular part of the probability function, the radial part of the probability function, and a probability contour. Draw simple sketches to illustrate. How would each of these be affected by a change in the principal quantum number, n?

1.11 Solution: The angular part of the probability function, $\theta^2\phi^2$, is a scaling factor which depends only on the angular orientation of a point in space with respect to a framework located on the nucleus. It is represented by a surface; the magnitude of $\theta^2\phi^2$ is the magnitude of a vector extending from the origin to the surface. It is independent of n, but does depend on l and m_l. For angular parts of the wavefunction that contain imaginary numbers, the probability function is obtained by multiplying the wavefunction by its complex conjugate—that is, by the function in which the sign of each term containing an imaginary number is multiplied by minus one.

The radial probability function, R^2, depends on the quantum numbers n and l. It varies with the distance from the nucleus, r, but not with angular orientation. It has zero values, or nodes at $r = 0$ and $r = \infty$ and $(n - l - 1)$ points in between.

The total probability function is the product of the angular probability function and the radial probability function, evaluated at every point in space. Contour lines are drawn by connecting points having the same preselected value of $R^2\theta^2\phi^2$.

Boundary surface diagrams are often depicted for orbitals. These enclose volumes within which a given probability (usually 90%) of finding the electron exists.

Sketches of each of these functions are shown below for a $3p_z$ orbital.

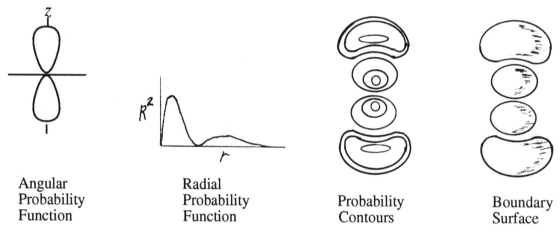

Angular Probability Function | Radial Probability Function | Probability Contours | Boundary Surface

Sketches of Functions for $3p_z$ orbital

1.12 Problem: Give the characteristic valence shell configuration for the following periodic groups (e.g., ns^1 for the alkali metals)

(a) Noble gases
(b) Halogens
(c) Coinage metals (Group 11)
(d) Ti family
(e) N family

1.12 Solution:
(a) ns^2np^6 except for He, $1s^2$
(b) ns^2np^5
(c) $(n-1)d^{10}ns^1$
(d) $(n-1)d^2ns^2$
(e) ns^2np^3

1.13 Problem: Write the electron configuration beyond a noble gas core for (for example, F, [He]$2s^22p^5$) Rb, La, Cr, Fe, Cu, Tl, Po, Gd, and Lu.

1.13 Solution: The periodic table provides at a glance a guide to electron configurations. The period structure is indicated in the figure below. With minor variations, the outermost electron configuration is determined by the group—main group elements having a number of outermost s and p electrons equal to their group number or group number minus ten (for Groups 13-18)—the noble gases completing the periods as indicated below. The first transition elements have a $4s^2$ configuration plus a number of d electrons equal to their number minus twenty (except Cr and Cu which have d^5s^1 and $d^{10}s^1$ configurations, respectively). The lanthanides begin filling the $4f$ after La which has a $6s^25d^1$ configuration.

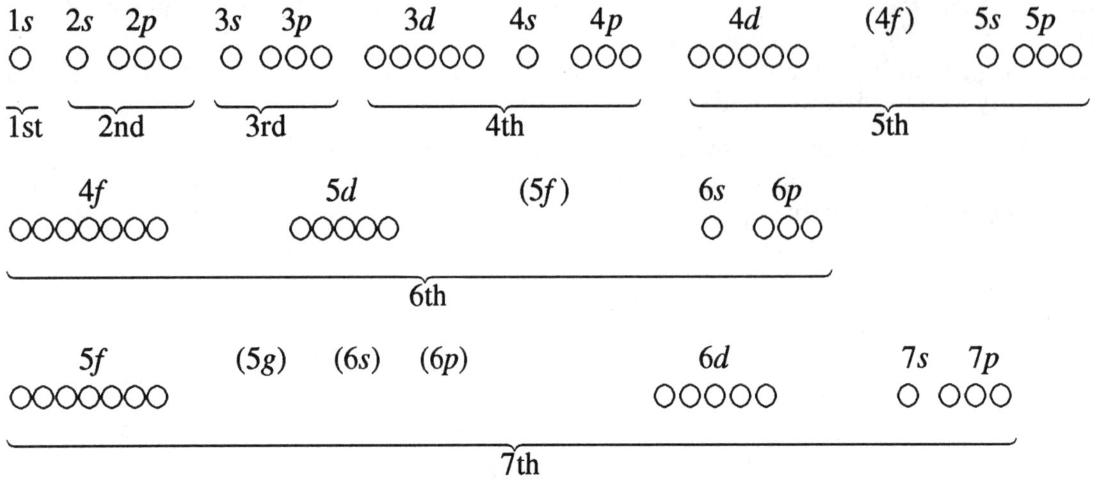

```
 1s   2s   2p      3s   3p     3d       4s   4p     4d         (4f)      5s   5p
 O    O   OOO      O   OOO   OOOOO      O   OOO   OOOOO                   O   OOO
 ‾‾   ‾‾   ‾‾‾‾‾‾‾‾‾   ‾‾‾‾‾‾‾‾‾‾‾‾‾‾‾‾‾‾‾‾‾‾‾‾‾   ‾‾‾‾‾‾‾‾‾‾‾‾‾‾‾‾‾‾‾‾‾‾‾‾‾‾‾‾‾‾‾
 1st  2nd       3rd               4th                          5th

    4f                 5d           (5f)         6s   6p
 OOOOOOO              OOOOO                       O   OOO
 ‾‾‾‾‾‾‾‾‾‾‾‾‾‾‾‾‾‾‾‾‾‾‾‾‾‾‾‾‾‾‾‾‾‾‾‾‾‾‾‾‾‾‾‾‾‾‾‾‾‾‾‾‾‾‾‾‾‾‾‾‾‾
                           6th

    5f          (5g)  (6s)  (6p)          6d           7s   7p
 OOOOOOO                                OOOOO          O   OOO
 ‾‾‾‾‾‾‾‾‾‾‾‾‾‾‾‾‾‾‾‾‾‾‾‾‾‾‾‾‾‾‾‾‾‾‾‾‾‾‾‾‾‾‾‾‾‾‾‾‾‾‾‾‾‾‾‾‾‾‾‾‾‾
                           7th
```

Makeup of the periods in the conventional long form of the periodic table

Rb, [Kr] $5s^1$ Tl, [Xe] $4f^{14}5d^{10}6s^26p^1$
La, [Xe] $5d^16s^2$ Po, [Xe] $4f^{14}5d^{10}6s^26p^4$
Cr, [Ar] $3d^54s^1$ Gd, [Xe] $4f^7\ 5d^1\ 6s^2$
Fe, [Ar] $3d^64s^2$ Lu, [Xe] $4f^{14}5d^1\ 6s^2$

1.14 Problem: Write the electron configuration beyond a noble gas core and give the number of unpaired electrons for (for example, F$^-$, [He] $2s^22p^6$):

K^+, Ti^{3+}, Cr^{3+}, Fe^{2+}, Cu^{2+}, Sb^{3+}, Se^{2-}, Sn^{4+}, Ce^{4+}, Eu^{2+}, and Lu^{3+}.

1.14 Solution: In forming ions, atoms will lose electrons from levels of highest principal quantum number, n; with levels of equal n, the electron will be lost from the orbital of highest l. Electrons in degenerate orbitals will occupy orbitals singly to the maximum permissible extent (Hund's rule of maximum multiplicity). These considerations lead to the following configurations.

Configurations	# Unpaired e	Configurations	# Unpaired e
K^+ [Ar]	(0)	Se^{2-} [Ar] $3d^{10}4s^24p^6$	(0)
Ti^{3+} [Ar] $3d^1$	(1)	Sn^{4+} [Kr] $3d^{10}$	(0)
Cr^{3+} [Ar] $3d^3$	(3)	Ce^{4+} [Xe]	(0)
Fe^{2+} [Ar] $3d^6$	(4)	Eu^{2+} [Xe] $4f^7$	(7)
Cu^{2+} [Ar] $3d^9$	(1)	Lu^{3+} [Xe] $4f^{14}$	(0)
Sb^{3+} [Kr] $4d^{10}5s^2$	(0)		

1.15 Problem: What is the spectroscopic notation for an atom in an energy state in which $L = 3$ and $S = 2$?

1.15 Solution: The term notation and the corresponding L values are

$$
\begin{array}{ccccccccc}
 & S & P & D & F & G & H & I & K & L \ldots \\
L = & 0 & 1 & 2 & 3 & 4 & 5 & 6 & 7 & 8 \ldots
\end{array}
$$

For $L = 3$ and $S = 2$ the term would be ^5F where the spin multiplicity 5 is obtained from the value of $2S + 1$.

1.16 Problem: What is the number of microstates for an f^3 configuration? Which of these is unique to the ground-state term arising from this configuration?

1.16 Solution: The number of microstates $= \frac{n!}{e!h!}$ where n is the number of spin orbitals (that is, twice the number of orbitals), e is the number of electrons, and h is the number of holes (equal to $n - e$). For an f^3 configuration

$$\text{number of microstates} = \frac{14!}{3!\,11!} = \frac{14 \times 13 \times 12}{3 \times 2} = 364$$

The microstates uniquely belonging to the ground state are represented below

$$
\begin{array}{c}
m_l \\
\begin{array}{ccccccc}
3 & 2 & 1 & 0 & -1 & -2 & -3
\end{array}
\end{array}
$$

These are just four of the $(2L + 1)(2S + 1)$ or 28 microstates of the array for the ground state ^4I term. The first microstate is written such that the $\Sigma m_l = L$ and $\Sigma m_s = S$ for the ground state term.

1.17 Problem: Derive the spectral terms for the d^7 configuration. Identify the ground-state term. Give all J values for the ground-state term and indicate which is lowest in energy.

1.17 Solution: The number of microstates is $\frac{10!}{7!\,3!} = 120$. Only the "top left" part of the table of microstates is reproduced here, the bottom being symmetrical across $M_L = 0$ and the right across $M_S = 1/2$. We generate this "core" of the array by tabulating the 35 microstates with positive M_L and M_S values. There are 25 microstates for $M_L = 1 \rightarrow 5$, 10 for $M_L = 0$, and 25 for $M_L = -1 \rightarrow -5$. There are also 60 corresponding microstates for $M_S = -1/2$ and $-3/2$.

	$m_l =$	+2	+1	0	−1	−2	+2	+1	0	−1	−2	$M_L = \sum m_L$	$M_S = \sum m_S$
1.		+	+	+	+	+	+	+				3	3/2
2.		+	+	+	+	+	+		+			2	3/2
3.		+	+	+	+	+	+			+		1	3/2
4.		+	+	+	+	+	+				+	0	3/2
5.		+	+	+	+	+		+	+			1	3/2
6.		+	+	+	+	+		+		+		0	3/2
7.		+	+	+	+		+	+	+			5	1/2
8.		+	+	+	+		+		+	+		3	1/2
9.		+	+	+	+		+	+		+		4	1/2
10.		+	+	+	+		+	+			+	3	1/2
11.		+	+	+	+			+	+	+		2	1/2
12.		+	+	+	+		+		+		+	2	1/2
13.		+	+	+	+		+			+	+	1	1/2
14.		+	+	+	+			+		+	+	0	1/2
15.		+	+	+	+				+	+	+	1	1/2
16.		+	+	+		+	+	+	+			4	1/2
17.		+	+	+		+	+		+	+		2	1/2
18.		+	+	+		+	+	+		+		3	1/2
19.		+	+	+		+	+	+			+	2	1/2
20.		+	+	+		+		+	+	+		1	1/2
21.		+	+	+		+	+		+		+	1	1/2
22.		+	+	+		+	+			+	+	0	1/2
23.		+	+	+		+		+	+		+	0	1/2
24.		+	+		+	+	+	+	+			3	1/2
25.		+	+		+	+	+		+	+		1	1/2
26.		+	+		+	+	+	+		+		2	1/2
27.		+	+		+	+	+	+			+	1	1/2
28.		+	+		+	+	+		+		+	0	1/2
29.		+	+		+	+		+	+	+		0	1/2
30.		+		+	+	+	+	+	+			2	1/2
31.		+		+	+	+	+			+	+	0	1/2
32.		+		+	+	+	+	+	+			1	1/2
33.		+		+	+	+	+	+			+	0	1/2
34.			+	+	+	+	+	+	+			1	1/2
35.			+	+	+	+	+	+		+		0	1/2

The table of microstates is

M_L \ M_S	+3/2	+1/2	−1/2	−3/2
5		1		
4		2		
3	1	4		
2	1	6		
1	2	8		
0	2	8		

The term structure is ^2H, ^2G, ^4F, ^2F, $2\,^2D$, ^4P and ^2P using the methods of Section 1.5.1, DMA, 3rd ed.

Microstate **1** is associated with the ground state. In general, the ground state is the one with maximum spin multiplicity and, within this restriction, the one with the largest L value, that is the ^4F term. This has J values running from $L + S$ through $L - S$ or $J = 9/2, 7/2, 5/2$ and $3/2$. Since the d-orbitals are more than half filled, the maximum J state is the ground state $^4J_{9/2}$.

1.18 Problem: Derive the spectral terms for the f^2 configuration. Identify the ground-state term. Give all J values for the ground-state term and indicate which is lowest in energy.

1.18 Solution: The total number of microstates is $\frac{14 \times 13}{2} = 91$.

Only the "top left" part of the table of microstates is reproduced here, the bottom being symmetrical across $M_L = 0$ and the right across $M_S = 0$.

	$m_l = +3$	$+2$	$+1$	0	-1	-2	-3	$+3$	$+2$	$+1$	0	-1	-2	-3	$M_L = \Sigma m_l$	$M_S = \Sigma m_s$
1.	++														5	1
2.	+	+													4	1
3.	+		+												3	1
4.	+			+											2	1
5.	+				+										1	1
6.	+					+									0	1
7.		++													3	1
8.		+	+												2	1
9.		+		+											1	1
10.		+			+										0	1
11.			++												1	1
12.			+	+											0	1
13.	+										+				6	0
14.	+										+				5	0
15.	+										+				4	0
16.	+											+			3	0
17.	+											+			2	0
18.	+												+		1	0

#	+3	+2	+1	0	−1	−2	−3	+3	+2	+1	0	−1	−2	−3	$M_L = \sum m_l$	$M_S = \sum m_s$
19.	+													+	0	0
20.		+						+							5	0
21.		+							+						4	0
22.		+								+					3	0
23.		+									+				2	0
24.		+										+			1	0
25.		+											+		0	0
26.			+					+							4	0
27.			+						+						3	0
28.			+							+					2	0
29.			+								+				1	0
30.			+									+			0	0
31.				+				+							3	0
32.				+					+						2	0
33.				+						+					1	0
34.				+							+				0	0
35.					+			+							2	0
36.					+				+						1	0
37.					+					+					0	0
38.						+		+							1	0
39.						+			+						0	0
40.							+	+							0	0

The table of microstates is

M_L \ M_S	1	0	−1
6		1	
5	1	2	
4	1	3	
3	2	4	
2	2	5	
1	3	6	
0	3	7	

The term structure is ^1I, ^3H, ^1G, ^3F, ^1D, ^3P, and ^1S using the methods of Section 1.5.1, DMA, 3rd ed.

^3H is the ground state term with $J = 4, 5, 6$. Since the orbitals are less than half filled, $^3H_4$ is lowest in energy.

1.19 Problem: What spectroscopic terms arise from the configuration $1s^2 2s^2 2p^6 3s^2 3p^5 3d^1$?

1.19 Solution: This problem asks for the spectroscopic terms for a configuration containing nonequivalent electrons—electrons in orbitals differing from each other in their n and/or l quantum numbers. In such cases the equivalent electrons may be treated individually to give "spectroscopic terms" which may then be multiplied together under the Clebsch-Gordon product rules already used to obtain J values from L and S values. A filled shell yields a ^1S state, one-electron occupancy (or one hole) gives a doublet state with $L = l$, and other occupancies may be worked out by techniques such as those in Problems 1.17 and 1.18.

For our configuration the terms from equivalent electrons would be:

$1s^2$	^1S	$S = 0, \ L = 0$
$2s^2$	^1S	$S = 0, \ L = 0$
$2p^6$	^1S	$S = 0, \ L = 0$
$3s^2$	^1S	$S = 0, \ L = 0$
$3p^5$	^2P	$S = \frac{1}{2}, L = 1$
$3d^1$	^2D	$S = \frac{1}{2}, L = 2$

The product of $L_1 \times L_2$ gives $(L_1 + L_2), (L_1 + L_2 - 1) \ldots |L_1 - L_2|$. The product of S_1 and S_2 gives $(S_1 + S_2), (S_1 + S_2 - 1), \ldots |S_1 - S_2|$. Both run in integer steps from the sum through to the difference in quantum numbers.

The $P \times D$ product would thus give L values of $(2 + 1), (2 + 1 - 1)$, and $(2 - 1)$ or 3, 2, and 1 or F, D, and P terms. The S values would be $(\frac{1}{2} + \frac{1}{2})$ and $(\frac{1}{2} - \frac{1}{2})$ or 1 and 0. The result of $^2P \times {}^2D$ is thus $^3F, {}^3D, {}^3P$ and $^1F, {}^1D, {}^1P$, where the superscript indicates the spin multiplicity $2S + 1$. The product of a ^1S with any other term leaves the other term unchanged.

1.20 Problem: What nodal planes might be expected for the $f_{x^3 - \frac{3}{5}xr^2}$ orbital?

1.20 Solution: Nodal surfaces arise when the function is zero,

$$x^3 - \tfrac{3}{5}xr^2 = 0$$

This occurs when $x = 0$, giving the yz plane as a nodal plane, and when $x = \sqrt{\frac{3}{5}}\, r$, a conical surface in which the generatrix makes an angle of 39.2° with the x axis.

1.21 Problem: Use Slater's rules (See DMA, 3rd ed., p. 40) to calculate the effective nuclear charge Z^* for the elements Li-F. How is the trend in Z^* reflected in the ionization energies (IE's) in this period?

1.21 Solution:

Li	$S = 2(0.85) = 1.70$	$Z^* = 3 - 1.70 = 1.30$
Be	$S = 2(0.85) + 1(0.35) = 2.05$	$Z^* = 4 - 2.05 = 1.95$
B	$S = 2(0.85) + 2(0.35) = 2.40$	$Z^* = 5 - 2.40 = 2.60$
C	$S = 2(0.85) + 3(0.35) = 2.75$	$Z^* = 6 - 2.75 = 3.25$
N	$S = 2(0.85) + 4(0.35) = 3.10$	$Z^* = 7 - 3.10 = 3.90$
O	$S = 2(0.85) + 5(0.35) = 3.45$	$Z^* = 8 - 3.45 = 4.55$
F	$S = 2(0.85) + 6(0.35) = 3.80$	$Z^* = 9 - 3.80 = 5.20$
Ne	$S = 2(0.85) + 7(0.35) = 4.15$	$Z^* = 10 - 4.15 = 5.85$

The values of Z^* increase across the period as we would expect. IE is predicted to be $13.6Z^{*2}/n^2$ eV. However, the monotonic increase in Z^* does not perfectly parallel the observed IE changes since it does not reproduce the small reversals which occur between Be and B and between N and O.

1.22 Problem: Use Slater's rules (See DMA, 3rd ed., p. 40) to calculate the effective nuclear charge for the alkaline earth elements. How is the trend in Z^* reflected in the IE's in this group?

1.22 Solution:

   Be $S = 2.05$ $Z^* = 1.95$
   Mg $S = 2(1.00) + 8(0.85) + 0.35 = 9.15$ $Z^* = 2.85$
   Ca $S = 10(1.00) + 8(0.85) + 0.35 = 17.15$ $Z^* = 2.85$
   Sr $S = 28(1.00) + 8(0.85) + 0.35 = 35.15$ $Z^* = 2.85$
   Ba $S = 46(1.00) + 8(0.85) + 0.35 = 53.15$ $Z^* = 2.85$
   Ra $S = 78(1.00) + 8(0.85) + 0.35 = 85.15$ $Z^* = 2.85$

Z^* increases in going from Be to Mg because of the eight elements lying between. It remains the same for Ca–Ra because of the "perfect" shielding of the intervening electrons, all of which are in the $(n-1)$ shell. We expect successive IE's in a family to be in the ratio

$$\frac{\text{IE}(n+1)}{\text{IE}(n)} = \frac{Z^*(n+1)}{Z^*(n)} \frac{n^2}{(n+1)^2}$$

The observed (Table 1.7 in DMA, 3rd ed.) monotonic decrease in IE from Be to Ba down the family can be rationalized if Z^* increases more slowly than n. This trend is duplicated using the above values for Z^*; however, the rate of decline in IEs is overestimated rather seriously. The small increase in IE at Ra because of the filling of the relatively non-shielding $4f$'s is not replicated by the value of Z^*.

1.23 Problem: Use Slater's rules (See DMA, 3rd ed., p. 40) to calculate the effective nuclear charges for K, Rb, and Cs. Compare these results with those for Cu, Ag, and Au. Comment on the IE's of these elements in light of these charges.

1.23 Solution:

   K $S = 10(1.00) + 8(0.85) \quad = 16.80$ $Z^* = 2.20$
   Rb $S = 28(1.00) + 8(0.85) \quad = 34.80$ $Z^* = 2.20$
   Cs $S = 46(1.00) + 8(0.85) \quad = 52.80$ $Z^* = 2.20$
   Cu $S = 10(1.00) + 18(0.85) = 25.30$ $Z^* = 3.70$
   Ag $S = 28(1.00) + 18(0.85) = 43.30$ $Z^* = 3.70$
   Au $S = 60(1.00) + 18(0.85) = 75.30$ $Z^* = 3.70$

The imperfect screening of the d electrons is reflected in the larger values of Z^* for the coinage metals. Each coinage metal is calculated to have an IE $(3.70)^2/(2.20)^2 = 2.8$ times as large as that of the alkali metal in the same period. The actual increase is less than this.

1.24 Problem: The trend in electron affinities (EA's) for Group 15 (VB) is opposite to that for 16 (VIB). Explain the trend for each family. (See Table 1.8, DMA, 3rd ed.)

1.24 Solution: At the outset it should be stated that there is no universally accepted answer to the question posed, but approaches to the question should cause us to look more closely at our model of the atom.

First we must find the trends alleged to exist in the question. Inspection of a table of electron affinities (EA's) shows the following orders of EA's (values quoted are in eV units).

O(1.461) < S(2.0771) > Se(2.0207) > Te(1.9708) > Po(1.9)   Series *a*
N(–0.07) < P(0.746) < As(0.81) < Sb(1.07) > Bi(0.946)   Series *b*

The difference in neighboring pairs shows a trend in (EA Group 16 – EA Group 15) as follows

Δ(EA) for O – N > S – P > Se – As > Te – Sb > Po – Bi   Series *c*

If the latter can be explained, then, perhaps, the initial question may be answered. This trend is similar to that observed for the differences in ionization energies (IE's) of the pairs isoelectronic to O^- and N^-, that is, F and O, Cl, and S, etc.

Δ(IE) for F – O >> Cl – S $\geq$ Br – Se > I – Te > At – Po   Series *d*

Here we recognize the "normal" increase in IE on crossing a period, the increase being greatest at the top of the table and decreasing as we descend a group. We attribute such trends to inefficient shielding of the nuclear charge within a set of electrons having the same *n* and *l* quantum numbers. We find a "normal" decrease in IE on descending, within a group of the table.

But we also recognize that O, S, and Se have lower IE's than that of either of their neighbors in the same period. This we may attribute to the particular stability of the half-filled *p* orbitals. The interaction involved here is sometimes referred to as spin-exchange interaction and decreases in importance with average separation of the electrons, that is, with the size of the atom.

Both of the above factors work in the same direction to explain Series *c*. To explain Series *b* we have to postulate that the spin exchange term which stabilizes the neutral N, P, etc., atom decreases faster than the "normal" decrease in the "zeroth" IE.

Finally, to explain the much smaller EA's of the later second period elements than that of the third period elements of similar groups we postulate that the small size of the atoms of the second group elements makes it much more difficult to crowd electrons around than for the larger atoms.

1.25 Problem: Rearrange the equations for *d* orbitals of well-defined m_l values (Problem 1.8) to express the "real" set of *d* orbitals.

1.25 Solution: Linear combinations of solutions of the wave equation will also be solutions of the wave equation. Imaginary coefficients are permissible. When the angular function for the *d* orbitals of well-defined m_l values are expanded, the following combinations will readily yield the "real" set of *d* orbitals.

$$d_{x^2-y^2} = \frac{1}{\sqrt{2}}(d_2 + d_{-2}) = \frac{\sqrt{3}}{2}\frac{(x^2-y^2)}{r^2} \qquad d_{xy} = -\frac{1}{\sqrt{2}}i(d_2 - d_{-2}) = \sqrt{3}\frac{xy}{r^2}$$

$$d_{z^2} = d_0 = \frac{1}{2}\frac{3z^2-r^2}{r^2}$$

$$d_{xz} = \frac{1}{\sqrt{2}}(d_{-1} - d_1) = \sqrt{3}\frac{xz}{r^2} \qquad d_{yz} = -\frac{1}{\sqrt{2}}i(d_{-1} + d_1) = \sqrt{3}\frac{yz}{r^2}$$

1.26 Problem: Use the data in Problems 1.7 and 1.28 (See Table 1.7 and 1.8 in DMA, 3rd ed., pp. 43-48) to predict whether the following hydrides should dissociate to H^+ and X^- or H^- and X^+ in the gas phase:
(a) LiH (b) CsH (c) HBr

1.26 Solution: For the reaction $B^+(g) + C^-(g) \leftarrow BC \rightarrow B^-(g) + C^+(g)$, the energy difference between the products on the right and those on the left is $(IE_C - EA_B) - (IE_B - EA_C) = (IE_C + EA_C) - (IE_B + EA_B) = 2(\chi_C - \chi_B)$—that is, just twice the difference in the absolute electronegativities.
(a) $2(\chi_H - \chi_{Li}) = 7.18 - 3.01 = 4.17$ eV. $Li^-(g) + H^+(g)$ lie 4.17 eV higher than $Li^+(g) + H^-$ which will therefore be the products.
(b) $2(\chi_H - \chi_{Cs}) = 7.18 - 2.18 = 5.00$ eV. $Cs^+(g)$ and $H^-(g)$ are the products.
(c) $2(\chi_{Br} - \chi_H) = 7.59 - 7.18 = 0.41$ eV. $H^+(g)$ and $Br^-(g)$ are the products.
Thus if the electronegativity of X is lower than that of H, hydride dissociation is favored; if the electronegativity of X is higher than that of H, proton dissociation is favored.
[Dissociation of neutral molecules to give ionic products is not a spontaneous process in the gas phase. The preceeding calculations should be interpreted as showing the relative expense of removing a proton versus removing a hydride ion during heterolytic bond cleavage in a gas phase ion-molecular reaction.]

1.27 Problem: Which of the following is softer?
(a) Cl or Br   (c) S or O   (e) Al^{3+} or Tl^{3+}
(b) S^{2-} or Se^{2-} (d) H or Na   (f) K^+ or Cu^+

1.27 Solution: Periodic trends may be noted and most chemists would simply select the correct response from the knowledge that softness increases with a descent in a group or on going from a noble gas electron configuration to a pseudo-noble-gas configuration. Quantitatively, softness is measured by the inverse of the hardness parameter: $\sigma = 1/\eta$. The softness orders are
(a) Br ($\sigma = 0.24$) > Cl (0.21)
(b) Se is softer than S and the order should be preserved for anions.
(c) S (0.24) > O (0.17)
(d) Na (0.44) > H (0.16)
(e) $Tl^{3+} > Al^{3+}$. The EA of M^{3+} is the IE for M^{2+}. Using data from Problems 1.7 and 1.28 (See Tables 1.7 and 1.8 DMA, 3rd ed., pp. 43-48) we calculate η for Al^{3+} as 45.8 eV and for Tl^{3+} as 10.4 eV.
(f) Proceeding as in (e) Cu^+ (0.16) > K^+ (0.07)

1.28 Problem: Rank the following in the order of increasing absolute electronegativity: F, Li, Ti^{4+}, P, H, C, Mg^{2+}, Li^+, using IE's in Problem 1.7 and EA values below:

	H	Li	C	F	Mg	P	Ti	Br	Cs
EA	0.7542	0.618	1.263	3.399	−0.4	0.746	0.079	3.365	0.4716

The EA for M^{n+} is the IE for $M^{(n-1)+}$.

1.28 Solution:

$$\chi_{abs} = \frac{IE + EA}{2}$$

$$\chi_{Li} = \frac{5.392 + 0.618}{2} = 3.00 \text{ eV} \quad \text{and} \quad \chi_{Li^+} = \frac{75.638 + 5.392}{2} = 40.51$$

We calculate the following values for absolute electronegativity.
Li (3.00 eV) < P (5.61) < C (6.24) < H (7.18) < F (10.41) < Li^+ (40.52) < Mg^{2+} (47.59) < Ti^{4+} (109.3)

1.29 Problem: The IE's of several group 4 elements are as follows:

	I	II	III	IV	V
Ti	6.82	13.58	27.49	43.27	99.22 eV
Zr	6.84	13.13	22.99	34.34	81.5
Pb	7.42	15.03	31.94	42.32	68.8

What stable oxidation states are expected for compounds of these elements?

1.29 Solution: Stable oxidation states are expected when the difference in successive IE's exceeds ~12 eV for valence electrons (See L. H. Ahrens, *J. Inorg. Nucl. Chem.* **1956**, *2*, 290, or DMA, 3rd ed., pp. 43-45). For Ti, Zr, and Pb, these differences are:

	Δ1,2	Δ2,3	Δ3,4	Δ4,5
Ti	6.8	13.9	15.8	56.0
Zr	6.3	9.9	11.4	47.2
Pb	7.6	16.9	10.4	26.5

From the above we expect to find compounds of Ti^{II}, Ti^{III}, and Ti^{IV}; Zr^{IV} only; and Pb^{II} and Pb^{IV}.

1.30 Problem: The IE's of second period atoms increase approximately linearly from B to N. A drop occurs between N and O. However, the energy necessary to remove an electron from the ground state ^3P oxygen to produce O$^+$ in the excited ^2P state lies on the extrapolated line of B, C, N IE's. Explain this observation. (See IE's in Problem 1.7.)

1.30 Solution: The IE's of B, C, and N all represent processes which remove a p electron from a singly occupied $2p$ orbital giving B$^+$, C$^+$, and N$^+$ in their ground states. For O which has the electron configuration $2s^22p^4$, the normal IE represents removal of an electron from the doubly occupied p orbital giving a ^4S O which is stabilized by exchange energy. Removal of a $2p$ electron from a singly occupied $2p$ orbital giving a ^2P state of O is the process comparable to that undergone by B, C, and N. Hence, its energy might be expected to fall on the same straight line when plotted vs. atomic number.

1.31 Problem: The emission spectrum of atomic Ca shows a transition from a ^3D state to a ^3P state. If the selection rule permits only $\Delta J = \pm 1$ or 0 (but not $J = 0$ to $J = 0$), how many lines will be observed for this transition?

1.31 Solution: For a ^3D state the S value must be 1 since $2S + 1 = 3$, and the L value is 2 for a D term. The J values run from $|L + S|$ through $|L - S|$ in interger steps, and would thus be 3, 2, and 1. For the ^3P state the J values would be 2, 1, and 0. From the chart below it may be seen that six transitions have $\Delta J = \pm 1$ or 0 and would be allowed.

		3	2	1	$J_{initial}$
	2	1	0	−1	
J_{final}	1	2	1	0	
	0	3	2	1	

I. SOME BASIC CONCEPTS
2
Molecular Models

2.1 Problem: Give the expected possible covalent MX_n compounds (where $X^- = Cl^-$ or Br^-) of Sn, At, and Be. Assuming only σ bonding, predict the geometry of each molecule.

2.1 Solution: To predict possible covalent states, start with the ground state electron configuration of the atom (see Problem 1.13) and obtain the simplest possible covalent state by utilizing only the initially unpaired electrons in covalent bond formation. More covalent bonds may be formed by promoting an initially paired electron into a low energy vacant orbital. Geometry may be predicted by counting both lone and bonding pairs as participating in the σ bond hybridization.

	Valence State Configuration	Hybridization	Geometry
Sn		sp^2	bent
		sp^3	tetrahedral
At		sp^3	linear
		sp^3d	T-shaped
		sp^3d^2	square pyramidal
		sp^3d^3	pentagonal bipyramidal
Ra			
		sd	bent

Note that the shape of the molecule is described by the location of atoms, and we predict the atom location in molecules containing lone pairs using the VSEPR model (see DMA, 3rd. ed., Section 2.3.3).

2.2 Problem: Assign formal charges and oxidation numbers, and evaluate the relative importance of three Lewis structures of OCN⁻, the cyanate ion. Compare these to the Lewis structures of the fulminate ion, CNO⁻ (encountered in explosive compounds).

2.2 Solution: Since the valence orbitals of C, O, and N are limited to those of the $2s$ and $2p$ orbitals, we expect the octet rule to be followed. We assign formal charges in the valence bond (VB) structures by dividing the electrons in the bonds equally between the participating atoms and calculating the charge on the pseudo-ion that is formed. Formal charges are useful in qualitatively evaluating the contribution of structures to a resonance hybrid. Oxidation numbers are found by assigning both of the electrons in a bond to the more electronegative atom. Oxidation numbers will be independent of our VB structure provided the same atoms are attached to each other. The structures in which the negative formal charge resides on the more electronegative atom, charges are minimized, and not highly separated, will contribute the most to the resonance hybrid.

$$[\ddot{\underset{..}{O}}=C=\ddot{N}]^- \quad [:\ddot{\underset{..}{O}}-C\equiv N:]^- \quad [:O\equiv C-\ddot{\underset{..}{N}}:]^-$$

(a)	(b)	(c)	
(0) (0) (−1)	(−1) (0) (0)	(+1) (0) (−2)	Formal charges
−II +IV −III	−II +IV −III	−II +IV −III	Oxidation numbers

Best charge distribution is in (*b*). Structure (*b*) should be most important, with structure (*a*) next. Structure (*c*) should be unimportant because of poor charge distribution. For fulminate ion (CNO⁻) the best structure is [C≡N—O]⁻ (formal charges −1, +1, −1). Note that the cyanate ion is stable, fulminate is explosive, and CON⁻ does not exist—a pattern consistent with increasingly poor charge distribution in these ions. The most electronegative element is not expected to be the central atom with the most shared electrons.

2.3 Problem: Write a reasonable electron dot structure and assign formal charges and oxidation numbers for each of the following: ClF, ClF₃, ICl₄⁻, HClO₃ (HOClO₂).

2.3 Solution:

:C̈l—F̈:
(0)   (0)
+I   −I

Cl (0) +III
each F (0) −I

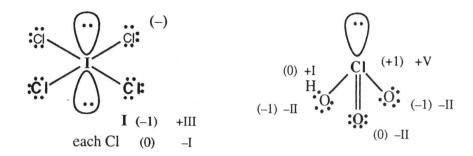

$$\begin{array}{lll} & \text{I } (-1) & +\text{III} \\ \text{each Cl} & (0) & -\text{I} \end{array}$$

2.4 Problem: Use Langmuir's formula given below (see DMA, 3rd ed., Section 2.4.5) to generate VB structures for BF_3, SO_3^{2-}, N_2F_2, and H_2CN_2 (consider CN_2^{2-} and then add $2H^+$).

$$\text{No. of covalent bonds} = \frac{8n - \text{No. of valence electrons}}{2}$$

2.4 Solution:

BF_3 $\frac{32-24}{2} = 4$ bonds

SO_3^{2-} $\frac{32-26}{2} = 3$ bonds

N_2F_2 $\frac{32-24}{2} = 4$ bonds

H_2CN_2 (Apply the formula to CN_2^{2-})

$\frac{24-16}{2} = 4$ bonds $\ddot{\text{N}}=\text{C}=\ddot{\text{N}}^{2-} + 2H^+ \longrightarrow \text{:N}\equiv\text{C}-\text{N}\begin{smallmatrix}H\\H\end{smallmatrix}$

One B=F bond, giving bond order $1\tfrac{1}{3}$ for BF_3, utilizes all low energy orbitals. Langmuir's formula gives the octet structure from which expanded octet structures can be generated. One S=O bond, using d orbitals, giving bond order $1\tfrac{1}{3}$ for SO_3^{2-}, gives better charge distribution.

2.5 Problem: Give the oxidation number, formal charge, and hybridization of the central atom in each of the following: NO_3^-, BF_4^-, $S_2O_3^{2-}$, ICl_2^+, ClO_3^-. What are the molecular shapes?

2.5 Solution: We proceed here as in Problem 2.1, except initially we assign extra electrons to the more electronegative atoms and, in the case of positively charged ions, remove electrons from the least electronegative atom. We recognize that only one bond between two atoms may be a σ bond and that further bonding must involve π (or more rarely δ) bonding.

	Oxidation Number	Number of Bonds	Hybrid	Formal Charge	Shape	Comments on Bonding
NO_3^-	+V	3	sp^2	+1	Planar	1 π bond to complete octet of N
BF_4^-	+III	4	sp^3	−1	Tetrahedral	No π bonding possible
$S_2O_3^{2-}$	+V[a]	4	sp^3	+1	Tetrahedral	Partial π bonding to lower charges
ICl_2^+	+III	2	sp^3	+1	Angular	π bonding is not important
ClO_3^-	+V	3	sp^3	+1	Pyramidal	Partial π bonding to lower charges

[a]The oxidation number of the central S of $S_2O_3^{2-}$ is sometimes given as +VI by analogy with SO_4^{2-}. However, for an S—S bond the shared electrons are divided equally, giving +V.

Only one resonance structure is shown in each case. The number of double bonds for $S_2O_3^{2-}$ and ClO_3^- corresponds to favorable charge distribution.

2.6 Problem: Select the reasonable electron-dot structures for each of the following compounds. Indicate what is wrong with each incorrect or unlikely structure.

(a) N=N=O :N—N≡O: :N≡N—O:
    N—N=O :N=N=O :N=N=O

(b) :S—C≡N:⁻ :S=C=N:⁻ :S=C=N:⁻ :S≡C—N:⁻

(c) [two cyclic P-N-Cl structures]

(d) [two cyclic B-N-R structures]

2.6 Solution:
(a) Structure 1 is wrong, 10 *e* on N, 18 *e* total

Structure 2 is unfavorable, adjacent (+) charges high (–2) charges on N

Structure 3 is OK, O has (–) charge, central N is (+)

Structure 4 is unfavorable, only 3 bonds

Structure 5 is *wrong*, 14 *e* total

Structure 6 is *wrong*, 10 *e* on O

(b) Structure 1 is unfavorable (–) on S, not the more electronegative N

Structure 2 is very unfavorable, (+) charge and only 6 *e* on N and (–2) charge on S

Struture 3 is best for charge distribution with (–1) on N

Structure 4 is very unfavorable, charges are widely separated, N is (–2), and π bonding is less important to S than to N

(c) Structure 1 has favorable charge distribution (all 0), but π bonding is less important for third period elements

Structure 2 has charge separation, but with (–1) on the more electronegative N

(d) Structure 1 is good, both B and N are neutral

Structure 2 has unfavorable charge separation, B (–1) and N (+1), but there are more strong bonds and the π bonding is delocalized around the ring

2.7 Problem: Give the expected hybridization of P, O, and Sb in $Cl_3P-O-SbCl_5$. The P—O—Sb bond angle is 165°.

2.7 Solution: The P has four σ bonds and is sp^3, Sb has six σ bonds and is d^2sp^3 (octahedral). The hybridization for O could be sp (180°), sp^2 (120°); or even sp^3 (109.5°). The 165° bond angle at O suggests sp to sp^2 hybridization with some degree of π bonding involving donation from the filled O p orbitals into the empty d orbitals of P.

2.8 Problem: Predict both the gross geometry (from the σ orbital hybridization) and the fine geometry (from bond-electron pair repulsion, etc.) of the following species: F_2SeO, $SnCl_2$, I_3^- and $IO_2F_2^-$.

2.8 Solution: F_2SeO has three σ bonds and a lone pair on Se. The sp^3 hybridization with one lone pair leads to a pyramidal molecule with bond angles less than 109° because of the lone-pair repulsion. The Se(F)(O) bond angle is expected to be larger than that of Se(F)(F) since π bonding to O is more favorable than to F because of the neutralization of charge for Se=O.

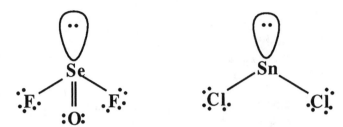

SnCl$_2$ has two σ bonds and one lone pair on Sn, leading to an angular molecule with the bond angle less than 120° because of lone-pair repulsion.

I$_3^-$ has two σ bonds and three lone pairs on I. The three lone pairs are expected to occupy the equatorial positions of a trigonal bipyramid, leading to a linear ion.

IO$_2$F$_2^-$ has four σ bonds and a lone pair on I. We expect the lone pair in an equatorial position of the trigonal bipyramid to give a "sawhorse". The F–I–F bonds are expected to be essentially linear because of the opposing effects of repulsion involving the lone pair and the double bonds to O. The I(=O)$_2$ angle is 100°, suggesting that the lone pair has a greater repulsive effect than the π bonds do.

2.9 Problem: Which of the following in each pair has the larger bond angle? Why?
(a) CH$_4$ and NH$_3$ (b) OF$_2$ and OCl$_2$ (c) NH$_3$ and NF$_3$ (d) PH$_3$ and NH$_3$.

2.9 Solution:
(a) The bond angle is smaller in NH$_3$ than in CH$_4$ because of the repulsion between the lone pair on N and the bonding pairs.
(b) The bond angle is greater in OCl$_2$ than in OF$_2$ because there is some π interaction in OCl$_2$ involving donation from the filled p orbitals on O and the empty d orbitals on Cl. No π bonding is possible for OF$_2$ because all orbitals are filled on both atoms.
(c) The bond angle in NH$_3$ is larger than that in NF$_3$ because the N—F bonds are longer than N—H bonds and the electron density is displaced toward the more electronegative F, both effects diminish the bonding pair–bonding pair repulsion.
(d) The bond angle is greater in NH$_3$ than in PH$_3$ because the P—H bonds are longer and the lower electronegativity of P permits electron density to be displaced toward H to a greater extent than in the case of NH$_3$. Both of these effects diminish the bonding pair–bonding pair repulsion.

2.10 Problem: Trimethylphosphine was reported by R. Holmes to react with SbCl$_3$ and SbCl$_5$ to form, respectively, (Me$_3$P)(SbCl$_3$) or (Me$_3$P)$_2$(SbCl$_3$) and (Me$_3$P)(SbCl$_5$) or (Me$_3$P)$_2$(SbCl$_5$). Suggest valence bond structures for each of these and indicate approximate bond angles around the Sb atom.

2.10 Solution: Me$_3$P→SbCl$_3$. There are five electron pairs on Sb, so the structure is based on a trigonal bipyramid with a lone pair and Me$_3$P in the equatorial plane. Lone pairs and groups of low electronegativity atoms prefer equatorial positions. The Sb⟨Cl/P angle should be less than 120° because of the lone–pair repulsion.

(Me$_3$P)$_2$SbCl$_3$. There are six electron pairs on Sb, so the structure is based on an octahedron with the PMe$_3$ *trans* to one another and *cis* to the lone pair. The PMe$_3$ groups are expected to bend away from the lone pair.

Me$_3$P→SbCl$_5$. There are six electron pairs on Sb giving an octahedron. The four equatorial Cl atoms might be expected to bend away from the PMe$_3$ slightly.

(Me$_3$P)$_2$SbCl$_5$. There are seven electron pairs on Sb giving a pentagonal bipyramid with both Me$_3$P groups in the less crowded axial positions. No distortion is expected for the symmetrical structure.

2.11 Problem: Sketch the following hybrid orbitals (indicate the sign of the amplitude of the wavefunction on your sketch):
(a) an *sd* hybrid (b) a *pd* hybrid.

2.11 Solution: (a) $s + d_{z^2}$—Combination of d_{z^2} with s increases the (+) lobes of d_{z^2} and diminishes the negative annulus. Angular hybrids result from combination of s with other d orbitals.

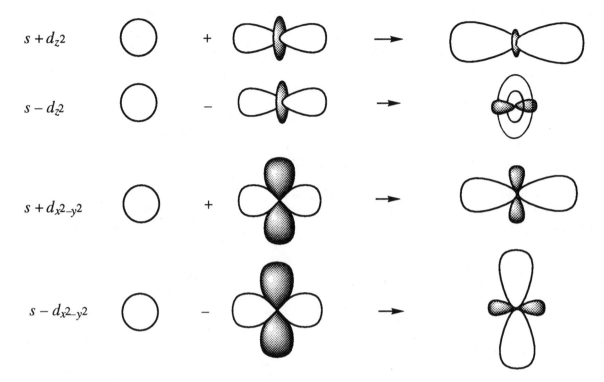

(b) $p_x \pm d_{z^2}$—Each combination increases one lobe of d_{z^2} and diminishes the other (only one combination is shown and the torus is omitted).

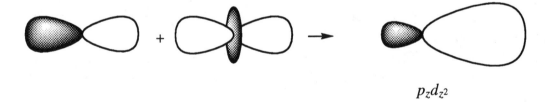

$p_z d_{z^2}$

2.12 Problem: The Ru—O—Ru bond angle in $(Cl_5Ru)_2O$ is 180°. What is the state of hybridization of the oxygen? Explain the reasons for the large bond angle. (See R. J. Gillespie, *J. Am. Chem. Soc.* **1960**, *82*, 5978.)

2.12 Solution: The 180° bond angles at O for $(Cl_5Ru)_2O$ suggests *sp* hybridization for O with the filled *p* orbitals on O involved in $p\pi d$ bonding with Ru.

2.13 Problem: Using H. A. Bent's isoelectronic groupings (see below), identify the species of list **2** (below) that are "isoelectronic" with each of those of list **1**. Each entry might have more than one "isoelectronic" partner in the other list.

1. $(H_2N)_2CO$ $HONO_2$ OCO CO $(CN)_2$ ClO_2^+ $Si_3O_9^{6-}$ $TeCl_2$ Bi_9^{5+}

2. H_2CCO $(HO)_2CO$ ONN H_2CNN H_3CNO_2 F_2CO BF B_2O_2 $(CH_3)_2CO$ CH_2CN^- SO_2 cyclic $(SO_3)_3$ ICl_2^+ Pb_9^{4-}

Isoelectronic groups (read horizontally):

| CO | CN$^-$ | NO$^+$ | NN | BF |

| —CH$_3$ | —N̈H$_2$ | —Ö̈H | —F̈: |

| >CH$_2$ | >N̈H | >Ö: |

| ≡CH | ≡N: | >Ö:$^{(+)}$ |

| =CH$_2$ | =N̈H | =Ö̈ |

2.13 Solution: For each horizontal line shown the molecules, ions, or groups underlined are "isoelectronic".

(1)	(2)
H**O**NO$_2$	**CH$_3$**NO$_2$
(**H$_2$N**)$_2$CO	(**HO**)$_2$CO, **F**$_2$CO, (**CH$_3$**)$_2$CO
O**CO**, HN**CO**	O**NN**, H$_2$C **NN**, CH$_2$ **CN**$^-$, H$_2$C**CO**
a b a	a b a b a b a b
CO	BF
(CN)$_2$	B$_2$O$_2$ (2 x 9 e)
ClO$_2^+$	SO$_2$ (18 e)
Si$_3$O$_9^{6-}$	(SO$_3$)$_3$ (3 x 24 e)
TeCl$_2$	ICl$_2^+$ (20 e)
Bi$_9^{5+}$	Pb$_9^{4-}$ (40 e)

2.14 Problem: Consult Pauling's electroneutrality principle (see DMA, 3rd ed., p. 62) and describe the reasonable charge distribution in NH$_4^+$ and in SO$_4^{2-}$.

2.14 Solution: For NH$_3$ the formal charge is +1 on N assuming equal sharing of electrons, but we expect the bonds to be polar (N is more electronegative than H), displacing charge toward N so that N becomes neutral and the positive charge is distributed equally among the four H atoms. For SO$_4^{2-}$ we expect the degree of π bonding and the polarity of the bonds to result in the distribution of the negative charge equally (–½ each) on the four oxygen atoms. The limiting cases involving zero and four double bonds are shown. Two double bonds make the formal charge on S zero, neglecting bond polarity.

2.15 Problem: The bromine atom in BrF$_5$ is below the plane of the base of the tetragonal pyramid. Explain.

2.15 Solution: In BrF$_5$ the important repulsion is between the lone pair and the bonding pairs of the four Br—F bonds in the base of the pyramid, displacing the F atoms upward and leaving the Br below the plane of the base of the pyramid.

2.16 Problem: H$_2$C=SF$_4$ gives the expected isomer with the double-bonded methylene group in an equatorial position. From the orientation of the π bond, would you expect the H atoms to be in the axial plane or the equatorial plane? (See K. O. Christie and H. Oberhammer, *Inorg. Chem.* **1981**, *20*, 297.)

2.16 Solution: π bonding can involve either d_{xz} (or d_{yz}), with the π lobes above and below the equatorial plane, or the d_{xy} orbital, with the π lobes in the equatorial plane. Since there is less crowding in the equatorial plane, the latter is expected. The sp^2 bonds to H will thus be perpendicular to the equatorial plane. The 97° S$<^F_F$ angle in the equatorial plane (see above reference) is consistent with this expectation. The σ hybridization for S is dsp^3, using d_{z^2}

2.17 Problem: The C—CH$_3$ bond distance in methyl acetylene is unusually short (146 pm) for a single bond. Show how this short bond can be rationalized in more than one way.

2.17 Solution: Since the strength of a bond involving some kind of s–p hybrid increases with the amount of s character (the lower-energy s orbitals lower the energy of the hybrids), the bond involving the sp hybrid orbitals for C–CH$_3$ is stronger and shorter than C–CH$_3$ bonds formed using sp^2 and sp^3 hybrids. Another explanation is that hyperconjugation involving "no bond" structures, H—C̈=C=C—H, increases the bond order of the C–CH$_3$ bond.

2.18 Problem: Calculate the dipole moment to be expected for the ionic structure H$^+$Cl$^-$ using the same internuclear separation as for the HCl molecule. Calculate the dipole moment for HCl assuming 19% ionic character. The bond distance for HCl is 127.4 pm.

2.18 Solution: Assuming +1 and –1 charges separated by 127 pm,

$$\mu = qd = 4.80 \times 10^{-10} \text{ esu} \times 127 \text{ pm} = 6.10 \text{ D}$$

This is for 100% ionic character; if the bond is only 19% ionic, the calculated dipole moment is 1.16 D, 0.19 x 6.10 = 1.16 D. The actual dipole moment for HCl is 1.03 D, corresponding to 17% ionic character.

2.19 Problem: Calculate the heats of formation (from electronegativities) for H_2S, H_2O, SCl_2, NF_3, and NCl_3. Calculate the heats of formation of these compounds from the bond energies for comparison. Compare the results with data from a handbook.

$$-\Delta H_f = 96.49n\, (\chi_A - \chi_B)^2$$

where n is the number of bonds (from the definition of Pauling's electronegativity scale).

	H	C	S	N	Cl	O	F
χ Values	2.2	2.6	2.6	3.0	3.2	3.4	4.0

	H–H	H–S	S–S	H–O	O–O	Cl–Cl	N–N	N–F	F–F
Av. Bond Energy	432	363	226	459	494	239	942	283	155

2.19 Solution:

H_2S $-\Delta H_f = 2(96.5)(2.6 - 2.2)^2 = 0.32(96.5) = -31$ kJ/mol

H_2O $-\Delta H_f = 2(96.5)(3.4 - 2.2)^2 = 2.88(96.5) = -278$ kJ/mol

SCl_2 $-\Delta H_f = 2(96.5)(3.2 - 2.6)^2 = -69$ kJ/mol

NF_3 $-\Delta H_f = 3(96.5)(4.0 - 3.0)^2 = -290$ kJ/mol

correcting for the extraordinary stability of the N≡N bond, requiring more energy (231 kJ/N) to break the bond,

$$-\Delta H_f = -290 + 231 = -59 \text{ kJ/mol}$$

H_2S	1/8 S_8	+ H_2	→	H_2S	
	+226	+432		–2 x 363	$\Delta H_f = -68$ kJ/mol
H_2O	H_2	+ ½ O_2	→	H_2O	
	+432	+ ½ x 494		–2 x 459	$\Delta H_f = -239$ kJ/mol
SCl_2	1/8 S_8	+ Cl_2	→	SCl_2	
	+226	+239		–2 x 255	$\Delta H_f = -45$ kJ/mol
NF_3	½ N_2	+ 3/2 F_2	→	NF_3	
	½ x 942	+ 3/2 x 155		–3 x 283	$\Delta H_f = -146$ kJ/mol

The values from bond energies are more reliable since the bond energies are obtained more directly from thermodynamic data. The agreement is poorest for NF, indicating that in this case the correction made for the N≡N bond is overdone.

2.20 Problem: Calculate the electronegativity differences from the bond energies for H—Cl, H—S, and S—Cl. Using Allred's electronegativity of H, compare the electronegativity values that can be obtained for S and Cl with those tabulated.

Average Bond energies
H–Cl	428 kJ/mol	H_2	432 kJ/mol
H–S	363	Cl_2	239
S–Cl	255	$(1/8) S_8$	226

$D(A-B) = \frac{1}{2}[D(A-A) + D(B-B)] + 96.5(\chi_A - \chi_B)^2$ and $\chi_H = 2.2$

2.20 Solution:

HCl $428 = \frac{1}{2}(432 + 239) + 96.5(\Delta\chi)^2$
$\Delta\chi = 1.0$

HS $363 = \frac{1}{2}(432 + 226) + 96.5(\Delta\chi)^2$
$\Delta\chi = 0.6$

SCl $255 = \frac{1}{2}(226 + 239) + 96.5(\Delta\chi)^2$
$\Delta\chi = 0.5$

Using $\chi_H = 2.2$, we get $\chi_{Cl} = 3.2$, and $\chi_S = 2.8$
Accepted values $\chi_{Cl} = 3.2$, and $\chi_S = 2.6$

2.21 Problem: Calculate the Lewis-Langmuir charges on carbon in
(a) CH_4 (b) C_2H_2 (c) HCN (d) CO_3^{2-}.

L. E. Allen proposed that Lewis-Langmuir (L.L.) charges (see Section 2.4.5, DMA, 3rd ed.) from the following equation (χ values are given in Problem 2.19).

(L.L. charge on A) = No. of valence e – No. of unshared e on A – $2 \sum_{bonds} \left(\dfrac{\chi_A}{\chi_A - \chi_B}\right)$

2.21 Solution:

(a) CH_4 L.L.(C) = $4 - 0 - 4 \times 2 \left(\dfrac{2.6}{2.6 + 2.2}\right)$ = -0.33

(b) HC≡CH L.L.(C) = $4 - 0 - 2 \left(\dfrac{2.6}{2.6 + 2.2}\right) - 3 \times 2 \left(\dfrac{2.6}{2.6 + 2.6}\right)$ = -0.08

(c) HC≡N L.L.(C) = $4 - 0 - 2 \left(\dfrac{2.6}{2.6 + 2.2}\right) - 3 \times 2 \left(\dfrac{2.6}{2.6 + 3.0}\right)$ = 0.13

(d)
```
    O  2-
    ‖
    C
   / \
  O   O
```
L.L.(C) = $4 - 0 - 2 \times 2 \left(\dfrac{2.6}{2.6 + 3.4}\right) - 2 \times 2 \left(\dfrac{2.6}{2.6 + 3.4}\right)$ = 0.53
(two C—O) (double bond)

2.22 Problem: Acetylene, unlike methane, has acidic properties and shows hydrogen bonding. Why do the Lewis-Langmuir charges calculated for CH_4 and C_2H_2 not show a greater negative charge for C in C_2H_2 than that in CH_4?

2.22 Solution: $\chi_C = 2.6$ is for sp^3 hybridization as in CH_4. For C_2H_2 the much higher value (3.3) for sp hybridization is appropriate.

$$L.L.(C) = 4 - 0 - 2\left(\frac{3.3}{3.3 + 2.2}\right) - 3 \times 2\left(\frac{3.3}{3.3 + 3.3}\right) = -0.2 \text{ for } C_2H_2.$$

2.23 Problem: Calculate the formal charges, oxidation numbers, and Lewis–Langmuir charges on S and F in S_2F_2, SF_2, SF_4, and SF_6. Which of these correlate well with the expected changes in bond polarity in the series?

2.23 Solution:
For L–L charges of S in FSSF:
$$L-L = 6 - 4 - 2\left(\frac{2.6}{2.6 + 4.0}\right) - 2\left(\frac{2.6}{2.6 + 2.6}\right) = 2 - 0.80 - 1 = 0.20$$

For L–L charges of S in
SF_2: $\quad L-L = 6 - 4 - 2 \times 2\left(\frac{2.6}{2.6 + 4.0}\right) = 0.40$
SF_4: $\quad L-L = 6 - 2 - 4 \times 2\left(\frac{2.6}{2.6 + 4.0}\right) = 0.80$
SF_6: $\quad L-L = 6 - 0 - 6 \times 2\left(\frac{2.6}{2.6 + 4.0}\right) = 1.20$

For F in all compounds
$$L-L = 7 - 6 - 2\left(\frac{4.0}{2.6 + 4.0}\right) = -0.20$$

S	S_2F_2	SF_2	SF_4	SF_6
Oxid. No.	+1	+2	+4	+6
L–L	+0.20	+0.40	+0.80	+1.20

The formal charges are zero on S and F in all of these compounds. For F the oxidation number is –1 and L–L is –0.20 for all of them. The trends for both oxidation numbers and L–L charges for S correlate with the expected increase in bond polarity with the S—F bonds through the series. The oxidation numbers here are adequate for comparisons. See DMA, 3rd ed., p. 782 for special considerations of bonding for sulfur fluorides.

2.24 Problem: Although ΔH_{vap} for HF is lower than that for H_2O, HF forms stronger H-bonds. Explain.

2.24 Solution: Each HF forms two H-bonds (one as an H-donor, one as an H-acceptor) whereas each H_2O molecule forms four H-bonds. HF vapor at the normal boiling point contains polymeric units up to $(HF)_6$ species—thus not all of the hydrogen bonds in HF are broken on vaporization, whereas water vapor at the boiling point of water consists essentially of monomeric H_2O molecules. Thus, four H-bonds are broken for every H_2O molecule vaporized, whereas fewer than two H-bonds are broken per HF vaporized.

2.25 Problem: The intensity of an infrared absorption is proportional to the change in the dipole moment occurring during the vibration. The asymmetric stretching in IHI^- is far more intense than that found for the stretching vibration of HI. Offer a reasonable explanation. Does your explanation also hold for the intensities of other H-bonded species?

2.25 Solution: In the asymmetric stretching vibration of IHI⁻, the negative charge may be viewed as moving from one iodine to the other for only a small displacement of the hydrogen (⁻I···HI going to IH···I⁻). Since the change in dipole moment with displacement along the vibrational coordinate is large, the intensity of the absorption is large. For HI, μ is very small since $\Delta\chi$ is small, and μ changes little during vibration, hence the IR absorption for this stretch is weak. Similar arguments hold for other hydrogen bonded species.

2.26 Problem: Explain why the anhydrous acid $HICl_4$ cannot be isolated, but the crystalline hydrate $HICl_4 \cdot 4H_2O$ may be obtained from ICl_3 in aqueous HCl.

2.26 Solution: $HICl_4$ has no good site for the proton. In the hydrate the proton is found in the $H_9O_4^+$ species.

2.27 Problem: List the substances in each of the following groups in order of increasing boiling points (bp's). (*Hint:* first group the substances according to the type of interaction involved.)
(a) LiF, LiBr, CCl_4, NH_3, CH_4, SiC, CsI.
(b) Xe, NaCl, NO, CaO, BrF, Al_2O_3, SiF_4.

2.27 Solution:
(a) LiF, LiBr, and CsI are ionic salts with very high bp's. Both melting points and bp's are expected to decrease as the distance between the centers of the ions increases. However, we find that radius ratio effects lead to high anion-anion repulsion in LiBr and consequent lower values of the melting point and bp than anticipated (See DMA, 3rd. ed., Section 5.7.2). SiC is a giant covalent molecule which remains solid at higher temperatures than any of the other compounds listed. It decomposes at temperatures above 2200°C. CH_4, NH_3, and CCl_4 contain discrete molecular units with covalent bonding. Only NH_3 has a non-zero dipole moment and the possibility of hydrogen bonding. We should know from experience that CH_4 and NH_3 are gases at room temperature whereas CCl_4 is a liquid. CCl_4 exposes more surface electrons and accordingly displays much greater London forces than CH_4 and NH_3. CH_4, with no dipole moment, and no surface electrons has the lowest bp in the set. Combining groups, the net order of bp's is thus:

$$CH_4 < NH_3 < CCl_4 < LiBr < CsI < LiF < SiC$$

(b) NaCl, CaO, and Al_2O_3 are salt-like solids with attraction between ions increasing with increasing charge on the ions. NO, BrF, and SiF_4 are molecular substances, all of which are gases at room temperature. As a first approximation we might expect the bp of NO to be approximately an average of the bp of N_2 and O_2, increased somewhat by the dipole moment (only $\Delta\chi = 0.5$). The actual bp is –151.8°C. For BrF the dipolar contribution should be much more significant ($\Delta\chi = 1.2$), and the bp of BrF is 20°C compared to that of –34°C for Cl_2 (the "average molar mass" of Br_2 and F_2).

That SiF_4 exists as a substance with discrete molecules, rather than as a network solid (as SiC_2 and AlF_3) is not a matter of any change from ionic to covalent bonding but rather that the coordination number of Si is satisfied by the number of halogens corresponding to a formula unit. The F atoms presented to the outside world by SiF_4 are not very polarizable and the intermolecular attractions are weak (sublimation point –95.7°C). The intermolecular interactions in the silicon tetrahalide have been the provocation for much speculation by chemists (see J. H. Hildebrand, *J.*

Chem. Phys. **1947**, *15*, 727).

The bp of Xe is expected to be higher than that of any of the noble gases of lower mass. We may recall that Ar is obtained from fractional distillation of liquid air and the Ar boils at a temperature between that of O_2 (–182.96°C) and N_2 (–195.8°C). Congeners in the 3rd, 4th, and 5th periods differ by 30 to 40 degrees in bp's (MeCl, –24°C; MeBr, 3.5°C; MeI, 42.4°C) so we might expect Xe to boil 70 to 80 degrees above the range for liquid air. The actual bp for Xe is –109.3°C. The order of increasing bp's is thus: NO < Xe < SiF_4 < BrF < NaCl < CaO < Al_2O_3.

2.28 Problem: When no chemical reaction occurs, the solubility of a gas in a liquid is proportional to the magnitude of the van der Waals interaction energy of the gas molecules (see DMA, 3rd ed., Table 2.11, p. 100). Indicate the relative solubility of O_2, N_2, Ar, and He in water. Why do deep-sea divers use a mixture of He and O_2 instead of air?

2.28 Solution: The van der Waals interaction leading to solution of these gases is expected to be the geometric mean of the van der Waals interaction between gas molecules and the van der Waals interaction between solvent molecules. The van der Waals interaction for the gases will parallel their bp's (the higher bp having the stronger van der Waals interaction). The bp order and hence that of increasing solubility is:

$$He < N_2 < Ar < O_2$$

Helium is used instead of N_2 in the "air" for deep-sea divers because of its much lower solubility in blood—thereby preventing the "bends", a painful condition resulting from the bubbles of N_2 released from the solution as the pressure decreases during the ascent.

2.29 Problem: Select from among the following radii and bond distances to calculate the expected S···S distance for SF_6 molecules in contact in the solid.

Covalent Bond Length		Crystal Radii		van der Waals Radii	
		Si^{IV}	43 pm	S	180 pm
S—F (in SF_6)	156 pm	F^-	119	F	147

2.29 Solution: The distance S—F···F—S is the sum of S—F covalent distance, the F···F van der Waals distance and the F—S distance, that is 156 + 2(147) + 156 = 606 pm.

2.30 Problem: Calculate the expected O···O separation in a system containing O—H–O if no hydrogen bonding were to occur and the O—H–O atoms are linearly arranged. Compare your results with the nearest O--H--O distance (275 pm) in ice I_h.

2.30 Solution: The van der Waals radius of O is 152 pm and that of H is 120 pm. The H—O bond distance is 95.7 pm. Thus, O···O contact for nonbonded atoms would be 304 pm. To calculate the O—H···O distance expected for no H-bonding, take the O—H covalent distance and add the H···O van der Waals distance.

$$r_{O-H(covalent)} + r_{H(vdW)} + r_{O(vdW)}$$
$$95.7 + 120 + 152 = 368 \text{ pm}$$

The observed distance of 275 pm is much less than the O—H···O distance calculated for a van der

2.31 Problem: For each of the following substances, in the liquid state, indicate the major type of interaction between the individual units (atoms, molecules, or ions).
  LiF, H_2O, CH_3NH_2, HCl, Xe, CCl_4

2.31 Solution:
LiF	$Li^+ \cdots F^-$	Ionic or coulombic attraction
H_2O and CH_3NH_2		Hydrogen bonding
HCl		Dipole-dipole attraction
Xe and CCl_4		Dispersion attraction

2.32 Problem: Why is $(CH_3)_2O$ quite soluble in H_2O and why is C_2H_6 very insoluble in H_2O?

2.32 Solution: The oxygen of dimethyl ether forms hydrogen bonds with H_2O. C_2H_6 cannot form hydrogen bonds to H_2O and separating H_2O molecules by C_2H_6 molecules breaks strong H_2O-H_2O hydrogen bonds.

2.33 Problem: Most substances expand when they are heated, leading to decreasing density with increasing temperature. However, $H_2O(l)$ has a maximum density at 4°C. How can you explain this?

2.33 Solution: In the structure of ice, water molecules are oriented so as to form the maximum number of H-bonds (four per O). This leaves considerable open space in the lattice (see the open channels in ice in Figure 2.21, p. 96 in DMA, 3rd ed.) As ice melts some H-bonds are broken and the open cage structure collapses, giving a decrease in volume. Structure is broken as temperature increases. Above 4°C, the normal volume expansion, with rising T, is more important than structure breaking and density decreases.

2.34 Problem: If a table of van der Waals radii is unavailable, what other source might be used for reasonable estimates of van der Waals radii for nonmetals? Why are these values reasonable?

2.34 Solution: Use ionic radii for anions as reasonable values for van der Waals radii. The values are similar since in each case an approaching atom encounters a completed octet of electrons and no bond is formed.

2.35 Problem: Compare the electronegativity of the N orbital involved in the N–H bond in each of the following species:

NH_3 NH_4^+ NH_2^- $C_5H_5\overset{+}{N}-H$ $CH_3C\equiv NH^+$

2.35 Solution: Electronegativity, as defined by Pauling, is the ability of an atom in a molecule to attract electrons. We expect the electronegativity to increase with increasing s character in the hybrid orbital, and with an increase in the positive formal charge. These considerations lead to an electronegativity order of N in the above series of

$\chi(CH_3CNH^+) > \chi(C_5H_5NH^+) > \chi(NH_4^+) > \chi(NH_3) > \chi(NH_2^-)$

2.36 Problem: What is the expected bond order of the Se–Se bonds in the cyclic Se_4^{2+} species?

2.36 Solution: The contributing structures to the resonance hybrid which are expected to be of major importance would be:

$$\left\{ \begin{array}{c} \ddot{S}e - \ddot{S}e \\ || \quad | \\ \ddot{S}e - \ddot{S}e \end{array}^{2+} \quad \begin{array}{c} :\ddot{S}e = \ddot{S}e: \\ | \quad | \\ :\ddot{S}e - \ddot{S}e: \end{array}^{2+} \quad \begin{array}{c} :\ddot{S}e - \ddot{S}e \\ | \quad || \\ :\ddot{S}e - \ddot{S}e \end{array}^{2+} \quad \begin{array}{c} :\ddot{S}e - \ddot{S}e: \\ | \quad | \\ :\ddot{S}e = \ddot{S}e: \end{array}^{2+} \right\}$$

From the above we expect the Se–Se bond order to be 1.25, that is, between any pair of Se atoms, there is one double bond in a sum of four contributing structures.

2.37 Problem: Which of the possible isomers of $(CH_3)_2PF_3$ is the one encountered? Why is there a site preference for the methyl group?

2.37 Solution: In $(CH_3)_2PF_3$ the methyl groups occupy the equatorial positions. In such cases the less electronegative substituents occupy equatorial positions in order to minimize bonding pair repulsion. There is greater displacement of electron density in the bonds toward the more electronegative substituent, decreasing the bonding pair–bonding pair repulsion. It is the bond angle between the methyl groups that is greater (124°) than the 120° angle in the trigonal plane of a regular trigonal bipyramid. The two axial F substituents bend away slightly from the CH_3 groups. (The bending is exaggerated to make it evident.)

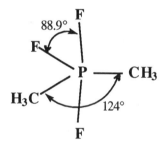

2.38 Problem: There are striking differences between the physical properties of the second period elements C and N and the corresponding elements of the third period, Si and P. Similarly, there are important differences between the oxides and oxoacids (just consider those with the highest oxidation states). What bonding tendencies account for the differences and why do they occur?

2.38 Solution: Carbon exists as diamond with each C bonded to four others by single bonds and as graphite where each C is bonded to three others in a planar hexagonal network—the latter occurs because of delocalized π bonding. Silicon gives ony the single bonded structure.

Nitrogen occurs as triple bonded N≡N while P_4 (only single bonds) is the stable form.

The oxides and oxoacids are:

CO_2	H_2CO_3	N_2O_5	HNO_3
SiO_2	H_4SiO_4	P_4O_{10}	H_3PO_4

In CO_2, carbon forms two σ bonds and two π bonds. In H_2CO_3, N_2O_5, and HNO_3, C or N form only three σ bonds and a π bond. In all of the Si and P oxides and oxoacids considered here Si or P form four σ bonds (sp^3 hybridization). Any π bonding must involve the d orbitals of the Si or P and occurs only to the extent needed to achieve low charges on all atoms.

Pi bonding occurs to a much greater extent for C and N than for Si and P because of the size of the atoms and of the orbitals. The sidewise π overlap is most favorable for small atoms (short X–X distance). Pi bonding is also unfavorable for large, diffuse orbitals because of poor overlap.

2.39 Problem: Explain why the valence bond structure of carbon monoxide is best described as :C⇐O:, featuring a dative bond from O to the less electronegative C.

(–)   (+)

2.39 Solution: The ground state electron configuration of C is $2s^2 2p^2$. We consider that the s orbital and the empty p orbital form two sp hybrids, one directed toward O and one directed 180° opposite. This contains the C lone pair, leaving the empty sp hybrid to interact with a filled O σ orbital. The two unhybridized singly occupied C $2p$ orbitals form π bonds by overlap with singly occupied O $2p$ orbitals.

2.40 Problem:
Compounds of the type $[i\text{-}(C_5H_{11})_4N]Cl \cdot y\, H_2O$ and $[i\text{-}(C_5H_{11})_4N]_2CrO_4 \cdot 2y\, H_2O$ (where y is approximately 40) are found to be isomorphous. It is unusual to find isomorphous compounds of the same cation with anions of different charge. What special features of these hydrated compounds might be responsible?

2.40 Solution: The structures of these very highly hydrated compounds are dominated by the large hydrogen-bonded cages. The anions fill cavities in the cages without altering the structure. See G. A. Jeffrey and R. K. McMullan, *Prog. Inorg. Chem.* **1967**, *8*, 43.

2.41 Problem: The dicyanamide ion, $C_2N_3^-$, is isoelectronic with carbon suboxide, C_3O_2.

(a) How many covalent bonds are expected for $C_2N_3^-$ or C_3O_2?
(b) Assuming the five atoms may be arranged as ABBBA or as BABAB, give a preferred Lewis structure for each. Give the reason for your choice. Give the expected shape of each.
(c) Comment on the statement that isoelectronic species will have the same shape.

2.41 Solution:
(a) Both species have 24 valence level electrons and 5 atoms, and each species is expected to have 8 bonds according to Langmuir's equation (see Problem 2.4).
(b) The linear ABBBA arrangement is expected for carbon suboxide and the bent BABAB for the dicyanamide ion based on the most favorable formal charge distribution in the competing structures.

$$\overset{-2}{\ddot{\underset{..}{C}}}=\overset{+2}{O}=C=\overset{+2}{O}=\overset{-2}{\underset{..}{\ddot{C}}}$$

best
$$\overset{..}{\underset{..}{\ddot{O}}}=C=C=C=\overset{..}{\underset{..}{\ddot{O}}}$$

$$\overset{-2}{\underset{..}{\ddot{C}}}=\overset{+1}{N}=\overset{+1}{N}=\overset{+1}{N}=\overset{-2}{\underset{..}{\ddot{C}}}$$

$$\overset{-1}{\underset{..}{\ddot{N}}}=C=\overset{+1}{N}=C=\overset{-1}{\underset{..}{\ddot{N}}}$$

(angular structures with C apex, charges shown; for dicyanamide the N-apex structure with ≡N−C−N−C≡N is labeled **best**)

(c) Isoelectronic species generally have the same shape *provided that the stoichiometry shows no changes in the ratio of electronegative to electropositive atoms.* The most electronegative atoms tend to be the terminal atoms, or at the apex of angular structures, because of formal charge considerations.

39

I. SOME BASIC CONCEPTS
3
Symmetry

3.1 PROBLEM: Use matrices to show that $\sigma_1\sigma_2 = C_4$ and $\sigma_2\sigma_1 = C_4^3$ for the kaleidoscope. Take the intersection of σ_1 and σ_2 as the z axis (that is, the axis perpendicular to the plane of the paper) and let the x axis be the intersection of σ_1 and the plane of the paper.

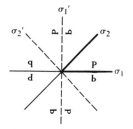

Reflections in a kaleidoscope

3.1 Solution: The mirror plane σ_1 carries **P** into **b**, or $x_2 = x_1$, $y_2 = -y_1$, and $z_2 = z_1$ (x and z are unchanged, y goes into the negative of itself). The matrix may be set up as:

New \ Orig.	x_1	y_1	z_1
x_2	1		
y_2		−1	
z_2			1

or

$$\sigma_1 = \begin{bmatrix} 1 & 0 & 0 \\ 0 & -1 & 0 \\ 0 & 0 & 1 \end{bmatrix} \begin{bmatrix} x \\ y \\ z \end{bmatrix} = \begin{bmatrix} x \\ -y \\ z \end{bmatrix}$$

The mirror plane σ_2 carries **P** into **ᴅ**, or $x_2 = y_1$, $y_2 = x_1$, and $z_2 = z_1$.

	x_1	y_1	z_1
x_2		1	
y_2	1		
z_2			1

or

$$\sigma_2 = \begin{bmatrix} 0 & 1 & 0 \\ 1 & 0 & 0 \\ 0 & 0 & 1 \end{bmatrix} \begin{bmatrix} x \\ y \\ z \end{bmatrix} = \begin{bmatrix} y \\ x \\ z \end{bmatrix}$$

The C_4 operation carries **P** into ⊐, or $x_2 = y_1$, $y_2 = -x_1$, and $z_2 = z_1$. Thus

	x_1	y_1	z_1
x_2		1	
y_2	-1		
z_2			1

or

$$C_4 = \begin{bmatrix} 0 & 1 & 0 \\ -1 & 0 & 0 \\ 0 & 0 & 1 \end{bmatrix} \begin{bmatrix} x \\ y \\ z \end{bmatrix} = \begin{bmatrix} y \\ -x \\ z \end{bmatrix}$$

The C_4^3 operation carries **P** into ⊐, or $x_2 = -y_1$, $y_2 = x_1$, and $z_2 = z_1$. Thus

	x_1	y_1	z_1
x_2		-1	
y_2	1		
z_2			1

or

$$C_4^3 = \begin{bmatrix} 0 & -1 & 0 \\ 1 & 0 & 0 \\ 0 & 0 & 1 \end{bmatrix} \begin{bmatrix} x \\ y \\ z \end{bmatrix} = \begin{bmatrix} -y \\ x \\ z \end{bmatrix}$$

We are now to show by matrix multiplication that $\sigma_1\sigma_2 = C_4$ and $\sigma_2\sigma_1 = C_4^3$. We will use the product rule for matrix multiplication.

$$p_{ik} = \sum_j a_{ij} b_{jk}$$

where p is an element in the *i*th row and the *j*th column of the product matrix, *a* and *b* are elements in matrices *A* and *B* and are located in the *i*th row and *j*th column and *j*th row and *k*th column, respectively. The sum of each p_{ik} element is taken over all *j* elements in the *i*th row of *A* with all *j* elements in the *k*th column of *B*. Labeling each matrix with subscripts to follow the multiplication more readily, we have

$$\begin{bmatrix} 1_{11} & 0_{12} & 0_{13} \\ 0_{21} & -1_{22} & 0_{23} \\ 0_{31} & 0_{32} & 1_{33} \end{bmatrix} \begin{bmatrix} 0_{11} & 1_{12} & 0_{13} \\ 1_{21} & 0_{22} & 0_{23} \\ 0_{31} & 0_{32} & 1_{33} \end{bmatrix} =$$

$$\quad\quad\quad \sigma_1 \quad\quad\quad\quad\quad\quad \sigma_2$$

$$\begin{bmatrix} (1_{11}\cdot 0_{11}+0_{12}\cdot 1_{21}+0_{13}\cdot 0_{31}) & (1_{11}\cdot 1_{12}+0_{12}\cdot 0_{22}+0_{13}\cdot 0_{32}) & (1_{11}\cdot 0_{13}+0_{12}\cdot 1_{23}+0_{13}\cdot 1_{33}) \\ (0_{21}\cdot 0_{11}-1_{22}\cdot 1_{21}+0_{23}\cdot 0_{31}) & (0_{21}\cdot 1_{12}-1_{22}\cdot 0_{22}+0_{23}\cdot 0_{32}) & (0_{21}\cdot 0_{13}-1_{22}\cdot 0_{23}+0_{23}\cdot 1_{33}) \\ (0_{31}\cdot 0_{11}+0_{32}\cdot 1_{21}+1_{33}\cdot 0_{31}) & (0_{31}\cdot 1_{12}+0_{32}\cdot 0_{22}+1_{33}\cdot 0_{32}) & (0_{31}\cdot 0_{13}+0_{32}\cdot 0_{23}+1_{33}\cdot 1_{33}) \end{bmatrix}$$

$$= \begin{bmatrix} 0 & 1 & 0 \\ -1 & 0 & 0 \\ 0 & 0 & 1 \end{bmatrix} \equiv C_4$$

And, with less fanfare, the product $\sigma_2 \sigma_1$ is

$$\underbrace{\begin{bmatrix} 0 & 1 & 0 \\ 1 & 0 & 0 \\ 0 & 0 & 1 \end{bmatrix}}_{\sigma_2} \underbrace{\begin{bmatrix} 1 & 0 & 0 \\ 0 & -1 & 0 \\ 0 & 0 & 1 \end{bmatrix}}_{\sigma_1} = \underbrace{\begin{bmatrix} 0 & -1 & 0 \\ 1 & 0 & 0 \\ 0 & 0 & 1 \end{bmatrix}}_{C_4^3}$$

Thus, σ_1 and σ_2 do not commute.

3.2 PROBLEM: Select the point group to which each of the species below belongs.

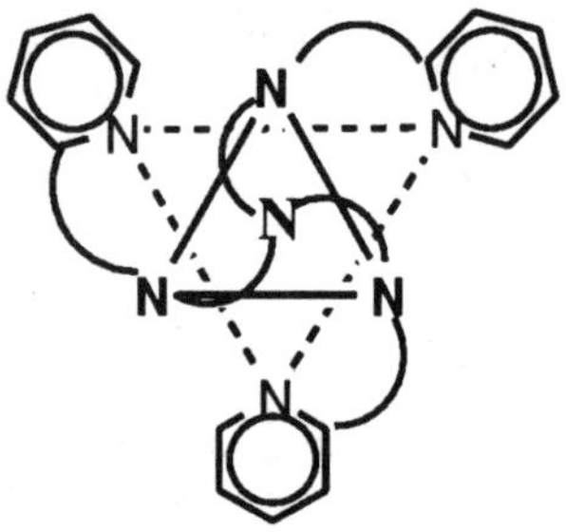

(a)

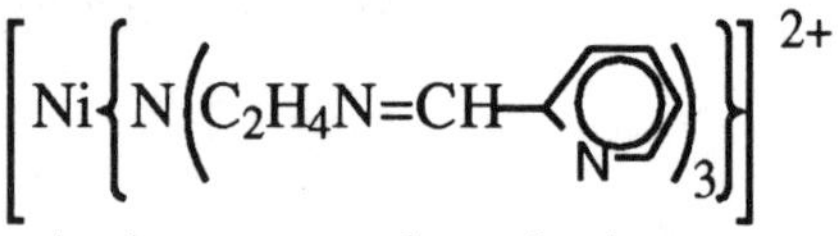

$\left[Ni\left\{ N(C_2H_4N=CH-\bigcirc_{N})_3 \right\} \right]^{2+}$

(a monocapped octahedron—
Ni at the octahedral center
is not shown)

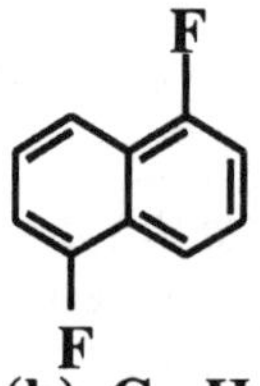

(b) $C_{10}H_6F_2$

(c) C_6H_6

(d) All of the tribromobenzenes
(1,2,3
1,2,4
1,3,5)

3.2 Solution: The "flow chart" below indicates the minimum number of generators which should be found to identify the point group to which a species belongs.

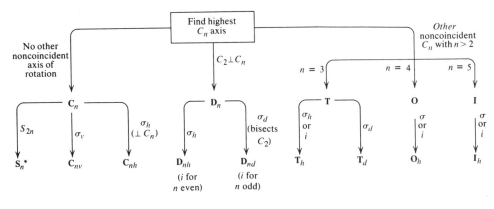

Flow chart for assignment of molecules to point groups.

(a) *The highest C_n is C_3.* Threre are no other C_3 and there are no $C_2 \perp C_3$. The rotation group is therefore C_3. The molecule is chiral—there are no σ, or other S_n, so the point group is C_3.
(b) Highest $C_n = C_2$; no other C_2 present; a $\sigma_h \perp C_2$ is present. Point group is C_{2h}.
(c) Highest $C_n = C_6$, $\perp C_2$, σ_h Point group is D_{6h}
(d) 1,2,3-isomer C_2, no $\perp C_2$, σ_v Point group is C_{2v}
    1,2,4-isomer C_1, σ Point group is C_s
    1,3,5-isomer C_3, $\perp C_2$, σ_h Point group is D_{3h}

3.3 PROBLEM: Assign the isomers of $[(gly)_2Co(OH)_2Co(gly)_2]$ depicted below to the appropriate point groups.

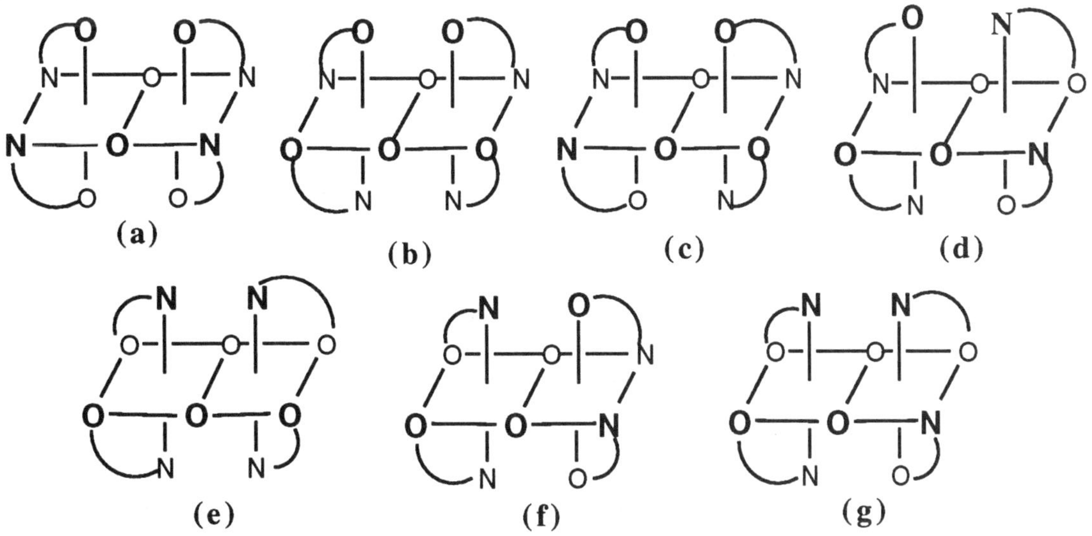

(gly = N⌒O = $NH_2CH_2CO_2^-$ and OH^- ions bridge the octahedra)

3.3 Solution:
(a) C_2, no $\perp C_2$, σ_h. Point group is $\mathbf{C_{2h}}$
(b) only σ $\quad\quad\quad\quad\quad\quad\quad\quad \mathbf{C_s}$
(c) only E $\quad\quad\quad\quad\quad\quad\quad\quad \mathbf{C_1}$
(d) only i $\quad\quad\quad\quad\quad\quad\quad\quad \mathbf{C_i}$
(e) C_2, no $\perp C_2$, σ_h $\quad\quad\quad\quad \mathbf{C_{2h}}$
(f) only C_2 $\quad\quad\quad\quad\quad\quad\quad\quad \mathbf{C_2}$
(g) only E $\quad\quad\quad\quad\quad\quad\quad\quad \mathbf{C_1}$

3.4 PROBLEM: Assign the species below to the appropriate point groups.

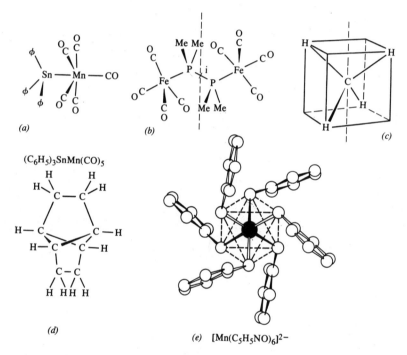

Improper rotational axes of symmetry. [Structure (a) is from H. P. Weber and R. F. Bryan, *Acta Cryst.* **1967**, *22*, 822. Structure (d) is after J. F. Chiang and S. H. Bauer, *Trans. Faraday Soc.* **1968**, *64*, 2248. Structure (e) is reprinted with permission from T. J. Bergendahl and J. S. Wood, *Inorg. Chem.* **1975**, *14*, 338; copyright © 1975, American chemical Society.]

3.4 Solution:

(a) only σ $\quad\quad\quad$ Point group is $\mathbf{C_s}$
(b) C_2, σ_h $\quad\quad\quad\quad\quad\quad\quad\quad \mathbf{C_{2h}}$
(c) C_3, + additional C_3, σ_d $\quad\quad \mathbf{T_d}$
(d) C_2, $\perp C_2$, σ_d $\quad\quad\quad\quad\quad \mathbf{D_{2d}}$
(e) C_3, no $\perp C_2$; S_6 $\quad\quad\quad\quad \mathbf{S_6}$

3.5 PROBLEM: A scalene triangle has three unequal sides. A regular scalene tetrahedron may be constructed by folding the pattern obtained by joining the midpoints of the sides of a scalene triangle having acute angles. To what point groups does the regular scalene tetrahedron belong?

3.5 Solution: A drawing of a regular scalene tetrahedron may be made by connecting a set of diagonally opposite corners of an orthorhombic prism as shown here. The three mutually perpendicular C_2 axes are readily apparent.
It belongs to a **D$_2$** point group.

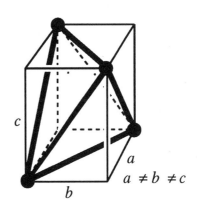

3.6 PROBLEM: Assign point groups to the following figures and their conjugates. What are the shapes of the conjugate figures?

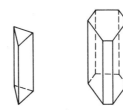

(a) Trigonal prism (b) Hexagonal prism with alternate sides of equal area (c) Square pyramid

3.6 Solution:
(a) C_3, $C_2 \perp C_3$, σ_h Point group is   **D$_{3h}$**
(b) C_3, $C_2 \perp C_3$, σ_h **D$_{3h}$**
(c) C_4, no $\perp C_2$, σ_v **C$_{4v}$**

The conjugate figures are obtained by connecting the centers of the faces and would be:
(a) a trigonal bipyramid.

(b) a hexagonal bipyramid—the hexagon having alternate interior angles of the same size.

(c) a square pyramid (inverted).

The conjugates would all have the same point groups as the figures to which they are conjugate.

3.7 PROBLEM: Determine the order of the **D$_{3h}$** group, the number of classes of operations, and the dimensions of the irreducible representations.

3.7 Solution: The order of the **C$_3$** subgroup is 3; adding a perpendicular **C$_2$** to give **D$_3$** doubles the order to give 6; adding σ_h again doubles the order to give 12 for **D$_{3h}$**.

The $3C_2$ axes belong to the same class and the $3\sigma_v$ belong to the same class, since they are interchanged by C_3, so the number of classes cannot be greater than 9 (without determining if C_3 and C_3^2, and S_3 and S_3^2 are in the same class). For any group, the dimensions of the representation are the same as the character under the identity operation—when these are squared and summed over all irreducible representations the order of the group is obtained. We find on assuming 8 irreducible representations, with one being two-dimensional, $2^2 + 1^2 + 1^2 + 1^2 + 1^2 + 1^2 + 1^2 + 1^2 = 11 \neq$ order of the group. Hence there must be at least 2 two-dimensional representations and 6 irreducible representations, since $2^2 + 2^2 + 1^2 + 1^2 + 1^2 + 1^2 = 12$, or 6 classes, 2 two-dimensional and 4 one-dimensional representations.

3.8 PROBLEM: Indicate how the slash mark on the "vee-bar" shown would be shifted under each of the operations of the $\mathbf{C}_{2v}$ group. Use your results as a guide to construct the multiplication table for the $\mathbf{C}_{2v}$ group.

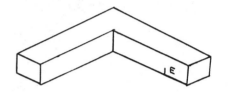

3.8 Solution: Let the z axis be the C_2 axis, let the y axis be in the plane of the "vee." Under the operations of the $\mathbf{C}_{2v}$ group, the slash mark is brought to the new position indicated on the figure below. The multiplication table for the $\mathbf{C}_{2v}$ group may be constructed by noting the position of the slash mark after two successive operations, and noting the single operation which would carry the slash mark to the same position. The operation carried out first is listed at the top of the multiplication table.

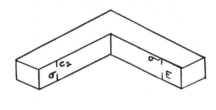

$\mathbf{C}_{2v}$	E	C_2	σ	σ'
E	E	C_2	σ	σ'
C_2	C_2	E	σ'	σ
σ	σ	σ'	E	C_2
σ'	σ'	σ	C_2	E

3.9 PROBLEM: Identify three symmetry operations that are their own inverses.

3.9 Solution: σ, i, C_2. $\sigma^2 = E$, $i^2 = E$, $C_2^2 = E$ (and, of course, $E^2 = E$.)

3.10 PROBLEM: Determine the independent operations of the $\mathbf{C}_{3h}$ group and construct the group multiplication table. Determine whether C_3 and C_3^2 are in the same class.

3.10 Solution: The operations of $\mathbf{C}_{3h}$ are $E, C_3, C_3^2, \sigma_h, S_3$, and S_3^5. Even powers of an S_n operation generate rotations, while odd powers yield further improper rotations.

$S_3 = \sigma_h C_3$

$S_3^2 = (\sigma_h C_3)(\sigma_h C_3) = C_3^2$

$S_3^3 = (\sigma_h C_3)C_3^2 = \sigma_h$

$S_3^4 = C_3^2 C_3^2 = C_3$

$S_3^5 = (\sigma_h C_3)C_3 = \sigma_h C_3^2$

$S_3^6 = E$

C_{3h}	$E = S_3^6$	S_3^1	S_3^2	S_3^3	S_3^4	S_3^5
E	E	S_3	S_3^2	S_3^3	S_3^4	S_3^5
S_3	S_3	S_3^2	S_3^3	S_3^4	S_3^5	E
S_3^2	S_3^2	S_3^3	S_3^4	S_3^5	E	S_3
S_3^3	S_3^3	S_3^4	S_3^5	E	S_3	S_3^2
S_3^4	S_3^4	S_3^5	E	S_3	S_3^2	S_3^3
S_3^5	S_3^5	E	S_3	S_3^2	S_3^3	S_3^4

In groups having a single generating operation (*Abelian* groups), all operations are in separate classes. Thus, the C_3 and C_3^2 (or C_3^{-1}) operations are not interchanged by σ_h, or any other operation of the group.

3.11 PROBLEM: Test the mutual orthogonality of the representations of the $\mathbf{D}_{3h}$ group.

$\mathbf{D}_{3h}$	E	$2C_3$	$3C_2$	σ_h	$2S_3$	$3\sigma_v$		
A_1'	1	1	1	1	1	1		x^2+y^2, z^2
A_2'	1	1	−1	1	1	−1	R_z	
E'	2	−1	0	2	−1	0	(x,y)	x^2-y^2, xy
A_1''	1	1	1	−1	−1	−1		
A_2''	1	1	−1	−1	−1	1	z	
E''	2	−1	0	−2	1	0	(R_x, R_y)	(xy, yz)

3.11 Solution: Orthogonality requires that for each pair of symmetry species given in the character table the sum of the product of the characters under *each* of the operations of the group should be zero. We will illustrate this test for orthogonality for the A_2'' and E'' symmetry species.

$$\begin{array}{cccccc} E & 2C_3 & 3C_2 & \sigma_h & 2S_3 & 3\sigma_v \\ (1)(2) + & 2(1)(-1) + & 3(-1)(0) + & (-1)(-2) + & 2(-1)(1) + & 3(1)(0) = 0 \end{array}$$

3.12 PROBLEM: Derive the $\mathbf{C}_{3v}$ character table.

3.12 Solution: The $\mathbf{C}_{3v}$ point group has the operations E, $2C_3$, and $3\sigma_v$. All of the σ_v's are in the same class since the C_3 operation carries one σ_v into another successively. Also, both C_3's are in the same class since reflection in a vertical mirror plane reverses the sense of the rotation. Since there are three classes of operations, there will be three distinct irreducible representations. Since the order of the group is six, one of the irreducible representations will be two-dimensional and the other two will be one-dimensional. Examining the transformation matrices for x, y, and z for one operation in each class (See Problem 3.1 for instructions on setting up transformation matrices and for matrices for rotations, see DMA, 3rd ed., p. 109.) we have

$$\begin{bmatrix} 1 & 0 & 0 \\ 0 & 1 & 0 \\ 0 & 0 & 1 \end{bmatrix} \quad \begin{bmatrix} -1/2 & \sqrt{3}/2 & 0 \\ \sqrt{3}/2 & -1/2 & 0 \\ 0 & 0 & 1 \end{bmatrix} \quad \begin{bmatrix} 1 & 0 & 0 \\ 0 & -1 & 0 \\ 0 & 0 & 1 \end{bmatrix}$$
$$E \qquad\qquad\qquad C_3 \qquad\qquad\qquad (\sigma_v)_{xz}$$

From the above it is readily apparent that z is invariant under all of the operations and gives an irreducible representation having a single unit element for each matrix of the representation. This representation is called the totally symmetric representation and given the Mulliken symbol A (or A_1 or A') and such a representation occurs for every group. The remaining two-dimensional rerpresentation is termed an E representation, and is the one to which x and y belong. The sum of the diagonal elements of these matrices gives the character of the representation. The characters for the remaining one-dimensional representation can be obtained from the orthogonality relationships. The character table is shown.

C_{3v}	E	$2C_3$	$3\sigma_v$
A_1	1	1	1
A_2	1	1	−1
E	2	−1	0

Alternatively, recognizing that there is always a totally symmetric representation with all characters = 1, the characters for A_2 can be obtained from othogonality relationships. Recognizing that the character under E must be 2 for the two-diminsional representation, the other two characters are obtained from orthogonality to both A_1 and A_2.

3.13 PROBLEM: Derive the character table for the C_3 group using the following equations, expressing the characters as simple and complex numbers.

$$\varepsilon = e^{(2\pi i/n)} \quad \text{and} \quad \varepsilon^{i\theta} = \cos\theta + i\sin\theta$$

3.13 Solution: Both C_n and S_n groups are cyclic (or Abelian) groups whose representations, and characters, may be developed by taking a single generator which may take as its value the n roots of 1 where n is the order of the group. (See DMA Section 3.5.5 or F. Theobald, *J. Chem. Educ.* **1982**, *59*, 277.) Successive powers of the value taken by the generator give the characters of the remaining elements of the group. The values taken by the generator give the characters of the remaining elements of the group. The values of the n roots of 1 are easily found by geometric construction locating n points regularly on a circle with coordinates chosen along the real number line and the imaginary number line, illustrated below for $n = 3$.

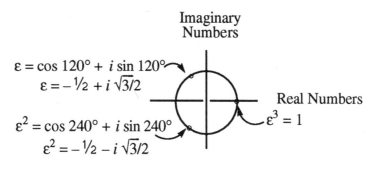

When $\chi(C_3) = 1$, $\chi(C_3^2) = 1$,   and $\chi(C_3^3) = \chi(E) = 1$
When $\chi(C_3) = \varepsilon$, $\chi(C_3^2) = \varepsilon^2$,   and $\chi(C_3^3) = \chi(E) = 1$
When $\chi(C_3) = \varepsilon^2$, $\chi(C_3^2) = \varepsilon$,   and $\chi(C_3^3) = \chi(E) = 1$
The character table for the C_3 group is thus:

C_3	E	C_3	C_3^2
A	1	1	1
E	$\begin{cases} 1 \\ 1 \end{cases}$	$\begin{matrix}(-1/2 + i\sqrt{3}/2) \\ (-1/2 - i\sqrt{3}/2)\end{matrix}$	$\begin{matrix}(-1/2 - i\sqrt{3}/2) \\ (-1/2 + i\sqrt{3}/2)\end{matrix}$

3.14 PROBLEM: Take the following direct products and find the irreducible representations contained.
(a) $E_g \times E_u$ in $\mathbf{D}_{3d}$ (b) $E_g \times T_{1g}$ in $\mathbf{O}_h$ (c) $E_g \times T_{2g}$ in $\mathbf{O}_h$
(d) $E_u \times E_u$ in $\mathbf{O}_h$ (See page 74 for the $\mathbf{O}_h$ character table, $\mathbf{D}_{3d}$ and $\mathbf{O}$ tables are below.)

3.14 Solution: The $\mathbf{D}_{3d}$ group is a subgroup of the $\mathbf{O}_h$ group. Its elements are E, $2C_3$, $3C_2$, i, $2S_6$, and $3\sigma_d$. With fewer classes of symmetry elements than the $\mathbf{O}_h$ group, it has fewer irreducible representations, it has no T representations. The $\mathbf{D}_3$ group is a subgroup of both $\mathbf{D}_{3d}$ and $\mathbf{O}_h$ groups. Since the $\mathbf{D}_3$ group has no inversion operations, its symmetry species do not carry g or u subscripts. In the following solution we make use of the fact that $g \times g = g$ and $u \times u = g$ whereas $g \times u = u$. Accordingly we need examine only products from the rotation groups alone.

(a) To find the irreducible representations contained in the direct product of E_g and E_u under $\mathbf{D}_{3d}$ symmetry we first obtain the characters for the reducible representation by multiplying together the characters under each class.

$\mathbf{D}_3$	E	$2C_3$	$3C_2$
E	2	-1	0
E	2	-1	0
Γ_R	4	1	0

$\mathbf{D}_{3d}$	E	$2C_3$	$3C_2$	i	$2S_6$	$3\sigma_d$
A_{1g}	1	1	1	1	1	1
A_{2g}	1	1	-1	1	1	-1
E_g	2	-1	0	2	-1	0
A_{1u}	1	1	1	-1	-1	-1
A_{2u}	1	1	-1	-1	-1	1
E_u	2	-1	0	-2	1	0

In this case inspection shows that the product is made up of $1A_1 + 1A_2 + 1E$. These will all be ungerade ($g \times u = u$), that is $1A_{1u} + 1A_{2u} + 1E_u$.

(b) In similar fashion

O	E	$6C_4$	$3C_2(=C_4^2)$	$8C_3$	$6C_2$
E	2	0	2	-1	0
T_1	3	1	-1	0	-1
Γ_R	6	0	-2	0	0

$\rightarrow T_1 + T_2$

or $E_g \times T_{1g} \rightarrow T_{1g} + T_{2g}$

O	E	$6C_4$	$3C_2(=C_4^2)$	$8C_3$	$6C_2$
A_1	1	1	1	1	1
A_2	1	-1	1	1	-1
E	2	0	2	-1	0
T_1	3	1	-1	0	-1
T_2	3	-1	-1	0	1

(c) Same results as in **(b)**

(d) Same results as in **(a)** except all products are gerade ($u \times u = g$).

In more complex cases it may be necessary to use the decomposition formula

$$n_\Gamma = \frac{1}{h} \sum_g n_g \chi_R \chi_\Gamma$$

where n_Γ is the number of times a particular irreducible representation is contained in a reducible representation, h is the order of the group, n_g is the number of operations in the class g, χ_R is the character of the reducible representation and χ_Γ that of the irreducible representation under the operations of class g, and the summation is taken over all classes. Using this formula we reduce the set of characters in part **(b)** as follows:

$$\begin{array}{lllllll}
 & E & 6C_4 & 3C_2 & 8C_3 & 6C_2 & \\
n(A_1) = \tfrac{1}{24} & [1(6)1 & + 6(0)1 & + 3(-2)1 & + 8(0)1 & + 6(0)1] & = 0 \\
n(A_2) = \tfrac{1}{24} & [1(6)1 & + 6(0)(-1) & + 3(-2)1 & + 8(0)1 & + 6(0)(-1)] & = 0 \\
n(E) = \tfrac{1}{24} & [1(6)2 & + 6(0)0 & + 3(-2)2 & + 8(0)(-1) & + 6(0)0] & = 0 \\
n(T_1) = \tfrac{1}{24} & [1(6)3 & + 6(0)1 & + 3(-2)(-1) & + 8(0)0 & + 6(0)(-1)] & = 1 \\
n(T_2) = \tfrac{1}{24} & [1(6)3 & + 6(0)(-1) & + 3(-2)(-1) & + 8(0)0 & + 6(0)1] & = 1 \\
\end{array}$$

3.15 PROBLEM: Give the symmetry labels for the p and d orbitals for the point groups of $[Co(NH_3)_6]^{3+}$, $[Co(en)_3]^{3+}$, $[Co(edta)]^-$, and *trans*–$[Cr(Cl)_2(NH_3)_4]^+$.

3.15 Solution: We must first assign the species to the appropriate point group (see Problems 3.2-3.4). We then find how the appropriate function describing the angular part of the wavefunction transforms in the given point group (see discussion in DMA, 3rd ed., Section 3.5.5). Most character tables list these functions at the far right of the table for the appropriate species. We will examine several ways of assigning labels to the orbitals in a given symmetry as an aid to understanding how the information included in character tables is derived.

$[Co(NH_3)_6]^{3+}$ This species has octahedral geometry, and, for freely rotating NH_3 ligands, would belong to the **O_h** point group. From the **O_h** character table on p. 74 we find that x, y, and z transform as T_{1u}; $2z^2 - x^2 - y^2$ and $x^2 - y^2$ as E_g; and xy, xz, and yz as T_{2g}. The symmetry labels become the orbital labels, lower case letters being used—t_{1u}, t_{2g}, and e_g.

[Co(en)$_3$]$^{3+}$ This species belongs to $\mathbf{D}_3$, a subgroup of $\mathbf{O}_h$. We can obtain the symmetry species for the p orbitals in $\mathbf{D}_3$ symmetry by reducing the characters of the T_{1u} symmetry for those classes that belong to $\mathbf{D}_3$.

$\mathbf{D}_3$	E	$2C_3$	$2C_2$		
Γ_p	3	0	-1	$\rightarrow$	$E + A_2$

By examining the behavior of the p_z orbital under $\mathbf{D}_3$ symmetry (it goes into itself under E or C_3 and inverts under C_2) we see that it belongs to A_2 symmetry. The p_x and p_y orbitals must belong to E symmetry species. We obtain the characters of the reducible representation for the d orbitals by adding the appropriate characters for the E_g and T_{2g} representation in $\mathbf{O}_h$ symmetry. Reduction under $\mathbf{D}_3$ symmetry gives the appropriate labels for $\mathbf{D}_3$.

$\mathbf{D}_3$	E	$2C_3$	$3C_2$
Γ_d	$(2+3)$	$(-1+0)$	$(0+1)$

$$n(A_1) = \tfrac{1}{6}[1(5)1 + 2(-1)1 + 3(1)1] = 1$$
$$n(A_2) = \tfrac{1}{6}[1(5)1 + 2(-1)1 + 3(1)(-1)] = 0$$
$$n(E) = \tfrac{1}{6}[1(5)2 + 2(-1)(-1) + 3(1)0] = 2 \quad (2E)$$

The d_{z^2} orbital is invariant under the $\mathbf{D}_3$ operations and belongs to the totally symmetric A_1 species. The d_{xy}, $d_{x^2-y^2}$, and d_{xz}, d_{yz} form two sets of E symmetry species.

[Co(edta)]$^-$ belongs to the $\mathbf{C}_2$ point group, a subgroup of the $\mathbf{O}_h$ group. There are only two symmetry species for this Abelian group of order two, the symmetric A species and the antisymmetric B species.

$\mathbf{C}_2$	E	C_2
A	1	1
B	1	-1

Letting the z axis be the C_2 axis, we examine each orbital for symmetric or antisymmetric behavior under the C_2 operation.

	Behavior under C_2	Symmetry Species
p_z	symmetric—goes into itself	A
p_y	antisymmetric—goes into its negative	B
p_x	antisymmetric—goes into its negative	B
d_{z^2}	symmetric—goes into itself	A
$d_{x^2-y^2}$	symmetric—goes into itself	A
d_{xy}	symmetric—goes into itself	A
d_{xz}	antisymmetric—goes into its negative	B
d_{yz}	antisymmetric—goes into its negative	B

trans-[$Cr(Cl)_2(NH_3)_4$]$^+$ belongs to the D_{4h} group, a subgroup of the O_h group. The correlation of symmetry species between the groups is not as simple as with the previous subgroups of O_h discussed in this problem. Thus, we will work directly with the D_4 character table given below.

D_4	E	$2C_4$	$2C_2(=C_4^2)$	$2C_2'$	$2C_2''$
A_1	1	1	1	1	1
A_2	1	1	1	−1	−1
B_1	1	−1	1	1	−1
B_2	1	−1	1	−1	1
E	2	0	−2	0	0

Here the C_4 axis is chosen to coincide with the z axis and the $2C_2'$ axes are chosen to coincide with the x and y axes. Let us assign symmetry labels to the p and d orbitals in D_{4h} symmetry. We know that the p orbitals are ungerade and the d orbitals are gerade, so we use the D_4 character table to obtain the desired information. Looking at the transformation of each orbital individually we find

D_4	E	C_4	C_4^2	C_2'	C_2''
p_x	1	0	−1	1	0
p_y	1	0	−1	−1	0

Here we have arbitrarily chosen a single C_2' coincident with the x axis. The zero under C_4 indicates that the orbital goes into neither itself nor its negative; "it goes into something else". We find that neither p_x nor the p_y orbital transforms as a symmetry species belonging to D_4, but together they transform as an E species—their characters sum to the characters of E. Accordingly p_x and p_y are labeled as e_u orbitals in D_{4h} symmetry. The remaining orbitals transform as follows

D_4	E	C_4	C_4^2	C_2'	C_2''	
p_z	1	1	1	−1	−1	$\rightarrow a_{2u}$
d_{z^2}	1	1	1	1	1	$\rightarrow a_{1g}$
$d_{x^2-y^2}$	1	−1	1	1	−1	$\rightarrow b_{1g}$
d_{xy}	1	−1	1	−1	1	$\rightarrow b_{2g}$
d_{xz}	1	0	−1	−1	0	$\left.\rule{0pt}{2.2ex}\right\} \rightarrow e_g$
d_{yz}	1	0	−1	1	0	

The reader interested in obtaining the labels of species of atomic orbitals and spectroscopic states by descent from spherical symmetry should consult R. L. DeKock, A. J. Kromminga, and T. S. Zwier, *J. Chem. Educ.* **1979**, *56*, 510.

3.16 PROBLEM: The representations for the same orbitals in centrosymmetric groups and closely related subgroups usually differ only in the dropping of the *g* and *u* subscripts. Explain why the *p* orbitals are T_{1u} in O_h and T_2 in T_d (subscripts 1 and 2).

3.16 Solution: The T_1 or T_2 distinction for **O** (or **O**$_h$) is made using the major axis, C_4, since both have characters of zero under the S_6 operations. Since C_4 is missing for **T**$_d$, the T_1 or T_2 distinction is made using S_4.

In similar vein we find the A_2 labeling in **O**$_h$ is based on the behavior under the S_6 operation. On descending to the **D**$_{4h}$ group the S_6 disappears so the C_4 operation defines the A and B labeling and the A_2 species in **O**$_h$ becomes a B_1 species when the symmetry is lowered to **D**$_{4h}$.

3.17 PROBLEM: Assign the point groups for the following water clusters. (See C. J. Tsai and K. D. Jordan, *J. Phys. Chem.*, **1993**, *97*, 5208.)

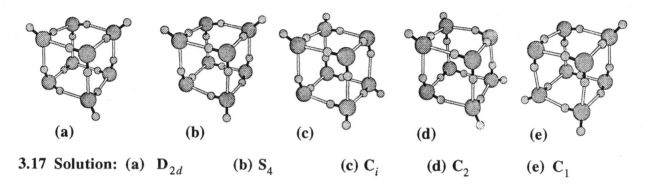

(a)   **(b)**   **(c)**   **(d)**   **(e)**

3.17 Solution: (a) **D**$_{2d}$ (b) **S**$_4$ (c) **C**$_i$ (d) **C**$_2$ (e) **C**$_1$

3.18 PROBLEM: Sketch the following vibrational modes for octahedral XeF$_6$ using vectors to show the directions of motion of atoms for the following stretching modes and identify the Xe orbitals having the corresponding sign patterns.

     (a) A_{1g} (b) E_g (one mode)   (c) T_{1u} (one mode)

3.18 Solution:

(a) A_{1g} (b) E_g (c) T_{1u}

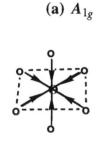 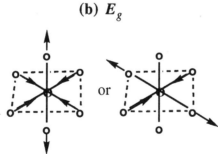 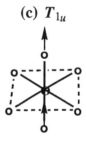

Totally symmetric,   Sign patterns as d_{z^2} and $d_{x^2-y^2}$ (or motion along x or y)
as for s Sign patterns as the p orbitals

3.19 PROBLEM: Assign the symmetry species (representations) for the vibrational modes shown for H$_2$O (C$_{2v}$). Arrows show the directions of motion of the atoms in the yz plane.

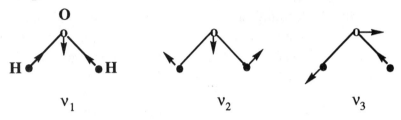

3.19 Solution: v_1 and v_2 are totally symmetric, A_1.
For v_3 vectors are reversed by C_2 and $\sigma_v(xz)$ and the vectors are unchanged by $\sigma_v(yz)$, the plane of the paper. This is the B_2 representation.

3.20 PROBLEM: Instead of using Cartesian coordinates x and y, the position of a point in the xy plane can be specified by the polar coordinates r and ϕ.

(a) What is the mathematical relation between x, y, and r, ϕ?

(b) Clockwise rotation around the z axis through an angle θ takes the point (x, y) into (x', y'). The new polar coordinates are $[r, (\phi - \theta)]$. Express mathematically the relation between (x', y') and the polar coordinates $[r, (\phi - \theta)]$. Convert the relationship into one between (x', y') and (x, y).

(c) Express the results of **b** in matrix notation, thereby obtaining the standard rotation matrix.

3.20 Solution:
(a) $r = \sqrt{x^2 - y^2}$
$\tan \phi = y/x \quad x = r \cos \phi \quad y = r \sin \phi$

(b) $x' = r \cos \phi' = r \cos (\phi - \theta)$
$y' = r \sin \phi' = r \sin (\phi - \theta)$
From the trigonometric relationships for cos and sin of $\phi - \theta$ we have
$x' = r[\cos \phi \cos \theta + \sin \phi \sin \theta]$
or $\quad x' = x \cos \theta + y \sin \theta$

$y' = r[\sin \phi \cos \theta - \cos \phi \sin \theta]$
$y' = y \cos \theta - x \sin \theta \quad$ or $\quad -x \sin \theta + y \cos \theta$

(c) Expressing the above in matrix form we have

$$\begin{bmatrix} x' \\ y' \end{bmatrix} = \begin{bmatrix} \cos \theta & \sin \theta \\ -\sin \theta & \cos \theta \end{bmatrix} \begin{bmatrix} x \\ y \end{bmatrix}$$

3.21 PROBLEM: Nuclei are termed chemically equivalent if permuted by a C_n operation, enantiopically equivalent if permuted only by an S_n operation, and magnetically equivalent if permuted by an operation which leaves all other magnetically active nuclei invariant. Indicate the types of equivalent sets of halogen nuclei found in the following species.

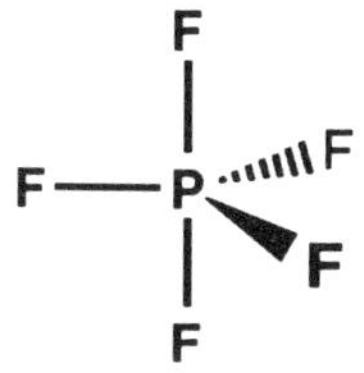

3.21 Solution: We first assign the species to the appropriate point group. Proceeding as in Problem 3.2, we obtain $\mathbf{D}_{3h}$ for PF$_5$, $\mathbf{O}_h$ for SF$_6$, $\mathbf{D}_{2d}$ for the W tetramer, and $\mathbf{C}_{2h}$ for the iodine dimer. We must examine the permutations possible under each operation of the group (see D. H. McDaniel, *J. Phys. Chem.* **1981**, *85*, 471), with the following results:

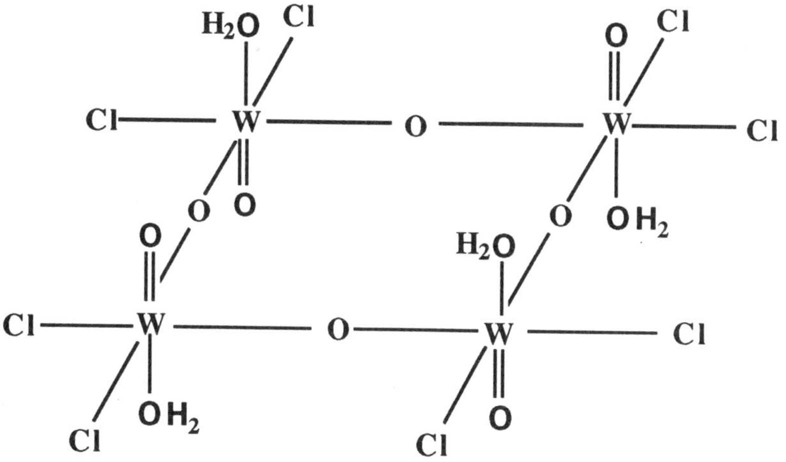

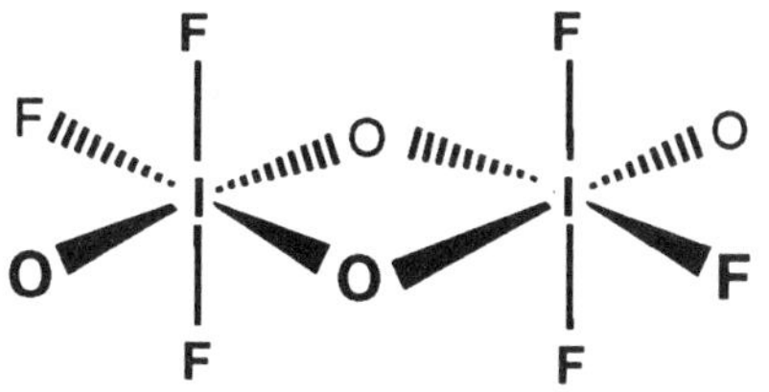

(a) In PF$_5$, the equatorial F are chemically and magnetically equivalent (C_3), the axial F are chemically equivalent (C_2) and magnetically equivalent (σ_h). The magnetic equivalence of the axial F would be removed in a chiral environment, that is, in an optically active solvent.

(b) In SF$_6$ all F are chemically equivalent ($6C_4$, etc.) and magnetically equivalent.

(c) Here the $3C_2$ operations show that there are two sets of chemically equivalent Cl (four Cl having parallel bond directions in each set); the σ_d's or S_4's scramble these two sets giving the result that all Cl atoms are enantiotopically equivalent; all Cl atoms are magnetically unique in a chiral environment, but equivalent in an achiral environment.

(d) There are three sets of F atoms which give chemically equivalent pairs under the C_2 operation; under the i and σ_h we find one set of four enantiotopically equivalent F atoms and one set of two F; all F atoms are magnetically unique.

3.22 PROBLEM: If A and B are symmetry elements belonging to the same class, then $X^{-1}AX = B$ for some member X and its inverse X^{-1}. Show that if all operations of a group commute, the group will have only one-dimensional representations. Give some examples of such groups.

3.22 Solution: If all operations commute, then $X^{-1}AX = X^{-1}XA = A$ for all operations of the group and thus each symmetry element is in a class by itself. Since there are as many different representations as there are classes, the number of representations must equal the number of symmetry elements, that is, the order of the group. Also, since the character under the identity operation for an irreducible representation is the same as the dimension of the matrix representation, and $\sum_\Gamma [\chi_\Gamma(E)]^2 = h$, each of the irreducible representations must be one-dimensional representations. Groups for which all operations commute are termed Abelian groups. All cyclic groups, groups generated by the successive application of one symmetry operation (such as S_n in the case of S_n and C_{nh} groups or C_n for C_n groups) are Abelian groups, but the converse is not necessarily true. Thus, the D_2 group is an Abelian group, but is not a cyclic group. In cases where an E representation is given in the character table for an Abelian group, it may be noted that it consists of a pair of one-dimensional representations.

3.23 PROBLEM: (a) For a reducible representation, demonstrate that the character under any operation R is $\chi(R) = \sum_\Gamma n_\Gamma(R)$ where the sum is over the Γ irreducible representations of the group, n_Γ is the number of times the Γth irreducible representation is contained in the reducible one and $\chi_\Gamma(R)$ is the character of the Γth irreducible representation under the operation R.

(b) Using the result of (a) and the orthogonality of irreducible representations, show that $n_\Gamma = \frac{1}{h}\sum_\Gamma \chi(R)\chi_\Gamma(R)$. This formula allows you to determine the number of times the Γth irreducible representation is contained in any reducible representation.

3.23 Solution: (a) The demonstration follows from the rules for matrix addition and multiplication of a matrix by a constant. For matrix addition:

$$\begin{bmatrix} a_{11} & a_{12} & \cdots & a_{1n} \\ a_{21} & a_{22} & & \\ \vdots & & \ddots & \\ a_{n1} & & & a_{nn} \end{bmatrix} + \begin{bmatrix} b_{11} & b_{12} & \cdots & b_{1n} \\ b_{21} & b_{22} & & \\ \vdots & & \ddots & \\ b_{n1} & & & b_{nn} \end{bmatrix} = \begin{bmatrix} a_{11}+b_{11} & a_{12}+b_{12} & \cdots & a_{1n}+b_{1n} \\ a_{21}+b_{21} & a_{22}+b_{22} & & \\ \vdots & & \ddots & \\ a_{n1}+b_{n1} & & & a_{nn}+b_{nn} \end{bmatrix}$$

and

$$p\begin{bmatrix} a_{11} & a_{12} & \cdots & a_{1n} \\ a_{21} & a_{22} & & \\ \vdots & & \ddots & \\ a_{n1} & \cdots & & a_{nn} \end{bmatrix} = \begin{bmatrix} pa_{11} & pa_{12} & \cdots & pa_{1n} \\ pa_{21} & pa_{22} & & \\ \vdots & & \ddots & \\ pa_{n1} & \cdots & & pa_{nn} \end{bmatrix}$$

We illustrate with a sum of three matrices which can be generalized to any number. Suppose the matrix for the operation R in the reducible representation is

$$\begin{bmatrix} d_{11} & d_{12} & \cdots & d_{1n} \\ d_{21} & d_{22} & & \\ & & & \\ d_{n1} & \cdots & & d_{nn} \end{bmatrix} = n_a \begin{bmatrix} a_{11} & a_{12} & \cdots & a_{1n} \\ a_{21} & a_{22} & & \\ & & & \\ a_{n1} & & & a_{nn} \end{bmatrix} + n_b \begin{bmatrix} b_{11} & b_{12} & \cdots & b_{1n} \\ b_{21} & b_{22} & & \\ & & & \\ b_{n1} & \cdots & & b_{nn} \end{bmatrix} + n_c \begin{bmatrix} c_{11} & c_{12} & \cdots & c_{1n} \\ c_{21} & c_{22} & & \\ & & & \\ c_{n1} & & & c_{nn} \end{bmatrix}$$

$$= \begin{bmatrix} n_a a_{11} + n_b b_{11} + n_c c_{11} & n_a a_{12} + n_b b_{12} + n_c c_{12} & \cdots \\ n_a a_{21} + n_b b_{21} + n_c c_{21} & n_a a_{22} + n_b b_{22} + n_c c_{22} & \\ \cdots & & & n_a a_{nn} + n_b b_{nn} + n_c c_{nn} \end{bmatrix}$$

The character of the matrix for the reducible representation is

$$d_{11} + d_{22} + d_{33} + \ldots + d_{nn} = \sum_{j=1}^{n} d_{jj} = \chi(R) =$$

$$\sum_{j=1}^{n} n_a a_{jj} + n_b b_{jj} + n_c c_{jj} = n_a \sum_{j=1}^{n} a_{jj} + n_b \sum_{j=1}^{n} b_{jj} + n_c \sum_{j=1}^{n} c_{jj} =$$

$$n_a \chi_a(R) + n_b \chi_b(R) + n_c \chi_c(R) = \sum n_\Gamma \chi(R)$$

This demonstration assumes all the matrices to be the same dimension and so is not a general proof. However, the general result is obvious since matrices of less than maximum dimension could be expressed as a matrix of maximum dimension containing zeros in some rows and columns for example $\begin{bmatrix} 1 & 1 \\ 1 & 1 \end{bmatrix}$ could be expressed as $\begin{bmatrix} 1 & 1 & 0 \\ 1 & 1 & 0 \\ 0 & 0 & 0 \end{bmatrix}$, etc.

(b) Since we know for each matrix in the reducible representation
$$\chi(R) = \sum_\Gamma n_\Gamma \chi_\Gamma(R),$$

We can multiply by $\chi_{\Gamma'}(R)$ and sum over all operations R in the group.

$$\sum_R \chi(R)\chi_{\Gamma'}(R) = \sum_\Gamma \sum_R n_\Gamma \chi_\Gamma(R) \chi_{\Gamma'}(R)$$

But, $\sum_R \chi_\Gamma(R) \chi_{\Gamma'}(R) = 0$ if $\Gamma \neq \Gamma'$

$$= h \text{ if } \Gamma = \Gamma'$$

Then $\sum_R \chi(R)\chi_\Gamma(R) = \sum_\Gamma n_\Gamma h = n_\Gamma h$

So, $n_\Gamma = \dfrac{1}{h} \sum_R \chi(R)\chi_\Gamma(R)$

3.24 PROBLEM: On page 60 there are figures which can be copied and used for the construction of a tetrahedron, an octahedron, and a cube. Identify the point groups for each, considering the shading of the faces.

3.24 Solution: The *regular* tetrahedron belongs to the T_d group since it has more than one noncoincident C_3 axis and a dihedral mirror plane that carries one C_3 axis into another. The *shaded* tetrahedron retains all of the proper rotational axes of the regular tetrahedron, but loses all of the improper axes of rotation. Thus, it belongs to the T group—a pure rotation group. In similar fashion we find that the shaded cube has several noncoincident C_4 axes, but no improper axes, and belongs to the O group. The shaded octahedron is somewhat more interesting since the shading reduces the number of both proper and improper elements. The highest rotational axis is now a C_3 axis, of which there are several, but a center of inversion is retained, making this an example of a T_h group.

3.25 PROBLEM: On page 61 there are figures which can be copied and used for the construction of a dodecahedron with trigonal faces, a dodecahedron with pentagonal faces, and an icosahedron. Determine the point groups for the polyhedra taking into account any shading of the faces. Ignoring the shading, which of the polyhedra are conjugate Platonic solids?

3.25 Solution: The trigonal dodecahedron is not a Platonic solid since all faces are not equivalent. It has three mutually perpendicular C_2 axes, and several dihedral mirror planes, so it belongs to the D_{2d} group. The icosahedron with shaded faces has no mirror planes or other S_n axes, but it does have several C_5 axes. It belongs to an I group. The pentagonal dodecahedron also has several C_5 axes, but with no shading it also has an inversion center, thus it belongs to the I_h group. The pentagonal dodecahedron and the regular icosahedron are conjugate solids, that is, one can be formed from the other by connecting the centers of the faces.

3.26 PROBLEM: The buckyball is shown in the figure here. Determine the point group. A figure on page 60 can be copied and cut out for construction of a buckyball.

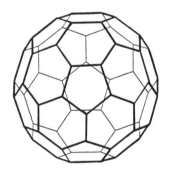

Diagram of the C_{60} molecule.

3.26 Solution: There are C_5 axes through the centers of all pentagons and C_3 axes through the centers of all hexagons giving the I rotation group. One of many symmetry planes is evident through the C—C bond shared by the two front hexagons. This identifies it as the I_h group.

3.27 PROBLEM:
(a) Connect the points at the centers of the faces of a tetrahedron. What is the conjugate of a tetrahedron?
(b) What is the conjugate of the pentagonal dodecahedron?

3.27 Solution:
(a) The tetrahedron is its own conjugate.
(b) The conjugate of the pentagonal dodecahedron is the icosahedron, both I_h. (We thank C. J. Tsai, University of Pittsburgh for the pentagonal dodecahedron and its conjugate.)

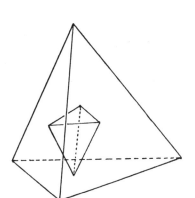

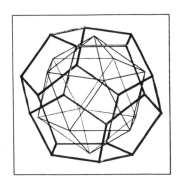

Dodecahedron

Icosahedron

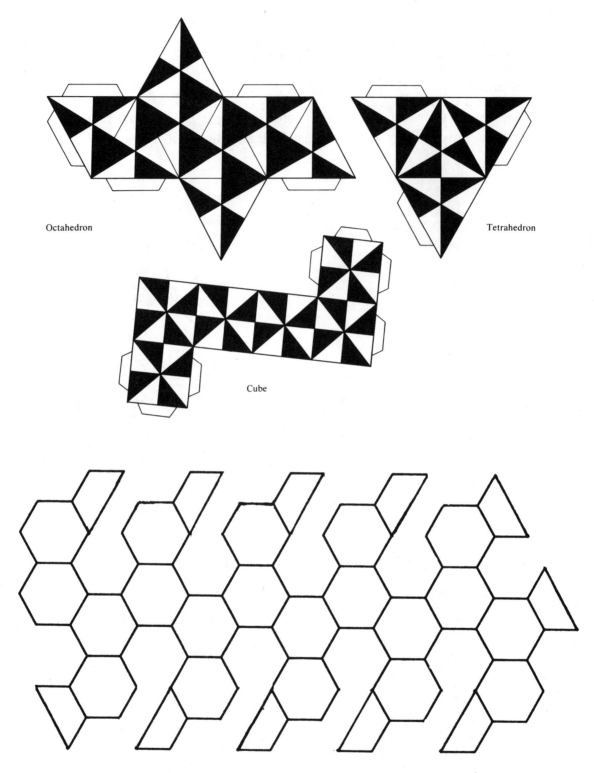

Octahedron

Tetrahedron

Cube

Buckyball (C_{60})

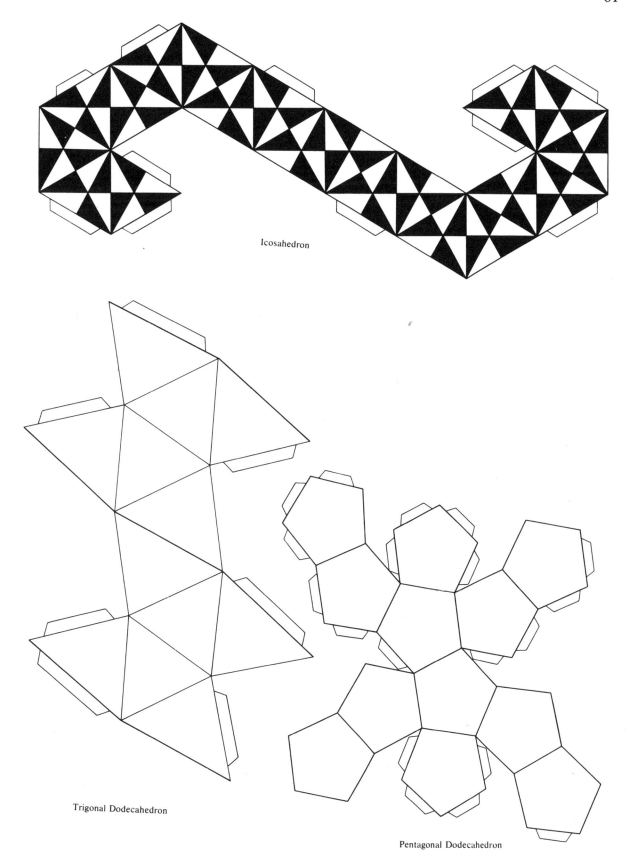

Icosahedron

Trigonal Dodecahedron

Pentagonal Dodecahedron

II. BONDING AND STRUCTURE
4
Discrete Molecules
Molecular Orbitals

4.1 Problem: Give the bond order and the number of unpaired electrons for Be_2^+, B_2^+, C_2^+, O_2, O_2^+, O_2^-, O_2^{2-}.

4.1 Solution: The molecular orbital (MO) energy-level diagram for diatomic molecules as a function of atomic number is shown below.

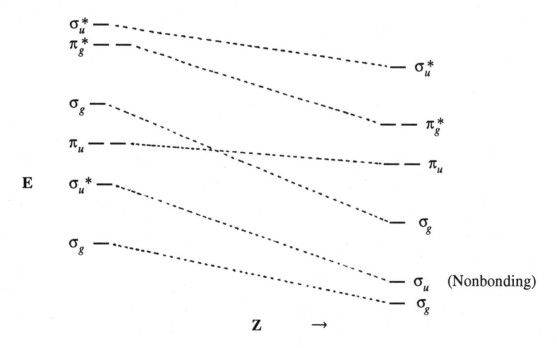

Configuration interaction (or, if you prefer, hybridization) mixes the σ_g orbitals of the lighter diatomic molecules as shown on the left—this ordering is shown through N_2. Electrons populate the orbitals in accordance with Hund's rules of maximum spin-multiplicity; the bond order is obtained by substracting the number of antibonding electrons from the number of bonding electrons and dividing by two. The answers for our problem are thus:

	MO Configuration	Bond Order	Number of unpaired e
Be_2^+	$\sigma_s^2 \sigma_s^{*1}$	0.5	1
B_2^+	$\sigma_s^2 \sigma_s^{*2} \pi^1$	0.5	1
C_2^+	$\sigma_s^2 \sigma_s^{*2} \pi^2 \pi^1$	1.5	1
O_2	$\sigma_s^2 \sigma_s^{*2} \sigma_p^2 \pi^4 \pi_x^{*1} \pi_y^{*1}$	2	2
O_2^+	$\sigma_s^2 \sigma_s^{*2} \sigma_p^2 \pi^4 \pi^{*1}$	2.5	1
O_2^-	$\sigma_s^2 \sigma_s^{*2} \sigma_p^2 \pi^4 \pi_x^{*2} \pi_y^{*1}$	1.5	1
O_2^{2-}	$\sigma_s^2 \sigma_s^{*2} \sigma_p^2 \pi^4 \pi^{*4}$	1	0

4.2 Problem: Sketch sigma bonding orbitals that result from combinations of the following orbitals on separate atoms: p_z and d_{z^2}, s and p_z, $d_{x^2-y^2}$ and $d_{x^2-y^2}$ (let the σ bond be along the x axis in the latter case.)

4.2 Solution:

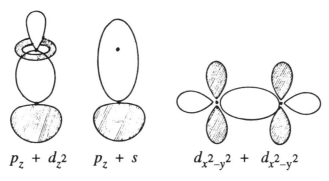

$p_z + d_{z^2}$ $p_z + s$ $d_{x^2-y^2} + d_{x^2-y^2}$

4.3 Problem: Sketch π bonding orbitals that result from combination of the following orbitals on separate atoms with σ bonding along with the z axis:
(a) p_x and p_x,   (b) p_x and d_{xz},   (c) d_{xz} and d_{xz}.

4.3 Solution:

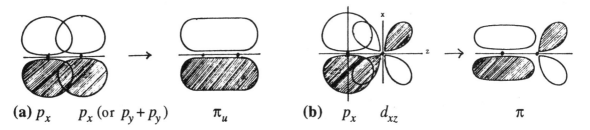

(a) p_x p_x (or $p_y + p_y$) π_u (b) p_x d_{xz} π

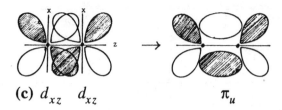

(c) d_{xz} d_{xz} → π_u

4.4 Problem: Sketch a delta bond for an X—Y molecule, identifying the atomic orbitals and showing the signs of the amplitudes of the wavefunctions (signs of the lobes).

4.4 Solution:

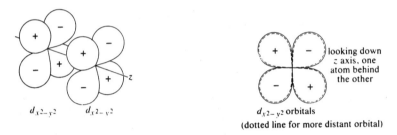

$d_{x^2-y^2}$ $d_{x^2-y^2}$

$d_{x^2-y^2}$ orbitals (dotted line for more distant orbital)

looking down z axis, one atom behind the other

4.5 Problem: Explain why the CO ligand donates the lone pair on carbon in bonding to metals, rather than the pair on oxygen.

4.5 Solution: The lone pair on the more electronegative O of CO has more s character and is lower in energy than the lone pair on C. The more loosely bound electron pair on C is the pair donated to metals in forming metal carbonyl complexes.

4.6 Problem: Write the molecular orbital (MO) configurations (e.g., σ_{1s}^2 for H_2) and give the bond orders of NO^+, NO, and NO^-. Which of these species should be paramagnetic?

4.6 Solution:

		Bond order		Isoelectronic with
NO^+	$\sigma_s^2 \sigma_s^{*2} \pi_x^2 \pi_y^2 \sigma^2(p_z)$	3	Diamagnetic	N_2
NO	$\sigma_s^2 \sigma_s^{*2} \pi_x^2 \pi_y^2 \sigma^2(p_z) \pi_x^{*1}$	2.5	Paramagnetic	O_2^+
NO^-	$\sigma_s^2 \sigma_s^{*2} \pi_x^2 \pi_y^2 \sigma^2(p_z) \pi_x^{*1} \pi_y^{*1}$	2.0	Paramagnetic	O_2

4.7 Problem: Why must atomic orbitals (AO's) have very similar energies to form MO's? (Consider the case of combining orbitals of two atoms having much different energies as indicated by ionization energies or electronegativities.)

4.7 Solution: If orbitals on atoms X and Y, differing greatly in energy, are occupied by one electron in each and the elements are combined, the electron pair will occupy the lower energy orbital. If X = Na and Y = F, in combination the electron of Na is transferred to the lower energy orbital of F, giving Na^+ and F^-, not a covalent bond

4.8 Problem:
(a) Draw an MO energy-level diagram for the MO's that would arise in a diatomic molecule from the combination of unhybridized d orbitals. Label the AO's being combined and the resulting MO's. Let the z axis lie along the σ bond.
(b) Sketch the shape of these orbitals.

4.8 Solution:

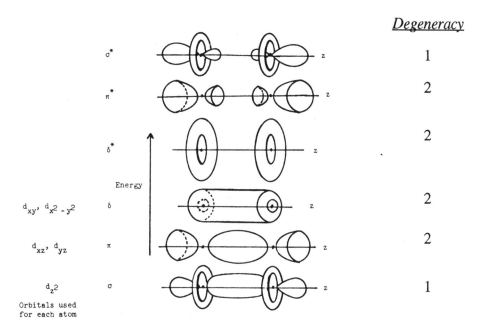

The σ bonding interaction should be most favorable (lowest in energy) because of ideal, directed overlap. Overlap is poorer for π interaction and poorer still for the parallel lobes giving δ bonding. The antibonding orbitals increase in energy in the reverse order $\delta^* < \pi^* < \sigma^*$.

4.9 Problem:
Account for the differences in dissociation energies and bond lengths on addition and removal of an electron from
(a) O_2 (b) N_2 (c) CO (d) NO

	O_2^+	O_2	O_2^-	N_2^+	N_2	N_2^-
Bond energy	642.9	493.6	395.0	840.7	941.7	765 kJ/mol
Bond length	111.6	120.8	(135)	109.8	111.6	119 pm

	CO$^+$	CO	CO$^-$	NO$^+$	NO	NO$^-$
Bond energy	804.5	1070.2	784	1046.9	626.9	487.8 kJ/mol
Bond length	111.5	112.8	—	106.3	115.1	125.8 pm

4.9 Solution: In general, we find that removing an electron (e^-) from a diatomic species tends to decrease interelectronic repulsion and thus to shorten the bond length. This factor may be opposed if the e^- being removed is being taken from a bonding orbital—as is the case on going from N_2 to N_2^+—and little effect is observed. Shorter bond lengths mean stronger interactions among all overlapping orbitals, not just the set involving the e^- being removed or added. The e^- being removed or added will come from the highest occupied molecular orbital (HOMO) or go into the lowest unoccupied MO (LUMO). These frontier orbitals control the chemistry of these species and are well worth looking at in detail.

(a) With O_2 both HOMO and the LUMO are identical, that is the π_g^* obritals. Since these orbitals are antibonding, removal of an e^- increases the bond strength whereas addition of an e^- decreases the bond strength, the former effect being slightly larger.

(b) With N_2 the HOMO is the σ_g orbital which is only slightly bonding. In valence bond terminology we would say there is strong mixing between the lone pairs and the σ bonding orbital. Removal of an e^- from this orbital decreases the bond energy somewhat and has little effect on the bond length. The LUMO is a π_g^* orbital. Adding an e^- here is equivalent to removing an e^- from a π_u bonding orbital, and results in a decrease in bond energy and an increase in bond length.

(c) CO is isoelectronic with N_2 and might be expected to behave in a smiliar fashion. It does so but the effects are larger. The σ orbital is somewhat more bonding in CO [compare ionization energies (IE's) of CO and C to those of N_2 and N]. Another complication is that in comparing dissociation energies we must consider the energy of the products. CO^+ will dissociate to C^+ and O, but CO^- will dissociate to C and O^-. Accordingly the antibonding character of the π^* orbital in CO must be gauged by comparing the electron affinity (EA) of CO with that of O. The data presented here allow the EA of CO to be calculated if we know that of O, and it turns out that it costs approximately 125 kJ/mol to force an electron onto C.

	C	CO	N	N_2	O
Ionization Energy (eV)	11.26	14.01	14.53	15.58	13.62
Electron Affinity (eV)	1.26	−1.5	−0.07	−1.90	1.46

(d) With NO we again have a case where the HOMO and LUMO are the same, the π^* orbitals. Here the asymmetry in the dissociation products of NO^+ and NO^- is particularly noticeable. The high IE of N leads to a high dissociation energy of NO^+, so we find that removing the π^* e^- of NO increases the dissociation energy much more than adding a π^* e^- decreases the dissociation energy.

4.10 Problem: The bond dissociation energy of C_2 (599 kJ/mol) decreases slightly on forming C_2^+ (513 kJ/mol) and increases greatly on forming C_2^- (818 kJ/mol). Why is the change much greater for the addition of an electron?

4.10 Solution: $C_2\ \sigma_{2s}^2\ \sigma_{2s}^{*2}\pi^4\sigma_{2p}^0$. Both the π and σ_{2p} orbitals are suitable for bonding, so the direction of change in bond energies are those anticipated. Ionization to form C_2^+ removes a π bonding electron. The electron added to form C_2^- is a σ bonding electron, giving a B.O. of 2.5. The addition of a σ bonding electron involving directed overlap has a greater effect than removal of a bonding π electron. (See Problem 4.9.)

4.11 Problem: Draw a Double Quartet representation (see DMA, 3rd ed., p. 163f) for ClO_2. Is resonance required for the representation?

4.11 Solution: Resonance is required to give equivalent Cl—O bonds.

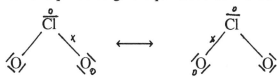

4.12 Problem: The formation of OF_2 from F_2 and O_2 is endothermic. Compare the Double Quartet descriptions of OF_2, F_2, and O_2 with respect to localization of electron pairs.

4.12 Solution: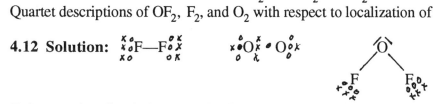

F_2 has one localized electron pair, O_2 has none, and OF_2 has four localized pairs (2 bonds and 2 lone pairs on O).

4.13 Problem: The BH^- group is isoelectronic with CH. What would be the effect on the π MO's and energy levels of $C_5H_5^-$ if one CH group were substituted by

(a) BH^- giving the planar borole $C_4BH_5^-$?
(b) NH^+ giving the planar pyrrole $C_4NH_5^+$?

4.13 Solution: In both cases the symmetry of C_5H_5 is reduced to C_{2v}. Inspection of the character table (Problem 4.18) reveals that there are no degenerate irreducible representations in this point group. Hence, the MO's of e symmetry of C_5H_5 will be split.

(a) Suppose B is placed at the 1 position. Since B is less electronegative than C, any MO which has a contribution from p_1 will be raised in energy since the B π orbital is less stable than a C π orbital. The C_2 axis lies in the plane of the molecule and passes through B. Since the π AO's are all perpendicular to the molecular plane, each AO changes sign by the C_2 operation; however, the linear combinations do not necessarily do so. If we choose the xz plane as the molecular plane, all the MO's of $C_4BH_5^-$ will change sign on reflection in this plane so that all orbitals will have either a_2 or b_2 symmetry. Hence, $\Psi(a_2'')$ becomes $\Psi(1b_2)$; $\Psi(e_{1a}'')$ transforms as $\Psi(1a_2)$ and $\Psi(e_{1b}'')$ is $\Psi(2b_2)$. Similarily, $\Psi(e_{2a}'')$ becomes $\Psi(3b_2)$ and $\Psi(e_{2b}'')$ is $\Psi(2a_2)$. The energy-level diagram looks as shown below. From this diagram, the aromatic species corresponding to $C_5H_5^-$ is $C_4BH_5^{2-}$.

(b) For N substitution, the N π orbital is more stable than that of C. Hence, the composition of the MO's is not changed, but the energy splittings for e's are in the opposite sense to that for the borole—i.e., $1b_2 < 2b_2 < 1a_2 < 3b_2 < 2a_2$. Mixing among the b_2's could conceivably reverse some of the orderings by stabilizing the low-energy members and destabilizing high-energy; this could also happen with the a_2's. In contrast, mixing among borole orbitals of the same symmetry would maintain the ordering shown above. In this case, the aromatic six π-electron species would be neutral C_4NH_5.

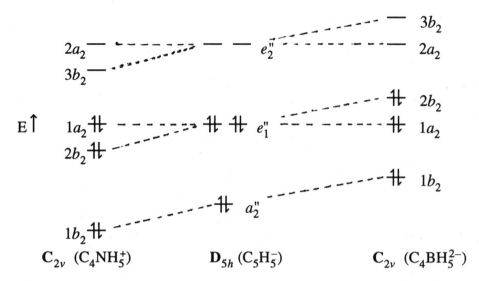

4.14 Problem: The Lewis structure for CO_2 shows four lone electron pairs of on the oxygen. The MO description has two nonbonding pairs. How do you reconcile the two descriptions?

4.14 Solution: The valence bond structure of CO_2 ($\ddot{\underset{..}{O}}=C=\ddot{\underset{..}{O}}$) shows electon pairs on individual atoms or shared between two atoms. We might consider the σ bond for each oxygen to use an sp orbital. The other sp orbital (using the p_z orbital with the major lobe of the hybrid directed away from the C) can accommodate a lone electron pair. One p orbital (p_x or p_y) of each oxygen is used to form a double bond. The other (p_x or p_y) orbital contains a lone pair.

In the MO description nonbonding pairs arise because orbitals are too low in energy for bonding, because the symmetry of the orbitals do not match those of the adjacent atoms, or there is insufficient overlap. There are two combinations of oxygen p_x and p_y orbitals. The e_{1u} combinations form the two π bonds. The e_{1g} combinations are nonbonding because there are no e_{1g} orbitals on C. These accommodate two lone pairs localized on the oxygens. The lower energy a_{1g} and a_{1u} electron pairs are nonbonding largely localized on oxygens. The higher energy a_{1g} and a_{1u} electron pairs form the bonds. The nonbonding a_{1g} and a_{1u} pairs are probably largely oxygen s electrons. They are nonbonding because of their very low energy and this results because of the very high electronegativity of O compared to C. Another possible interpretation is similar to the valence bond description, using oxygen sp hybrids to form the σ bonds. These hybrids used for bonding have the major lobes directed toward C ("in"). The other hybrids have the major lobes directed away from C ("out") and accommodate lone pairs. The MO energy-level diagram for CO_2 shown in Figure 4.14 in DMA, 3rd ed., p. 169 uses the unhybridized description.

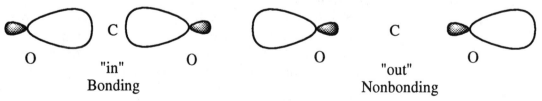

4.15 Problem: The preferred electron dot structure for NO_2 shows five lone pairs on the oxygens and one unshared electron on N. Show how this description is consistent with the MO description.

4.15 Solution: The bonding orbitals for NO_2 are $2a_1$ and $2b_2$ (σ) and b_1 (π). (See Figure below.) The $\cdot N \begin{smallmatrix} \diagup O \\ \diagdown O \end{smallmatrix}$ very low energy a_1 and b_2 (from s) orbitals are largely localized on the oxygens because of their low energy. The nonbonding a_2, $3a_1$, and $3b_2$ orbitals (from p_y) are localized on the oxygens. These are the orbitals for the five lone pairs on O's. The $4a_1$ nonbonding orbital is localized largely on N to accomodate the unshared e^-. Combinations of AO's are shown below. Combinations of AO's are shown below, see the energy-level diagram in Problem 4.18 for NO_2^-.

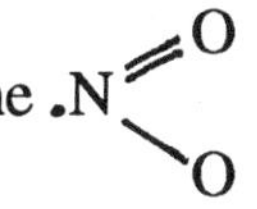

4.16 Problem: In forming NO_2^+ from NO_2, from which MO is the electron lost? On which atom is this electron localized? To which orbital is an electron added in forming NO_2^-? What are the consequences, in terms of molecular shape, of the loss and gain of an electron?

4.16 Solution: In forming NO_2^+ from NO_2 the e^- is lost from the nonbonding HOMO a_1 orbital, largely localized on N (see Problem 4.15). To form NO_2^- an e^- is added to this orbital. NO_2^+ is linear (CO_2 structure) and NO_2^- is angular with a smaller bond angle because the directed nonbonding a_1 orbital has two e^-, rather than one as in NO_2.

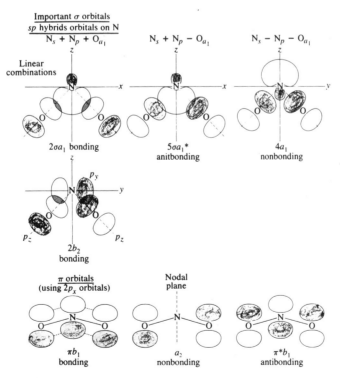

Molecular orbitals for NO_2

4.17 Problem: Why is the first excited state of BeH_2 bent, whereas that of BH_2 is linear? (Consider the Walsh diagram.)

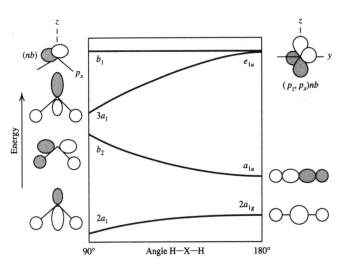

Walsh correlation diagram for XH_2

4.17 Solution: The lowest energy for any population of orbitals will determine the geometry. H:Be:H is linear with the configuration $2a_{1g}^2 a_{1u}^2$. Bending the molecule causes a faster increase in the a_{1u} orbital (relabeled b_2 in C_{2v} geometry; as the H atoms come closer their interaction is antibonding) than a decrease in the a_{1g} orbital (which loses the g label in C_{2v}). On excitation of BeH_2, the energy of the electron in e_{1u} orbital is lowered on bending by going into $3a_1$ (e_{1u} splits into a_1 and b_1 orbitals in C_{2v}). The decrease in energy of $3a_1^1$ and $2a_1^2$ more than compensates for the increase in the a_u^1 orbital that becomes b_2^1 on bending, so the excited state of BeH_2 is bent. For BH_2 the five electrons give $2a_1^2 b_2^2 \, 3a_1^1$ for a bent molecule. Promotion of the $3a_1$ electron goes into b_1. The b_1 orbital is insensitive to bond angle and the greater decrease in energy of b_2 (bent) $\rightarrow a_{1u}$ (180°) than the increase for $2a_1$ (bent) $\rightarrow a_{1g}$ (180°) causes the excited BH_2 to be linear.

4.18 Problem: Apply the group theoretical treatment to obtain the bonding description for NO_2^-.

4.18 Solution: The NO_2^- ion has C_{2v} symmetry.

C_{2v}	E	C_2	$\sigma_v(xz)$	$\sigma_v'(yz)$
A_1	1	1	1	1
A_2	1	1	−1	−1
B_1	1	−1	1	−1
B_2	1	−1	−1	1

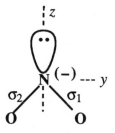

Examining the σ orbitals unchanged by the group operations, we find they transform as $A_1 + B_2$:

	E	C_2	σ_{xz}	σ_{yz}
Γ_σ	2	0	0	2

$= A_1 + B_2$

For N, s and p_z (represented in sketches on p. 69 as an *sp* hybrid) belong to a_1 and can combine with the O a_1 group orbital to give bonding (σ), nonbonding, and antibonding MO's. The N p_y orbital (b_2) combines with the O b_2 (p_z) group orbital to form the second σ bond. The O p_y group orbitals transform in the same way as σ_1 and σ_2 ($A_1 + B_2$), but these are nonbonding because of poor overlap with the N orbitals. The O p_x group orbitals belong to A_2 and B_1:

	E	C_2	σ_{xz}	σ_{yz}
Γ_π	2	0	0	-2

$= A_2 + B_1$

The a_2 group orbital is nonbonding since there is no a_2 orbital for N. The π bonding MO is b_1. The low energy oxygen s orbitals are nonbonding. Sketches of the orbitals are given on p. 69 and the energy-level diagram is presented below.

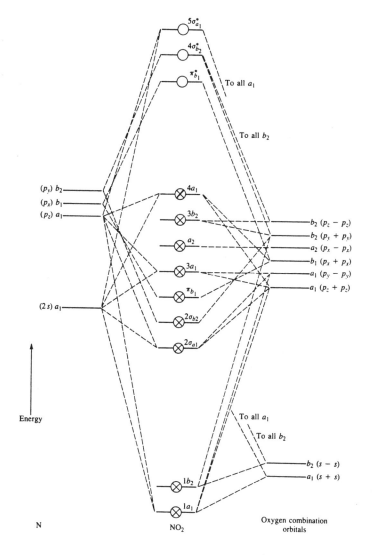

Qualitative MO energy-level diagram for NO_2^-

4.19 Problem: Apply the group theoretical treatment to obtain the MO description for σ bonding in PF_5 (D_{3h}).

4.19 Solution: The σ orbitals are in two nonequivalent sets—2 axial and 3 equatorial. Those within a set are interchanged by symmetry operations of the D_{3h} group, but those of one set are not interchanged with those of the other set by any operation of the group. The D_{3h} character table is given on p. 47.

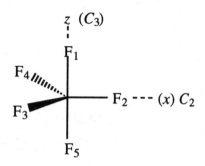

The σ orbitals unchanged by group operations are:

D_{3h}	E	C_3	C_2	σ_h	S_3	σ_v	
$\Gamma\sigma_{ax}$	2	2	0	0	0	2	$= A_1' + A_2''$
$\Gamma\sigma_{eq}$	3	0	1	3	0	1	$= A_1' + E'$

These are the representations for the F group orbitals and the P σ bonding orbitals. The atomic orbitals on P belonging to these representations can be identified from the D_{3h} character table.

a_1'	a_2''	e'
s	p_z	(p_x, p_y)
d_{z^2}		$(d_{xy}, d_{x^2-y^2})$

We need both s and d_{z^2} (they are $2a_1'$), $p_z(a_2'')$ and either (p_x, p_y) or $(d_{xy}, d_{x^2-y^2})$, or some combination of the two e' pairs. Since the p orbitals are much lower in energy, we shall use (p_x, p_y). We do not need to apply the projection operator method to obtain the axial group orbitals, since one is totally symmetric ($\sigma_1 + \sigma_5$) and the other ($\sigma_1 - \sigma_5$) is (a_2''), changing sign for σ_h, C_2, and S_3. Applying the **projection operator** method (see DMA, 3rd ed., p. 179f) for the equatorial orbitals, we have:

D_{3h}	E	C_3	C_3'	C_2	C_2'	C_2''	σ_h	S_3	S_3'	σ_v	σ_v'	σ_v''
σ_2	σ_2	σ_3	σ_4	σ_2	σ_4	σ_3	σ_2	σ_3	σ_4	σ_2	σ_4	σ_3

Multiplying by characters for a_1' (all +1), summing, simplifying, and normalizing we get:
$$a_1' = \frac{1}{\sqrt{3}}(\sigma_2 + \sigma_3 + \sigma_4)$$

For e', we get: $e' = (2\sigma_2 - \sigma_3 - \sigma_4)$ (one of the e' pair)

Performing the C_3 operation on this orbital gives $(-\sigma_2 + 2\sigma_3 - \sigma_4)$, but we have only interchanged superscripts. Multiplying this by 2 and adding to e' above, we get:

$$
\begin{array}{ll}
(e')_a & 2\sigma_2 - \sigma_3 - \sigma_4 \\
 & -2\sigma_2 + 4\sigma_3 - 2\sigma_4 \\
\hline
(e')_b & 3\sigma_3 - 3\sigma_4 \quad \text{or} \quad \sigma_3 - \sigma_4
\end{array}
$$

Normalizing the 2 LCAO,

$$(e')_a = \frac{1}{\sqrt{6}}(2\sigma_2 - \sigma_3 - \sigma_4) \qquad (e')_b = \frac{1}{\sqrt{2}}(\sigma_3 - \sigma_4)$$

The equatorial orbitals are the same as the σ orbitals for BF_3. They are sketched below. The P a_1' orbital for bonding to the axial F is $(s + d_{z^2})$, giving better overlap. The a_2'' group orbital overlaps with p_z. The e' group orbitals overlap with p_x and p_y.

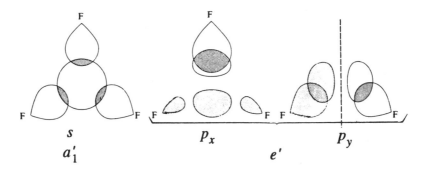

Bonding σ molecular orbitals for the equatorial bonds

4.20 Problem: Apply the group theoretical treatment to obtain the MO bonding description for σ bonding for SF_6 (O_h) using the s, p, and d orbitals on S.

4.20 Solution: We can look at the transformation properties of the central atom orbitals to identify their representations, but we know that the s orbital is totally symmetric (A_{1g}), the p orbitals are T_{1u}, and the d orbitals are E_g and T_{2g} (O_h character table below). We obtain the representations of the LGO's by examining the effects of all classes of operations on the individual ligand orbitals. The result for the vectors numbered as shown is given in the table for the O group. We can add the subscripts g or u by examining the effects of i and/or planes of symmetry. The reducible representation for the σ vectors reduces to $A_1 + E + T_1$ or, for O_h, $A_{1g} + E_g + T_{1u}$. Using the corresponding sulfur orbitals as templates, we can sketch the LGO's as **shown**. The sulfur t_{2g} orbitals are nonbonding.

74

Orientation of the ligand σ orbitals for SF$_6$

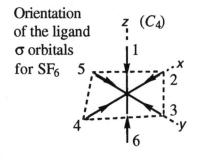

O_h	E	$8C_3$	$6C_2$	$6C_4$	$3C_2(=C_4^2)$	i	$6S_4$	$8S_6$	$3\sigma_h$	$6\sigma_d$	
A_{1g}	1	1	1	1	1	1	1	1	1	1	$x^2+y^2+z^2$
A_{2g}	1	1	−1	−1	1	1	−1	1	1	−1	
E_g	2	−1	0	0	2	2	0	−1	2	0	$(2z^2-x^2-y^2, x^2-y^2)$
T_{1g}	3	0	−1	1	−1	3	1	0	−1	−1	
T_{2g}	3	0	1	−1	−1	3	−1	0	−1	1	(xz, yz, xy)
A_{1u}	1	1	1	1	1	−1	−1	−1	−1	−1	
A_{2u}	1	1	−1	−1	1	−1	1	−1	−1	1	
E_u	2	−1	0	0	2	−2	0	1	−2	0	
T_{1u}	3	0	−1	1	−1	−3	−1	0	1	1	(x, y, z)
T_{2u}	3	0	1	−1	−1	−3	1	0	1	−1	

Effects of the Symmetry Operations of the **O** Point Group on the Ligand Orbitals of an Octahedral Complex.

O	E	C_4^z	$C_2^z(C_4^2)$	C_3	C_2'	i (for O_h)
σ_1	1	1	1	0	0	0
σ_2	1	0	0	0	0	0
σ_3	1	0	0	0	0	0
σ_4	1	0	0	0	0	0
σ_5	1	0	0	0	0	0
σ_6	1	1	1	0	0	0
Γ_σ	6	2	2	0	0	0 $= A_1 + E + T$

We can write these characters for Γ_σ more simply as the number of σ bonds unchanged by one operation of each class. The character for i is 0. There is always an totally symmetric representation, A_{1g} with character +1 for i. Hence, for $\Gamma_\sigma(i) = 0$, the symmetry species for Γ_σ are $A_{1g} + E_g + T_{2u}$.

The systematic way to obtain the detailed description of the LGO's and their LCAO wavefunctions is to apply the **projection operator** approach (see DMA, 3rd ed., p. 179). Here we tabulate all of the symmetry operations and write down the orbital obtained as a result of performing each operation on this orbital. For example, choosing σ_1, this is still σ_1 after the identity operation, it becomes σ_2 after one C_3 operation, σ_3 after another C_3 operation, etc.

Results of Applying *All* Symmetry Operations to σ_1

E	$C_4(1)$	$C_4(2)$	$C_4(3)$	$C_4(4)$	$C_4(5)$	$C_4(6)$	$C_2(1)$	$C_2(2)$	$C_2(3)$
σ_1	σ_1	σ_5	σ_2	σ_3	σ_4	σ_1	σ_1	σ_6	σ_6

$C_3(1)$	$C_3(2)$	$C_3(3)$	$C_3(4)$	$C_3(5)$	$C_3(6)$	$C_3(7)$	$C_3(8)$
σ_2	σ_3	σ_4	σ_5	σ_5	σ_2	σ_3	σ_4

$C_2'(1)$	$C_2'(2)$	$C_2'(3)$	$C_2'(4)$	$C_2'(5)$	$C_2'(6)$
σ_6	σ_6	σ_2	σ_3	σ_5	σ_4

Next we multiply each orbital generated by the character for the corresponding symmetry operation (for the A_1 representation all +1), and sum. In this we get $a_1 = 4\sigma_1 + 4\sigma_2 + 4\sigma_3 + 4\sigma_4 + 4\sigma_5 + 4\sigma_6$, or simplifying, $a_1 = \sigma_1 + \sigma_2 + \sigma_3 + \sigma_4 + \sigma_5 + \sigma_6$. This is normalized to give the LCAO wavefunction by dividing by the square root of the sum of the squares of the coefficients.

$$\Psi(a_1) = \frac{1}{\sqrt{6}} (\sigma_1 + \sigma_2 + \sigma_3 + \sigma_4 + \sigma_5 + \sigma_6)$$

Multiplying the orbitals in the table by the characters for the representation E and summing gives:

$$e(1) = 4\sigma_1 + 4\sigma_6 - 2\sigma_2 - 2\sigma_3 - 2\sigma_4 - 2\sigma_5$$

or

$$2\sigma_1 + 2\sigma_6 - \sigma_2 - \sigma_3 - \sigma_4 - \sigma_5$$

and $\Psi[e(1)] = \frac{1}{\sqrt{12}} (2\sigma_1 + 2\sigma_6 - \sigma_2 - \sigma_3 - \sigma_4 - \sigma_5)$. This is one of the pair of e LGO's. This one matches perfectly with the sulfur d_{z^2} orbital. We can get the other one by performing an operation which interchanges ligands along z with those in the xy plane. Using a C_2 axis such that $\sigma_1 \rightarrow \sigma_2$ and $\sigma_6 \rightarrow \sigma_4$, we get $2\sigma_2 + 2\sigma_4 - \sigma_1 - \sigma_3 - \sigma_5 - \sigma_6$. This is equivalent to the LGO already obtained, it corresponds to a different numbering scheme. Adding it to the one obtained just gives another equivalent combinations. If we keep in mind that the other e orbital is $d_{x^2-y^2}$, we recognize that we want an LGO entirely contained in the xy plane with σ_1 and σ_6 not participating. Multiplying the second LGO obtained above by 2 and adding to the first LGO, we get the second e orbital.

$$3\sigma_2 + 3\sigma_4 - 3\sigma_3 - 3\sigma_5 \quad \text{and} \quad \Psi[e(2)] = \frac{1}{2} (\sigma_2 - \sigma_3 + \sigma_4 - \sigma_5).$$

The t_1 LGO can be obtained similarily, recognizing that for each we need a pair of orbitals of opposite sign along one axis.

$$t_1(1) = \frac{1}{\sqrt{2}} (\sigma_1 - \sigma_6)$$

$$t_1(2) = \frac{1}{\sqrt{2}} (\sigma_3 - \sigma_5) \qquad t_1(3) = \frac{1}{\sqrt{2}} (\sigma_2 - \sigma_4)$$

In this case we can get $t_1(2)$ and $t_1(3)$ by operating on $t_1(1)$ with C_4 along x and y, respectively, giving us the independent LGO's directly. The t_1 LGO's match the p orbitals (t_{1u}).

The S—F bond length (156 pm) in SF_6 is shorter than expected for a single bond. We have used the equivalent of sp^3d^2 hybridization, with the possibility of some degree of S—F $p\pi d$ bonding to account for the short bond length. There is controversy concerning the participation of d orbitals in bonding. Zare[1] prefers a MO description using p orbitals on S for 3-center bonding as for XeF_2 (see DMA, 3rd ed., p. 176). In SF_6 there are three equivalent 3-center, 4-electron bonds, using p orbitals. Although the bond order is 0.5 for each S—F bond of the 3-center bond, there is an ionic contribution. Presumably this is sufficiently important to enhance the strength of the S—F bond, with a covalent bond order of 0.5, to account for the bond length somewhat shorter than that

---

(1) T. Kiang and R. N. Zare, *J. Am. Chem. Soc.* **1980**, *102*, 4024.

expected for a single bond.

4.21 Problem: The H_3^+ ion is the simplest case of a three-center bond. The three H atoms are in a triangular arrangement (D_{3h}). Sketch the three MO's or (LCAO) and give a qualitative energy-level diagram.

4.21 Solution: There are three LCAO. The bonding combination is the sum of the three $1s$ orbitals. There are just two other LCAO and these must involve a change in sign and, hence, a nodal plane. From the D_{3h} character table (See Problem 3.11) we see that the only representations that change sign upon rotation by C_3 are E' and E''; since two MO's are degenerate, each involves one nodal plane. The representation must be E' since for s orbitals there is no change in sign for the σ_h operation. The two equivalent LCAO correspond to the σ group orbitals for BF_3 (see DMA, 3rd ed., p. 177f), except that here we are using only s orbitals.

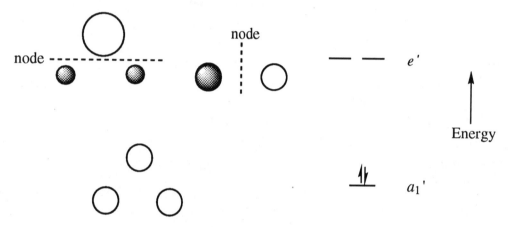

We can find the representations for the MO's more systematically by seeing how many AO's are left unchanged by each symmetry operation of the group.

D_{3h}	E	C_3	C_2	σ_h	S_3	σ_v
Γ_H	3	0	1	3	0	1

Γ_H can be reduced by inspection to give $A' + E'$. We can apply the projection operator method exactly as for the σ LCAO for BF_3 (see DMA, 3rd ed., p. 179f) to obtain the result given above pictorially.

4.22 Problem: XeF_4 is square planar (D_{4h}). The usual VB description assumes participation of the Xe d orbitals. Assuming that the d orbitals are too high in energy for bonding and that the low energy $5s$ electrons are nonbonding, give a qualitative bonding description using only the Xe p_x and p_y orbitals for bonding.

4.22 Solution: Along the x axis Xe uses the p_x orbital for bonding to 2F. There are three LCAO, the bonding orbital where the signs match perfectly, the antibonding orbital where the lobes touching have opposite signs, and a nonbonding orbital that does not involve participation of the Xe orbital. There is an identical three-center bond along the y axis. Since each three-center MO involves one electron pair for the two Xe—F bonds, each Xe—F bond has a bond order of 0.5. (see DMA, 3rd ed., p. 176 for a description of similar bonding in XeF_2 and K. A. R. Mitchell, *The Use of Outer d Orbitals in Bonding, Chem. Rev.* **1969**, *69*, 157.)

4.23 Problem: Diatomic molecules can be assigned term symbols by analogy with atoms. Angular momentum (λ) is quantized around the internuclear axis and the individual quantum numbers are given as follows:

MO	λ	m_l
σ	0	0
π	1	± 1
δ	2	± 2

Proceeding by analogy to Russell-Saunders coupling and using Hund's rule, give term symbols for the ground state of H_2, B_2, and O_2. The symbols used are the representations of the $D_{\infty h}$ group, Σ, Π, and Δ, corresponding to Λ (for L) = 0, 1, and 2, respectively. (For an approach using group theory, see D. I. Ford, *J. Chem. Educ.* **1972**, *49*, 336.)

4.23 Solution: For H_2 the electron configuration is σ_g^2. Since both electrons are in the same orbital we must have the quantum numbers $m_l = 0$, $m_s = -\frac{1}{2}$ and $m_l = 0$, $m_s = \frac{1}{2}$. Hence $M_L = 0 + 0$ and $\Lambda = 0$; $M_S = \frac{1}{2} - \frac{1}{2} = 0$ and $S = 0$. The ground state is $^1\Sigma$. Since the wavefunction for each electron is *g*, the total wavefunction is also *g*, giving $^1\Sigma_g$.

For B_2 the electron configuration is $\sigma_g^2\sigma_u^{*2}\pi_u^2$. From the Pauli principle, the following sets of quantum numbers are possible for the π_u electrons. (Filled shells contribute $M_L = 0$, $M_S = 0$.)

	$m_s = +\frac{1}{2}$		$m_s = -\frac{1}{2}$		$M_L = \Sigma m_l$	$M_S = \Sigma m_s$
$m_l =$	+1	−1	+1	−1		
	↑	↑			0	1
	↑			↓	2	0
	↑			↓	0	0
		↑	↓		0	0
		↑		↓	−2	0
			↓	↓	0	−1

The number of individual microstates corresponding to various M_L and M_S values is:

$$\begin{array}{c|ccc} M_L & & & \\ 2 & & 1 & \\ 0 & 1 & 2 & 1 \\ -2 & & 1 & \\ \hline & -1 & 0 & 1 \\ & & M_S & \end{array}$$

Hence, the possible values of Λ are 2 and 0 and S = 1, 0, giving the terms $^1\Sigma$, $^1\Delta$, and $^3\Sigma$. Because both electrons are *u* (*u* × *u* = *g*), the possible terms become $^1\Sigma_g$, $^1\Delta_g$, $^3\Sigma_g$. From Hund's rule, the state of maximum spin multiplicity, $^3\Sigma_g$, is the predicted ground state.

For O_2 the electron configuration is $\sigma_g^2\,\sigma_u^{*2}\,\sigma_g^2\,\pi_u^4\pi_g^{*2}$. The same array of microstates as for B_2 will be generated and the ground state is again $^3\Sigma_g$. (see DMA, 3rd ed., p. 159)

4.24 Problem: Develop a qualitative MO energy-level diagram for acetylene. Label the orbitals.

4.24 Solution: Both C atoms are equivalent; group orbitals may be constructed as the sum and the difference of the AO's, or preferably, of the sp hybridized orbitals. Symmetry labels may be assigned readily by considering the symmetry of the MO relative to a plane through the σ bond and perpendicular to it (see DMA. 3rd ed., pp. 154-155). Group orbitals for the H atoms are formed from the sum and difference of their $1s$ orbitals. MO's result from the combination of the H group orbitals with the C group orbitals of the same symmetry. The resulting MO energy-level diagram is given below.

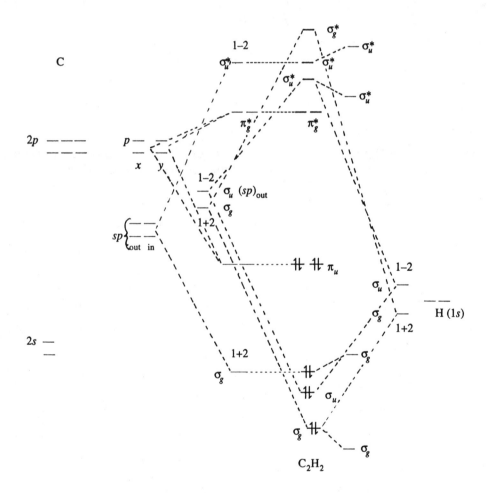

The shifts in the energies of the two σ_g and σ_u^* hydrogen group orbitals result from configuration interaction.

4.25 PROBLEM: Given below is a Walsh diagram for AH_3 molecules correlating the MO energy levels of pyramidal geometry with those of planar geometry. Compare the predictions of geometry based on the Walsh diagram with those based on the Gillespie VSEPR approach and on the Pauling hybridization approach for AH_3 molecules having 6, 7, 8, and 10 valence electrons.

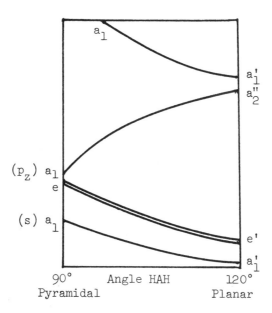

4.25 Solution: The geometry predicted from the Walsh diagram will be the one of lowest energy for the particular electron count. The VSEPR and the hybridization approach are discussed in DMA, 3rd ed., Sections 2.2 and 2.3, and touched on in Problems 2.1, 2.5, 2.8, and 2.9. Resulting geometries predicted by these various approaches are given below.

	WALSH	GILLESPIE	PAULING
6 e^-	trigonal planar	trigonal planar	sp^2 trigonal planar
7 e^-	HAH angle ~115° pyramidal	shallow pyramid (less repulsion from single electron than bond pairs.)	sp^3 pyramidal (with more s character in bonding orbitals giving HAH angle greater than 109°)
8 e^-	HAH angle 90° pyramidal	pyramidal with HAH angle less than 109° due to e^- pair–bond pair repulsion being greater than bond pair–bond pair repulsion.	sp^3 pyramidal
10 e^-	trigonal planar	"T"-shaped	must include 5 orbitals from A, sp^3d pseudo-trigonal bipyramidal

4.26 Problem: The Langmuir equation (see Problem 2.4) gives the same net bonding for polyatomic molecules as an MO approach which uses only s and p valence orbitals and averages five valence electrons or more per atom. Verify this for the oxygen species in Problem 4.1 and the NO species in Problem 4.6.

4.26 Solution: Number of bonds = $(8n - VE)/2$ where n = number of atoms using s and p valence orbitals and VE = sum of the valence electrons. For diatomic species the number of bonds will be $(16 - VE)/2$. For O_2 species VE = 2 (6) – (charge), giving the bond orders O_2 2, O_2^+ 2.5, O_2^- 1.5, and O_2^{2-} 1; for the NO species VE = 11 – (charge), giving the bond orders NO 2.5, NO$^+$ 3, and NO$^-$ 2.

4.27 Problem:
(a) Develop an equation similar to the Langmuir equation which will give the number of shared electrons of a given spin set in a Linnett Double Quartet (DQ) treatment of a polyatomic species.
(b) Use this equation as an aid to obtain the DQ representation of NO_2, PF_5, XeF_4, and SF_6.

4.27 Solution:
(a) Shared electrons (of one spin set) = $4N + H - e$, where N is the number of atoms using s and p orbitals, H is the number of hydrogen atoms, and e is the number of electrons in a given spin set.
(b) For NO_2 there are 17 valence electrons so there should be sets of 8 and 9 electrons which we will label x and o, respectively.
  # of x shared = 4 x 3 – 8 = 4
  # of o shared = 4 x 3 – 9 = 3

We note that for the structure for x's, a linear shape is "preferred" based on electron repulsion (no unshared x on N), while a bent shape is "preferred" for the structure for o's (one unshared o on N). This is resolved by a slightly bent molecule (larger angle than for NO_2^-). The NO_2 molecule dimerizes readily to form an additional bond. For the dimer both spin sets correspond to bent NO_2's.

For PF_5 VE = 40, or 20 x's and 20 o's. This description omits any d participation.
  # of unshared x (or o) = 4 x 6 – 20 = 4

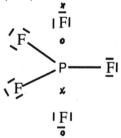

Note that this model predicts weaker axial bonding than equatorial bonding. Although the geometry can be obtained by considering the bipyramid to consist of two tetrahedral of opposed spin sets sharing a face, it is perhaps more fruitful to consider this as sp^2 hybrid equatorial bonds with "split p" axial bonds. See this type of bonding for XeF_2 as an MO description.

In similar fashion XeF_4 and SF_6 can be considered to have split p bonds (these would be considered to be two electron-three center bonds on a limited MO picture).

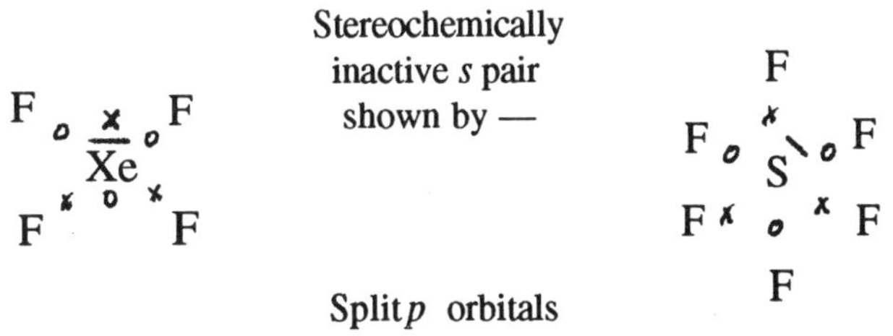

Stereochemically inactive s pair shown by —

F o×o F
 Xe
F ×o× F

F
F o S\o F
F× o ×F
F

Split p orbitals

4.28 Problem: Write Linnett-like electron dot pictures for the lowest $^3\Sigma$ and $^5\Sigma$ states of B_2.

4.28 Solution: The MO energy-level diagram of B_2 is shown below along with the electron population which gives the leading microstate giving rise to the triplet and quintet Σ states. Population of both the bonding and antibonding orbitals of a given type by electrons of the same spin gives a nonbonding electron of that spin on each atom, while population of a bonding orbital only gives an e of that spin as a bonding e shown between atoms. The resulting dot pictures are shown beneath the MO diagram.

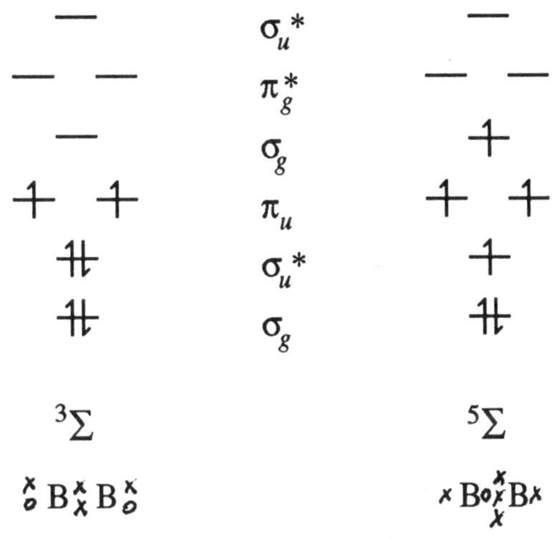

$^3\Sigma$ $^5\Sigma$

$^{\times}_{\circ}B^{\times}_{\times}B^{\times}_{\circ}$ $\times B\!\circ^{\times}_{\times}\!B\times$

4.29 Problem: Give DQ structures for O_3, O_3^+, and O_3^-.

4.29 Solution: Dividing the spin sets as 9 & 9, 9 & 8, and 9 & 10 gives the following numbers for the shared e (see Problem 4.27)): 3 & 3, 4 & 3, and 2 & 3, respectively.

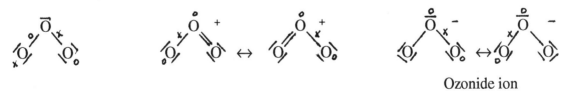

Ozonide ion

Ozonides are known for the heavier alkali metals, KO_3, etc. MO calculations give a bond order of $1\frac{1}{4}$, identical to the DQ result.

4.30 Problem: O_2F dimerizes at low temperature. Give DQ descriptions of both species.

4.30 Solution:
O_2F # VE = 19, giving spin sets of 9 and 10.
For 10 e, shared electrons = (3 x 4) – 10 = 2

Tetrahedral geometry for the electrons of one set is preferred.

For 9 e, shared electrons = 3 x 4 – 9 = 3

Trigonal geometry preferred.

The DQ formulation for O_2F is

For O_4F_2 # of shared electrons per spin set = 6 x 4 – 19 = 5
The DQ formulation is equivalent to the valence description.

II. BONDING AND STRUCTURE
5
Inorganic Solids
The Ionic Model

5.1 Problem: Which of each of the following pairs might be expected to be more ionic?
(a) $CaCl_2$ or $MgCl_2$
(b) NaCl or $CaCl_2$ (similar radii)
(c) NaCl or CuCl (similar radii)
(d) $TiCl_3$ or $TiCl_4$

5.1 Solution: Covalent character increases with polarization (charge displacement).
(a) $CaCl_2$ is expected to be more ionic than $MgCl_2$ because the larger Ca^{2+} cation is less polarizing.
(b) NaCl is expected to be more ionic than $CaCl_2$ because the lower cation charge of Na^+ causes less polarization of the Cl^-.
(c) NaCl is more ionic than CuCl because Cu^+ is an 18-electron cation and causes more polarization than Na^+ (8-e shell) since the d electrons are less efficient in screening the higher nuclear charge.
(d) $TiCl_3$ is more ionic than $TiCl_4$ (a liquid) because Ti^{3+} has lower charge and larger radius than Ti^{IV}—both favor less polarization.

5.2 Problem: For which of the following pairs indicate which substance is expected to be:
(a) More covalent (Fajans' rules):
 $MgCl_2$ or $BeCl_2$ $SnCl_2$ or $SnCl_4$
 $CaCl_2$ or $ZnCl_2$ $CdCl_2$ or CdI_2
 $CaCl_2$ or $CdCl_2$ ZnO or ZnS
          NaF or CaO

(b) Harder:
 NaF or NaBr
 Al_2O_3 or Ga_2O_3
 MgF_2 or TiO_2

5.2 Solution: (a) Covalent character increases with increasing polarization. $BeCl_2$ is more covalent than $MgCl_2$ because of the smaller radius of Be^{2+}. $ZnCl_2$ is more covalent than $CaCl_2$ because of the smaller radius of Zn^{2+} and the fact that Zn^{2+} has an 18-e configuration. $CdCl_2$ is more covalent than $CaCl_2$ because Cd^{2+} has an 18-e configuration. $SnCl_4$ is more covalent than $SnCl_2$ because of the smaller radius and higher charge of Sn^{IV}. CdI_2 is more covalent than $CdCl_2$ because the larger I^- is more easily polarized. ZnS is more covalent than ZnO because the larger S^{2-} is more easily polarized. CaO is more covalent than NaF because the higher ionic charges increase polarization.

(b) NaF is harder than NaBr because the lattice energy is greater for small ions. Al_2O_3 is harder than Ga_2O_3. The cation radii are almost the same, but Ga^{3+} is an 18-e ion, increasing polarization and "softening" the lattice. TiO_2 is harder than MgF_2 because of the higher ionic charges, increasing the lattice energy.

5.3 Problem: Variations in mechanical hardness for ionic substances correlate well with what one thermodynamic property?

5.3 Solution: Hardness (mechanical, not absolute hardness) is determined by the resistance to deformation of the lattice and this correlates well with lattice energy.

5.4 Problem: Which of the following are *not* possible close-packing schemes?
(a) ABCABC . . .   (d) ABCBC . . .
(b) ABAC . . .   (e) ABBA . . .
(c) ABABC . . .   (f) ABCCAB . . .

5.4 Solution: If the positions of the spheres in the first layer of a close-packed arrangement are designated A, those of adjacent layers can only be B or C, with the spheres of the adjacent layer in the indentations of the first layer. Arrangements **(e)** and **(f)** violate this requirement, repeating the same positions (BB and CC). These are not close-packed arrangements.

5.5 Problem: Calculate the number of formula units in the unit cells for the following:
(a) CsCl   (b) ZnS (zinc blende)   (c) CaF_2 (d) $CaTiO_3$ (See Figures on the next page.)

5.5 Solution: (a) For CsCl there is 1 Cs^+ at the center of the cube and 8 Cl^- at the corners:

$$1\ Cs^+$$
$$8 \times \tfrac{1}{8} = \underline{1\ Cl^-}$$
$$1\ CsCl/\text{unit cell}$$

(b) For ZnS (zinc blende) there are four Zn^{2+} within the cube, $8\,S^{2-}$ at the corners and 6 S^{2-} in the faces:

$$4\ Zn^{2+}$$
$$8 \times \tfrac{1}{8} = \quad 1\ S^{2-}\ \text{at corners}$$
$$6 \times \tfrac{1}{2} = \quad \underline{3\ S^{2-}\ \text{in faces}}$$
$$4\ ZnS/\text{unit cell}$$

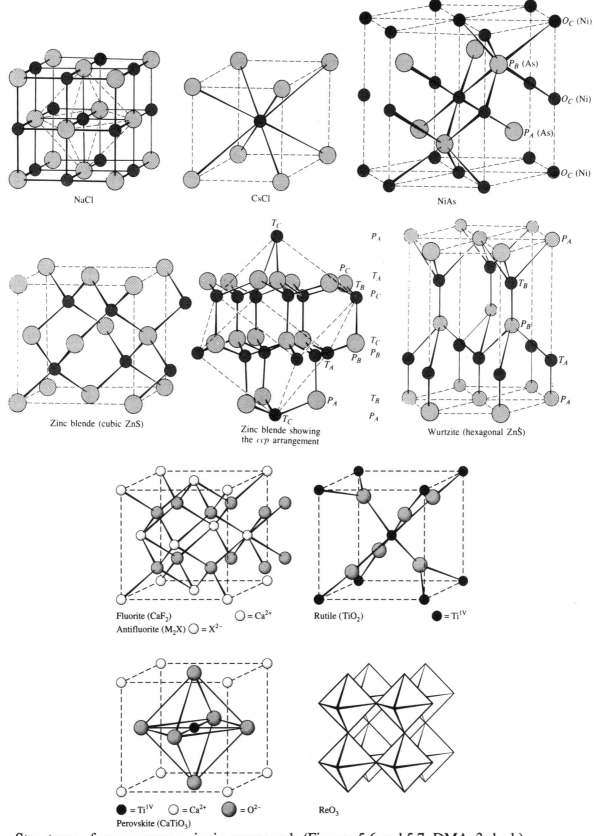

Structures of some common ionic compounds (Figures 5.6 and 5.7, DMA, 3rd ed.)

(c) For CaF$_2$ there are eight F$^-$ within the cube, 8 Ca^{2+} at the corners and 6 Ca^{2+} in the faces:

$$\begin{array}{ll} & 8\text{ F}^- \\ 8 \times \tfrac{1}{8} = & 1\text{ Ca}^{2+}\text{ at corners} \\ 6 \times \tfrac{1}{2} = & \underline{3\text{ Ca}^{2+}\text{ in faces}} \\ & 4\text{CaF}_2/\text{unit cell} \end{array}$$

(d) For CaTiO$_3$ there is one Ti^{IV} at the center of the cube, eight Ca^{2+} at the corners, and six O^{2-} in the faces:

$$\begin{array}{ll} & 1\text{ Ti}^{IV} \\ 8 \times \tfrac{1}{8} = & 1\text{ Ca}^{2+} \\ 6 \times \tfrac{1}{2} = & \underline{3\text{ O}^{2-}} \\ & 1\text{ CaTiO}_3/\text{unit cell} \end{array}$$

5.6 Problem: The wurtzite (*hcp* S^{2-}) structure of ZnS has open channels along the packing direction, but the zinc blende (*ccp* S^{2-}) structure does not. Explain.

5.6 Solution: Wurtzite has *hcp* S^{2-} with Zn^{2+} in *T* sites. For *hcp* the packing positions *A* and *B* are occupied by S^{2-}. The *T* sites occur only at *A* and *B* positions since each *T* site has a packing position at the apex directly above or below the *T* site. The octahedral sites at *C* positions are empty forming open channels. The *ccp* structure has packing *and* tetrahedral sites occurring at *A*, *B*, and *C* positions with no open channels.

5.7 Problem: Many MX type compounds with C.N. 6 have the NaCl (Cl$^-$ *ccp*) structure, whereas few have the NiAs (As *hcp*) structure. What features of the NiAs structure limit its occurrence? What characteristics of the compound are necessary for the NiAs structure? What unusual physical property of NiAs results from its structure?

5.7 Solution: NiAs has As in *hcp A* and *B* packing positions with Ni in all octahedral sites at *C* positions. There is *no* shielding between adjacent Ni atoms. This is tolerable only if there is a considerable amount of polarization (covalence) to diminish repulsion. There is metallic conduction along one direction of the NiAs crystal because of the Ni "wires" running in this direction. The more ionic MX compounds adopt the NaCl structure. The Cl$^-$ ions are in *ccp* (*ABC*) packing positions with Na$^+$ in all octahedral (*ABC*) positions.

5.8 Problem: What do the following pairs or groups of structures have in common, and how do they differ? Give the formula type (MX, MX$_2$, etc.) for each. (See DMA, 3rd ed., Section 5.6.)

(a) 3·2*PO*, 3·2*PT*, and 2·2*PT*
(b) 3·3*PT*$_{1/2}$*T*$_{1/2}$ and 3·2*PT*
(c) 3·3*PTT* and 2·2*PO*$_{1/2}$
(d) 2($\tfrac{3}{2}$)*PPO* and 2·2*PO*$_{1/2}$

5.8 Solution: (a) All are MX (1:1) compounds since the number of packing sites is the same as the number of octahedral sites or the number of tetrahedral sites in one layer. 3·2*PO* and 3·2*PT* are *ccp* (the first index is 3 for repeating *ABC* positions). In 3·2*PO* the cations fill all *O* sites. In 3·2*PT* and 2·2*PT* the cations fill all *T* sites in one *T* layer, the other *T* layer is vacant. 2·2*PT* has anions in *hcp* positions.

(b) Both are MX (1:1) compounds using P and 1/2 of the T sites and both are *ccp* (3P layers per repeating unit). In $3 \cdot 3PT_{1/2}T_{1/2}$, 1/2 of the T sites in each T layer are used. In $3 \cdot 2PT$ all T sites in one layer are used.

(c) Both are MX_2 or M_2X compounds since there are twice as many sites of one kind occupied as of the other kind. $3 \cdot 3PTT$ is *ccp* with (for MX_2) X in all T sites. The C.N. of M is 8 and the C.N. of X is 4. For $2 \cdot 2PO_{1/2}$ (for MX_2) M is in 1/2 of the octahedral sites. The C.N. of M is 6 and the C.N. of X is 3 (rutile). For each case the roles are reversed for M_2X compounds.

(d) For $2(\frac{3}{2})PPO$ there are $2 \times \frac{3}{2} = 3$ layers in the repeating unit, $P_A P_B O$. This is *hcp*. For $2 \cdot 2PO_{1/2}$ there are 4 layers in the repeating unit, $P_A O_{1/2} P_B O_{1/2}$, which is also *hcp*. In each case half of the octahedral sites are occupied—all of them in alternate layers for PPO (a layer structure) and half of them in *each* layer for $PO_{1/2}$ (rutile).

5.9 Problem: Why are layer structures such as those of $CdCl_2$ and CdI_2 usually not encountered for metal fluorides or compounds of the most active metals?

5.9 Solution: Since layer structures have adjacent layers of anions (Cs_2O is an unusual case in having adjacent *ccp* layers of *cations*), the repulsion would prohibit the structure unless there were a considerable amount of polarization (covalent). Metal fluorides and compounds of the most active metals are found in the more typical "ionic" structures such as those of NaCl and CaF_2.

5.10 Problem: Rubidium chloride assumes the CsCl structure at high pressures. Calculate the Rb—Cl distance in the CsCl structure from that for the NaCl structure (from ionic radii for C.N. 6, $r_{Rb^+} = 166$ pm, $r_{Cl^-} = 167$ pm). Compare with the Rb—Cl distance from radii for C.N. 8.

$$\frac{r_8}{r_6} = \left(\frac{8A_6}{6A_8}\right)^{1/(n-1)}$$

where r_8 and r_6 are the radii (or internuclear distances) for C.N. 8 and 6, respectively, n is the Bohr exponent {n is 10 for Rb^+ and 9 for Cl^-, or we can use an average value (9.5) for RbCl}. The Madulung constant (A) is 1.75 for NaCl and 1.76 for CsCl.

5.10 Solution: The internuclear distance expected for C.N. 6 is $166 + 167 = 333$ pm. For C.N. 8 we calculate

$$d(\text{C.N.8}) = 333\left(\frac{8A_6}{6A_6}\right)^{1/(9.5-1)} = 333\left(\frac{8 \times 1.75}{6 \times 1.76}\right)^{0.118} = 343 \text{ pm}$$

$$r_{Cl^-}(\text{C.N. 8}) = 167\left(\frac{8 \times 1.75}{6 \times 1.76}\right)^{1/(9-1)} = 173 \text{ pm, giving } d_{RbCl} = 348 \text{ as the sum of the}$$

C.N. 8 radii.

5.11 Problem: Calculate the cation/anion radius ratio (by using plane geometry) for a triangular arrangement of anions in which the cation is in contact with the anions but does not push them apart.

5.11 Solution: The lines joining the centers of the anions form an equilateral (60°) triangle

$$\cos 30° = \frac{b}{a} = \frac{r_X}{r_X + r_M}$$

$$r_X = 0.866\, r_X + 0.866\, r_M$$

$$\frac{r_M}{r_X} = 0.155$$

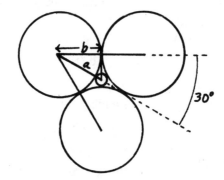

5.12 Problem: Why are the dashed lines in the sorting maps (shown below) more successful than radius rules indicated by solid lines in sorting structure types?

5.12 Solution: The radius ratio rules are based on the packing of hard spheres. In real crystals the spheres are somewhat compressible. Sizes of ions are important, but the limitations of the radius ratio rules are too restrictive.

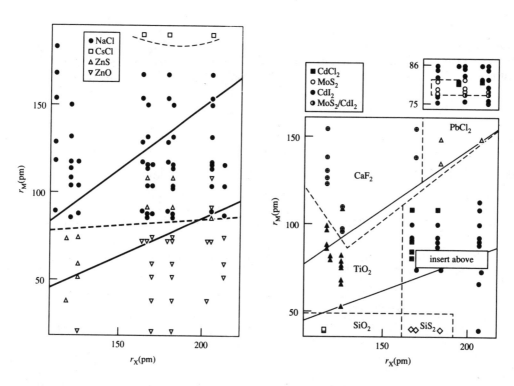

Plots of r_M and r_X to show the sorting of structures for MX and MX_2 compounds using ratio rules (solid rules) and by areas of predominant structures (dashed lines). Figures 5.12 and 5.13, DMA, 3rd ed.(Adapted with permission from J. K. Burdett, S. L. Price, and G. D. Price, *Solid State Commun.* **1981**, *40*, 923. Copyright © 1981, Pergamon Press, Ltd., Hedington Hill Hall, Oxford OX3 OBW UK.)

5.13 Problem: Estimate the density of MgO (NaCl structure) and zinc blende (ZnS) using radii to determine the cell dimensions and the number of formula units per unit cell.

5.13 Solution: For MgO the cubic cell edge equals

$$2r_{Mg^{2+}} + 2r_{O^{2-}} = 424 \text{ pm} = a$$

The volume of the unit cell = a^3 = 7.62 x 10^{-23} cm^3

Since there are 4 MgO per unit cell, we have $\dfrac{4 \times 40.3}{6.02 \times 10^{23}}$ g/unit cell.

The density is given by dividing the mass of the unit cell by the volume of the cell:

$$\dfrac{161.2}{6.02 \times 10^{23} \times 7.62 \times 10^{-23}} = 3.51 \text{g/cm}^3. \text{ Handbook gives 3.58 g/cm}^3.$$

For ZnS the S^{2-} are *ccp* (or face-centered cubic) with Zn^{2+} in T sites within the cell. The diagonal of a face of the cubic cell is $4r_{S^{2-}}$ = 680 pm (using the radius for C.N. 6). This is the hypotenuse of a right triangle with sides equal to a, or

$$(680)^2 = 2a^2 \qquad a = 482 \text{ pm} \qquad a^3 = 1.12 \times 10^{-22} \text{ cm}^3$$

4 ZnS/unit cell or $\dfrac{390}{6.02 \times 10^{23}}$ g/unit cell

$$\dfrac{390}{6.02 \times 10^{23} \times 11.2 \times 10^{-23}} = 5.78 \text{g/cm}^3. \text{ The handbook gives 4.10 g/cm}^3 \text{ for the density.}$$

The cell length for zinc blende is 541 pm. Thus, the effective radius for S^{2-} is 191 pm from this right triangle. ZnS is appreciably covalent so the ionic radius is a poorer value to use than the van der Waals radius (180 pm). Correcting the ionic radius for C.N. 6 to get a value C.N. 4 gives a smaller radius (163 pm) and poorer agreement with the observed density.

5.14 Problem: NaSbF$_6$ has the NaCl structure. The density is 4.37 g/cm^3. Calculate the radius of SbF$_6^-$ using the radius of Na$^+$, the density, the formula weight, and the number of formula units per unit cell. (r_{Na^+} = 116 pm)

5.14 Solution: For the NaCl structure there are 4 formula units per unit cell and the cell length is $2r(Na^+) + 2r(SbF_6^-)$.

Mass of a unit cell = 4 x 259/6.02 x 10^{23} = 1.72 x 10^{-21} g/unit cell. The density is the mass of the unit cell divided by its volume, or

$$\text{density} = 4.37 = \dfrac{1.72 \times 10^{-21}}{a^3 \text{ cm}^3} \text{g} \qquad a^3 = \dfrac{1.72 \times 10^{-21}}{4.37} = 3.94 \times 10^{-22} \text{cm}^3$$

a = 7.33 x 10^{-8} cm = 733 pm = $2r(Na^+) + 2r(SbF_6^-)$ = 2(116) + $2r(SbF_6^-)$

$$r(SbF_6^-) = \dfrac{501}{2} = 250 \text{ pm}$$

5.15 Problem: Compare the crystal (C.N. 6) radii of Na^+ (116 pm) and Mg^{2+} (86 pm), and those of S^{2-} (170 pm) and Cl^- (167 pm). Explain the differences within each pair.

5.15 Solution: There is a large decrease in ionic radius on going from Na^+ to Mg^{2+} but little difference on going from S^{2-} to Cl^-. An increase in nuclear charge pulls in the electrons of isoelectronic ions; an increase in ionic charge compresses ions because of increased attraction between cations and anions. With Na^+ and Mg^{2+} the two effects work in the same direction; with S^{2-} and Cl^- the two effects oppose each other.

5.16 Problem: Compare the thermochemical data for the alkali halides. What factors are important in establishing the order of increasing heats of formation within each group of halides (fluorides, chlorides, etc.)?

Thermochemical Data for Alkali Metal Halides							
				F	Cl	Br	I
			EA	328.0	349.0	324.7	295.4
S	IE		½D	77.2	119.5	95.0	74.3
160.7	520.5	Li	$-\Delta H_f$	616.9	408.3	350.2	270.1
			U_0	1049	862	819	763
107.8	495.4	Na	$-\Delta H_f$	573.6	411.1	361.4	288
			U_0	928	787	752	703
89.2	418.4	K	$-\Delta H_f$	567.4	436.7	393.8	327.9
			U_0	826	717	689	647
82.0	402.9	Rb	$-\Delta H_f$	553.1	430.5	389	328
			U_0	789	688	661	625
77.6	375.3	Cs	$-\Delta H_f$	554.7	442.8	395	337
			U_0	758	668	635	602

Data in kJ/mol at 298 K from DMA, 3rd ed, p. 231 except for ½D at 0 K
Electron affinity (EA), lattice energy (U_0), and heat of formation (ΔH_f) are thermodynamically negative (energy released) for formation of MX.
Sublimation energy (S), ionization energy (IE), and dissociation energy (½D) are thermodynamically positive (energy required).

5.16 Solution: For MCl, MBr, and MI ΔH_f becomes more negative with increasing size of M, a trend established by the ionization and sublimation energies of M. For MF the order is reversed because of the much greater in lattice energy from LiF to CsF.

5.17 Problem: Calculate the heat of formation of NaF$_2$ and the heat of reaction to produce NaF + ½ F$_2$. (Assume the rutile structure, A = 2.408, r_{F^-} (C.N. 3) = 116 pm and estimate $r_{Na^{2+}}$).

5.17 Solution: Assume $r_{Na^{2+}} \simeq r_{Mg^{2+}} = 86$ pm and $n = 7$ (Born exponent for Ne configuration)

$$U_0 = -\frac{1(-2)(4.80 \times 10^{-10} \sqrt{\text{dyne cm}})^2 (6.02 \times 10^{23}) 2.41}{2.02 \times 10^{-8} \text{cm}} \left(1 - \frac{1}{7}\right)$$

$= 2.83 \times 10^{13}$ erg $= 2,830$ kJ/mol

$$\begin{array}{c}
\text{Na}(s) + \text{F}_2(g) \xrightarrow{\Delta H_f} \text{NaF}_2(s) \\
\downarrow S \quad \quad D \downarrow \quad \quad \nearrow \\
\text{Na}(g) \quad \quad 2\text{F} \quad -U_0 \\
\text{IE }(1+2) \downarrow \quad 2\text{EA} \downarrow \\
\text{Na}^{2+}(g) + 2\text{F}^-
\end{array}$$

$\Delta H_f = -U_0 + D + 2\text{EA} + S + \text{IE}$
$= -2830 + 155 - 656 + 108 + 5058$
$= 1835$

NaF$_2$(s) → NaF(s) + ½ F$_2$ $\Delta H = -574 - 1835 = -2409$ kJ

5.18 Problem: Calculate the heats of formation of Ne$^+$F$^-$ and Na$^+$Ne$^-$, estimating radii of Ne$^+$ and Ne$^-$. What factors prohibit the formation of these compounds in spite of favorable lattice energies?

5.18 Solution: Estimate r_{Ne^-} (slightly larger than F$^-$) and r_{Ne^+} (slightly smaller than Na$^+$). Hence U_0 is approximately that of NaF (928) for NeF or NaNe.

$\Delta H_f (\text{NeF}) = -U_0 + \frac{1}{2} D_{F_2} + \text{EA} + \text{IE} = -930 + 77 - 328 + 2081$
$= 900$ kJ/mol (High IE prohibits formation)

Estimate EA for Ne as zero

$\Delta H_f (\text{NaNe}) = U_0 + S + \text{IE} = -920 + 108 + 496 = -316$ kJ/mol
NaNe should be thermodynamically stable from these assumptions, but EA is –120 and does not bind an electron.

5.19 Problem: From spectral data the dissociation energy of ClF has been determined to be 253 kJ/mol. The ΔH_f^0 of ClF(g) is –50.6 kJ/mol. The dissociation energy of Cl$_2$ is 239 kJ/mol. Calculate the dissociation energy of F$_2$.

5.19 Solution:

			ΔH
	$ClF(g)$	$\to Cl(g) + F(g)$	$+253$ kJ/mol
(1)	$\tfrac{1}{2} Cl_2(g)$	$\to Cl(g)$	119.5
(2)	$\tfrac{1}{2} F_2(g)$	$\to F(g)$	$\tfrac{1}{2} D(F_2)$
(3)	$Cl(g) + F(g)$	$\to ClF(g)$	-253
(4)	$\tfrac{1}{2} Cl_2(g) + \tfrac{1}{2} F_2(g)$	$\to ClF(g)$	$-50.6 = \Delta H_f^0$

Since equation (4) is the sum of (1), (2), and (3),
-50.6 kJ $= [-253 + 119.5 + \tfrac{1}{2} D(F_2)]$ kJ
$\tfrac{1}{2} D(F_2) = 82.9$ kJ $D(F_2) = 166$ kJ/mol

5.20 Problem: Although the electron affinity of F is lower than that of Cl, F_2 is much more reactive than Cl_2. Account for the reactivity of F_2 **(a)** with respect to the formation of solid halides MX or MX_2 and **(b)** with respect to the formation of aqueous solutions of MX or MX_2.

5.20 Solution:
(a) Less energy is gained for $F + e \to F^-$ (compared to Cl), but less energy is required to break the F—F bond compared to Cl—Cl, and U_0 is high for the small F^-, making the formation of metal fluorides very favorable.
(b) The same considerations for the EAs and dissociation energies cited in **(a)** apply. Additional energy is gained from the much higher hydration energy of F^- compared to Cl^-.

5.21 Problem: The $CuCl_5^{3-}$ is not very stable. What would be a suitable cation to use for the isolation of $CuCl_5^{3-}$?

5.21 Solution: Cs^+, $N(CH_3)_4^+$, or another large cation gives the best chance to isolate $CuCl_5^{3-}$. A small cation such as Li^+ would favor the dissociation of the complex anion and formation of solid LiCl because of its high lattice energy.

5.22 Problem: Draw the net of hexagons of a layer of graphite showing the single and double bonds.

5.22 Solution: There is a network of alternating single and bonds (conjugation). Each C has sp^2 hybridization, forming three trigonal σ bonds and one π bond.

```
    \   /       \   /       \
     C = C       C = C       C =
    /   \       /   \       /
= C       C = C       C = C
    \   /       \   /       \
     C = C       C = C       C =
    /   \       /   \       /
= C       C = C       C = C
    \   /       \   /       \
     C = C       C = C       C =
    /   \       /   \       /
```

5.23 Problem: Suggest a solvent, other than H_2O, for studies of ionic substances.

5.23 Solution: A solvent with a high dielectric constant and a high dipole moment can provide high solvation energy and lower attraction between cations and anions, both required to overcome the high coulombic attraction between cations and anions. HF, HCN, and NH_3 are reasonable solvents for ionic substances, but not as good as H_2O.

5.24 Problem: Synthetic gems, including emerald, sapphire, and ruby, have the same chemical composition as natural gems. Why are the natural gems much more expensive?

5.24 Solution: The scarcity of high quality natural gems makes them expensive. Natural gems can be identified under magnification by the presence of occulsions or other imperfections.

5.25 Problem: What is the significance of the term "molecular weight" with respect to diamond or SiO_2 (considering a perfect crystal of each)?

5.25 Solution: A crystal is a molecule for diamond (C) and SiO_2. More properly, we should use *formula weight* (or mass), as for ionic compounds, since "molecular weight" (or molar mass) has more specific meaning.

5.26 Problem: The triclinic feldspars form an isomorphous series involving replacement of Ca^{2+} for Na^+ from albite, $Na[AlSi_3O_8]$ to anorthite, $Ca[Al_2Si_2O_8]$. How can the series be isomorphous with changing ratio of Si/Al? Why do not K^+ and Ba^{2+} occur in this series?

5.26 Solution: The framework in each case consists of $[(Al,Si)_4O_8]$. The changing proportions of Al and Si are compensated by replacement of Ca^{2+} for Na^+. K^+ and Ba^{2+} are too large to replace Na^+ and Ca^{2+}. They form a separate series of feldspars.

5.27 Problem: Demonstrate that the formulas of repeating units can be produced from condensation of SiO_4^{4-} for the following minerals.

(a) SiO_2 (b) SiO_3^{2-} in pyroxenes   (c) $Si_4O_{10}^{4-}$ in talc   (d) $Si_4O_{11}^{6-}$ in amphiboles

5.27 Solution:

(a) $SiO_4^{4-} \rightarrow SiO_2 + 2O^{2-}$ SiO_2 contains SiO_4 tetrahedra with each O^{2-} shared between two tetrahedra.

(b) $SiO_4^{4-} \rightarrow SiO_3^{2-} + O^{2-}$ The chains contain SiO_4 tetrahedra with two O^{2-} unshared and two O^{2-} shared between two tetrahedra.

(c) $4\,SiO_4^{4-} \rightarrow Si_4O_{10}^{4-} + 6O^{2-}$ In the sheets of rings of 6 SiO_4 each SiO_4 has one O^{2-} unshared.

(d) $4\,SiO_4^{4-} \rightarrow Si_4O_{11}^{6-} + 5O^{2-}$ In the double chains of rings of 6 SiO_4 half of the SiO_4 (the ones of the rings joined in chains) have one O^{2-} unshared and the other half of the SiO_4 have two O^{2-} unshared.

5.28 Problem: Metals that are very malleable (can be beaten or rolled into sheets) and ductile (can be drawn into wire) have the *ccp* structure. Why are these characteristics favored for *ccp* rather than *hcp*?

5.28 Solution: For *ccp* there are slip planes along any cube diagonal. These occur along one direction only for *hcp*. Thus, for the *ccp* structure there are several ways to slide one layer over another without altering the structure.

5.29 Problem: Soft metals such as Cu become "work-hardened." Explain this and how the softness is restored by heating the metal.

5.29 Solution: Working (bending, etc.) disrupts the slip planes, particularly when puckered layers (not corresponding to slip planes) slide over one another. The original structure can be restored by heating (annealing) below the melting point.

5.30 Problem: Discuss the possible effects of extreme pressure on a metal (assume *bcc* at low pressure) and on a solid nonmetal.

5.30 Solution: The *bcc* structure is not close packed and only about 90% as dense as *ccp* and *hcp* structures. At high pressure the *bcc* structure would be expected to convert to a more dense close-packed structure (LeChatelier's Principle). Most solid nonmetals have crystal structures that accommodate specific bonding interactions. These are not likely to be altered by pressure effects, at least in some general predictable way. Some nonmetals form molecular crystals (e.g., N_2, O_2, and S_8). In these cases commonly, the molecules are close packed and a structure change with increasing pressure is not expected. At sufficiently high pressures, everything behaves as a metal.

5.31 Problem: Consider the sliding of one layer of a crystal over another until an e q u i v a l e n t arrangement is achieved. What are the consequences with respect to hardness, brittleness, and malleability for (1) a metal, (2) an ionic crystal, (3) a covalent crystal, and (4) a molecular crystal. In a crude way, how do these considerations relate to melting point?

5.31 Solution: Ionic crystals and covalent crystals (e.g., diamond and quartz) have high melting points because of strong "bonding". Molecular crystals are low melting because of weak interaction between molecules. Their melting points depend roughly on molar masses since van der Waals interactions increase with increasing number of electrons. Metals vary from low melting points (Hg is a liquid) to extremely high melting points for metals such as W. There are great variations in bond strengths for metals.

(1) **Metal**—all atoms are the same and the close-packed layers or slip planes can slide over one another. Bonding is weakened as the atoms slide from the original packing sites, but the same situation is restored by 1 atom displacement. Metals are malleable, particularly those with *ccp* structures (see Problem 5.28). Hardness and melting points vary with bond strength.

(2) **Ionic Crystal**—hard but brittle. Great force is required to separate cations from anions, but displacement brings anions into contact and cations into contact, causing the crystal to cleave or shatter.

(3) **Covalent Crystal**— hard and brittle. Displacement requires the breaking of covalent bonds that are not easily broken or restored.

(4) **Molecular Crystal**—soft and malleable. There is weak bonding between molecules so they can slide over one another. There is no strong repulsion during deformation.

5.32 Problem:
(a) NiAs and CdI_2 use *hcp* packing and octahedral sites. How do the structures differ?

(b) Several pairs of elements (e.g., Ni, Te and Co, Te) form a series of nonstoichiometric compounds in which the limiting composition MX has the NaAs structure and that of limiting composition MX_2 has the CdI_2 structure. Explain this observation.

5.32 Solution:
(a) The NiAs structure has As in *hcp* packing positions with octahedral holes filled by Ni. This gives a trigonal arrangement of Ni atoms around As. For CdI_2 there is an *hcp* array of I with half of the octahedral holes filled by Cd. That is, every other layer of octahedral sites is vacant and cleavage planes occur between the layers of iodine which face each other.

(b) When the composition is intermediate between MX and MX_2, the octahedral holes empty in the CdI_2 structure can be occupied by the "extra" M atoms. When all are occupied, the stoichiometry will be MX and the NiAs structure is reached.

II. BONDING AND STRUCTURE
6
Solid State Chemistry

6.1 Problem: Place a dot anywhere within the set of stick figures below in **(a)** and **(b)** and then form a 2–D lattice by putting dots at all of the sites in an identical environment.
(a) **(b)**

6.1 Solution:
An infinite number of sites could be selected for the first lattice point. Each subsequent lattice point must have an identical environment as the first one chosen. Thus the spacing of the lattice points will be the same as that given below. Usually lattice points are placed at the intersection of the maximum number of symmetry elements. In figure **(a)** this would be either the vertical mirror plane bisecting the stick figure or on the mirror plane between the stick figures. The stick figures in **(b)** are asymmetric, so any point may be selected as the first point. Note, however, that the inverted stick figures may be distinguished from the others, so lattice points can *not* reside on both inverted and standing figures. The dots mark identical points.

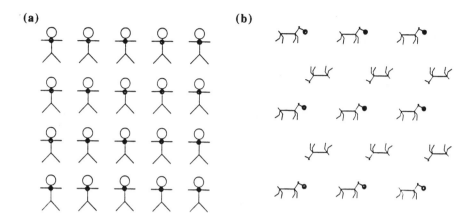

6.2 Problem: (a) Draw a planar set of lattice points and mark in at least four primitive unit cells.
(b) Demonstrate that these all have the same area.
(c) Mark in a multiply primitive cell (one containing more than one lattice point). What is its area compared to the primitive cell?

6.2 Solution:
(a)

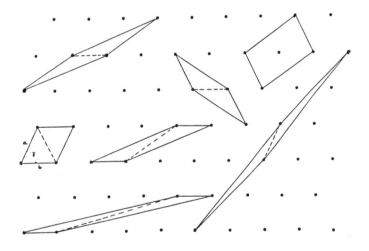

(b) Divide the primitive unit cells into two triangles of equal area. Every such triangle has either $\vec{a}$ or $\vec{b}$ as a base. The altitude of the triangle will be d_{10} or d_{01} and will be equal to $a \sin \gamma$ or $b \sin \gamma$. The area of all of the primitive unit cells will thus be $\frac{1}{2} ab \sin \gamma + \frac{1}{2} ab \sin \gamma = ab \sin \gamma$.

(c) For the multiply primitive unit cell chosen above the area is $(2b)a \sin \gamma$ and the cell contains two lattice points. For any cell which you may draw, the area will be proportional to the number of lattice points contained in the cell. For a three-dimensional lattice, the volume of any unit cell must be a simple multiple of the volume of a primitive cell.

6.3 Problem: Give the number of lattice points belonging to a unit cell in each of the following:
(a) Primitive hexagonal
(b) End-centered tetragonal
(c) I_{432}

6.3 Solution: (a) A primitive cell of any type contains *one* lattice point.
(b) An end-centered cell contains two lattice points, *i. e.*, $\frac{1}{8}(8) + \frac{1}{2}(2) = 2$.
(c) I_{432} denotes a body-centered (I) cubic (432) space lattice so the unit cell contains two lattice points.

6.4 Problem: Show that, for a cubic lattice, the vector $\vec{1}\vec{1}\vec{1}$ is perpendicular to the 111 crystal face.

6.4 Solution: The figure below shows a 111 plane in a unit cube bisected by a plane containing the points 1,1,0; 1,1,1; and the origin. A cross section is shown at the right with the intersection of the two planes giving the line labeled ℓ and the 111 ray labeled r.

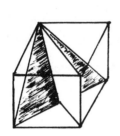
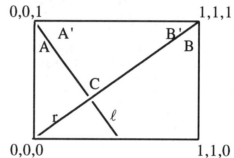

The angle between ℓ and r, labeled C, is to be found.
$$C = 180° - A' - B' = 180° - (90° - A) - (90° - B) = A + B$$
$$A = \tan^{-1}\sqrt{2}/2 = 35.26° \qquad B = \tan^{-1}\sqrt{2} = 54.74° \qquad C = 90°$$

Because of the cubic symmetry, the 1,1,0 point can be replaced with 1,0,1 or 0,1,1 with similar results. [More generally, the *hkl* plane will have Cartesian intercepts of $1/h$, $1/k$, and $1/l$, respectively, (in units of a) in the cubic system and the equation for the plane may be written as
$$hx + ky + lz = 1]$$
The distance of a point at x_1, y_1, z_1 from the plane is given by
$$(hx_1 + ky_1 + lz_1 - 1)p$$
where p is the distance of the plane from the origin, *i.e.*,
$$p = \frac{1}{\sqrt{h^2 + k^2 + l^2}}$$
The sum of the distance from the origin to the *hkl* plane and to the point x_1, y_1, z_1 is thus
$$d = (hx_1 + ky_1 + lz_1)p = \frac{hx_1 + ky_1 + lz_1}{\sqrt{h^2 + k^2 + l^2}}$$
For $x_1 = h$, $y_1 = k$, and $z_1 = l$, the above formula reduces to
$$d = \sqrt{h^2 + k^2 + l^2}$$
This is also the distance formula between the origin and a point with the coordinates h, k, and l. Thus *in the cubic system the hkl ray is normal to the hkl plane.*

6.5 Problem: Consider a tetragonal crystal with $c = 2a = 2b$. Calculate the angle between the 111 vector and the 111 face.

6.5 Solution: Proceeding in the same manner as in Problem 6.4, we draw a tetragonal cell with the 111 plane marked in and a plane perpendicular through the 1,1,1 point. On the cross section we mark in all of the coordinates in terms of *a* so that we may use normal distance relationships.

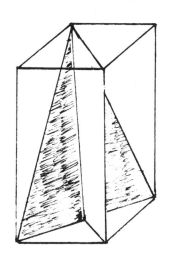
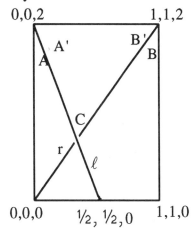

Here we find that

$A = \tan^{-1} \sqrt{2}/4 = 19.47°$ $\qquad B = \tan^{-1} \sqrt{2}/2 = 35.26°$ $\qquad C = A + B = 54.74°$

In noncubic systems, the *hkl* ray is usually *not* perpendicular to the *hkl* plane.

6.6 Problem: A crystal structure belongs to the space group $P4_2/n$.
(a) What is the point group of the space group in Schönflies notation?
(b) To what crystal system does the crystal belong?
(c) How many lattice points may be said to belong to the unit cell?

6.6 Solution: (a) The P simply tells us that the unit cell is primitive. The 4_2 indicates a four-fold screw axis, the *n* a glide plane, and the slash indicates that the glide plane is perpendicular to the four-fold screw axis. Setting translations equal to zero, the 4_2 becomes a four-fold rotation axis and the glide plane becomes a horizontal perpendicular mirror plane. The Hermann-Mauguin notation for the point group of the crystal is 4/m and the Schönflies notation is C_{4h}.
(b) The crystal belongs to the tetragonal system since the C_4 axis is the only rotational symmetry present.
(c) All primitive unit cells contain only one lattice point.

6.7 Problem:. With the aid of a stereogram for the D_{2h} point group, develop the group multiplication for this group.

6.7 Solution: A stereogram for the C_{2h} point group is shown below. On it we have marked one point, a solid dot, as the identity operation, and we have marked the rest with the symmetry operation which would carry the solid dot into the other position. The solid dots are labeled below and the open circles above the symbols. To obtain the product of two operations we find the first on the stereogram and go to the place it would be shifted by the second operation. The result is the product. Thus the product $C_2(x)C_2(z)$ is found by locating $C_2(z)$ on the stereogram (the solid dot at the upper right) and finding where it goes on carrying out a $C_2(x)$ operation (the open circle at the upper left). The result carries the label $C_2(y)$ and is the product sought. The results of all binary multiplications in this group are given in the $\mathbf{D}_{2h}$ multiplication table.

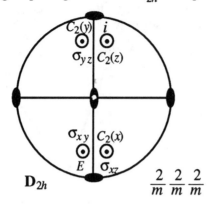

$\mathbf{D}_{2h}$	E	$C_2(z)$	$C_2(y)$	$C_2(x)$	i	σ_{xy}	σ_{xz}	σ_{yz}
E	E	$C_2(z)$	$C_2(y)$	$C_2(x)$	i	σ_{xy}	σ_{xz}	σ_{yz}
$C_2(z)$	$C_2(z)$	E	$C_2(x)$	$C_2(y)$	σ_{xy}	i	σ_{yz}	σ_{xz}
$C_2(y)$	$C_2(y)$	$C_2(x)$	E	$C_2(z)$	σ_{xz}	σ_{yz}	i	σ_{xy}
$C_2(x)$	$C_2(x)$	$C_2(y)$	$C_2(z)$	E	σ_{yz}	σ_{xz}	σ_{xy}	i
i	i	σ_{xy}	σ_{xz}	σ_{yz}	E	$C_2(z)$	$C_2(y)$	$C_2(x)$
σ_{xy}	σ_{xy}	i	σ_{yz}	σ_{xz}	$C_2(z)$	E	$C_2(x)$	$C_2(y)$
σ_{xz}	σ_{xz}	σ_{yz}	i	σ_{xy}	$C_2(y)$	$C_2(x)$	E	$C_2(z)$
σ_{yz}	σ_{yz}	σ_{xz}	σ_{xy}	i	$C_2(x)$	$C_2(y)$	$C_2(z)$	E

6.8 Problem: List the symmetry elements of the $\mathbf{C}_{3v}$ group and develop a stereogram for the group.

6.8 Solution: The $\mathbf{C}_3$ group is generated by repetition of the C_3 operation and reflection in a vertical plane. The rotation group is

$$C_3, C_3^2, C_3^3 (= E)$$

The full symmetry group is given by adjoining a vertical reflection to these operations, *i.e.*,

$$\sigma E = \sigma \qquad \sigma C_3 = \sigma' \qquad \sigma C_3^2 = \sigma''$$

The stereogram is developed by starting with a dashed circle (there is no horizontal mirror plane, which a solid circle would indicate) and placing a solid triangle in the center to represent the C_3 generator. Solid lines are drawn which bisect the triangle and represent the vertical mirror planes. A solid dot is placed beside one of the lines and the C_3 operation carries it around (like it is being pushed by a paddle wheel) to give the three dots representing the rotational operations. Reflection of this set of dots by any of the vertical mirrors produces the remainder of the dots. We may, of course, produce these by operating on a single dot in the rotation set with each of the three vertical mirrors separately.

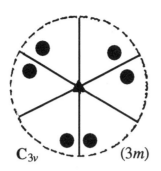

C_{3v} (3m)

6.9 Problem: A unit cell for perovskite is shown at the right.
   (a) How many Ca atoms are there per unit cell?
   (b) How many Ti/unit cell?
   (c) How many O/unit cell?
   (d) What is the chemical formula?
   (e) Into how many parts does a set of 222 planes divide the unit cell?
  (f) Sketch in this set of planes.

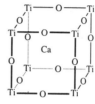

6.9 Solution: The number of atoms in a paralellepiped unit cell may be counted by summing 1/8 of the atoms at the corners, 1/4 of the edge atoms, 1/2 of the face atoms, and all interior atoms.
(a) One interior Ca atom/cell = 1 Ca atom per unit cell.
(b) (8 corner Ti atoms) x1/8 = 1 Ti/unit cell.
(c) (12 edge O atoms) x 1/4 = 3 O atoms/unit cell.
(d) From the numbers of atoms per unit cell, the chemical formula of perovskite is $CaTiO_3$.

(e) A set of *hkl* planes divides a cell into $h + k + l$ volumes between the planes, dividing the *a* edge into *h* equal lengths, the *b* edge into *k* equal lengths, and the *c* edge into *l* equal lengths. The 222 planes divide each edge of the unit cell into two equal lengths.

(f) The 222 planes thus intersect the unit cell at $0.5a$, $0.5b$, and $0.5c$ with parallel planes at 0 (or 1) on each axis. Since every lattice point is identical, each corner may be considered to be an origin (000), We show several views for greater clarity, the one on the left is the conventional orientation of the crystallographic axes.

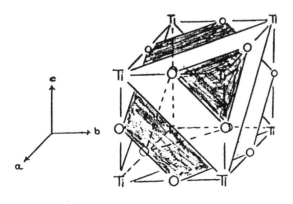
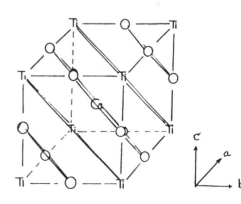

6.10 Problem: Derive a general formula for the distance between hkl planes in a cubic lattice. (*Hint::* What is the distance in reciprocal space from the origin to the point hkl ?)

6.10 Solution:
$$(d^*_{hkl})^2 = (ha^*)^2 + (kb^*)^2 + (lc^*)^2$$
$$= (h/a)^2 + (k/b)^2 + (l/c)^2$$
Since $a = b = c$
$$(d_{hkl})^2 = 1/(d^*_{hkl})^2 = a^2/(h^2 + k^2 + l^2)$$

$$d_{hkl} = \frac{a}{\sqrt{h^2 + k^2 + l^2}}$$

For an orthorhombic system, where $a \neq b \neq c$, the distance between hkl planes becomes
$$d_{hkl} = \frac{1}{\sqrt{(h/a)^2 + (k/b)^2 + (l/c)^2}}$$

6.11 Problem: (a) Under what circumstances with regard to relative sizes of ions and degree of nonpolar character are Frenkel and Schottky defects likely?
(b) The phenomenon of "half-melting" of solid electrolytes is related to which type of defect?

6.11 Solution: (a) Schottky defects, involving stoichiometric cation and anion vacancies, are likely for ionic compounds and ions of comparable size. Schottky defects minimize mechanical distortions and charge displacement. Frenkel defects, involving the migration of ions to interstitial sites, are more likely for more covalent compounds and for large (and highly polarizable) anions. Large, polarizable anions make it easier to accommodate the displaced cations and minimize repulsive effects.
(b) Solid electrolytes commonly have Frenkel defects, providing mobile cations. Half-melting refers to a temperature dependent transition, above which the cations are free to migrate.

6.12 Problem: Consider a linear carbyne structure with alternating short and long C—C bond distances as indicated below, with the z axis as given.

$$—C\equiv C—C\equiv C—C\equiv C—C\equiv C—C\equiv C—C\equiv C— \quad \rightarrow z$$

(a) Sketch a set of bonding and antibonding orbitals for the p_x orbitals of a C_2 unit in the above and label the $\pi_{bonding}$ and π^* orbitals.
(b) Sketch a portion of the wavefunction formed from $\pi_{bonding}$ at $k = 0$.
(c) Sketch a portion of the wavefunction formed from $\pi_{bonding}$ at $k = 0.5\vec{a}*$.
(d) Sketch a portion of the wavefunction formed from π^* at $k = 0$.
(e) Sketch a portion of the wavefunction formed from π^* at $k = 0.5\vec{a}*$.
(f) Show band structure (the energy versus k) for the bands formed from the $\pi_{bonding}$ and the π^* orbitals for a carbyne chain.

6.12 Solution: (a) These are similar to the bonding and antibonding π orbitals in ethylene and are obtained as the sum and difference of the p_x orbitals on the neighboring carbons.

b) $k = 0$ is the center of the first Brillouin zone and there is no change in the sign of the wavefunction with translation from one unit cell to the next.

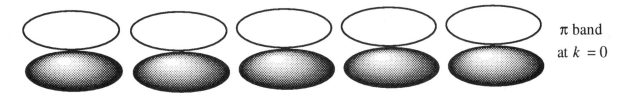

π band at $k = 0$

(c) $k = 0.5\vec{a}*$ is at the edge of the first Brillouin zone, and the sign of the wavefunction alternates with translation from one unit cell to the next.

π band at $k = 0.5\vec{a}*$

(d & e) This is similar to parts **b** and **c** above, but we use the antibonding orbital as our motif.

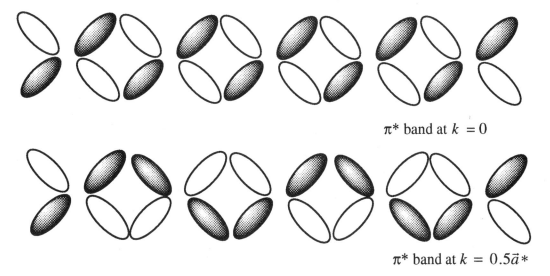

π* band at $k = 0$

π* band at $k = 0.5\vec{a}*$

(f) The energy is lowest with the π bonding band at the zone center (b) and increases as we move toward band edge (c). For the π* band, the energy is highest at the band center (d) and decreases as we approach the band edge. If the C—C distances were equal, the energies would be the same at the band edge (and, in fact, we should have chosen a single C for our motif). With alternating bond lengths, the π bonding orbital would be lower and the π* higher in energy at the band edge. Close-packed strands of carbynes have been proposed by A. G. Whittaker to account for the high-temperature, high-pressure phases of carbon (*Science* **1978**, *200*, 763). Krotos and Smalley were looking for C_nN_2 structures when they discovered C_{60}.

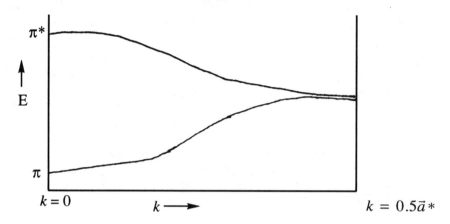

6.13 Problem: If Ge is added to GaAs, the Ge is about equally distributed between Ga and As sites. Which sites would Ge prefer if Se is added also? Would GaAs doped with Se be an *n*-type or a *p*-type semiconductor?

6.13 Solution: Ge prefers Ga sites if Se is added, since Se occupies As sites, as Se^+ plus an electron, giving an *n*-type semiconductor.

6.14 Problem: Sketch the curves for the distribution of energy states and their electron populations for a metallic conductor, an insulator, and a semiconductor.

6.14 Solution: For metals a conducting band is partially filled or a filled band overlaps an empty band. The filled and empty bands are well separated for an insulator, but the energy levels are close enough for promotion in the case of a semi-conductor. The densities of states (DOS) are sketched below.

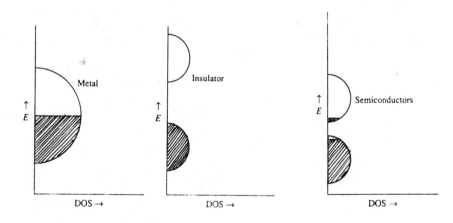

6.15 Problem: In a solid electrolyte what are the advantages of an *hcp* array of anions over a *ccp* array of anions? Assume that the metal ions are in octahedral holes.

6.15 Solution: With the anions in close-packed positions, the octahedral sites are line up at C positions for *hcp*, but they are staggered (A, B, and C positions) for *ccp*. The motion of cations would be less impeded in the open C channels for *hcp*.

6.16 Problem: (a) Explain how you might determine whether graphite forms intercalation compounds (GIC's) when placed in contact with various vapors such as $SnCl_4$, $SOCl_2$, HNO_3, CF_3CO_2H, etc. (b) If no reaction is found to occur, propose a different route to a possible compound. (c) What type of reaction, if any, do you expect?

6.16 Solution: (a) In many cases, visual inspection of the graphite will show the appreciable swelling typical of the graphite layers being pushed apart to accommodate the guest molecules. More quantitatively, the 001 spacing could be determined from the X-ray diffraction pattern. Other possible methods of study include observation of a possible weight increase (by suspending a graphite crystal on a quartz spring and noting the extension during reaction), measurement of the resistivity (the intercalate should be a good conductor), observation of color changes, etc.
(b) GIC's are normally formed by either electron donors or electron acceptors. None of the potential guest compounds listed is a candidate for electron donation, and only HNO_3 is a good oxidizing agent, *i. e.*, electron acceptor. Accordingly, a good oxidizer such as Cl_2 could be added. On accepting electrons a complex such as $[SnCl_6]^{2-}$ or $[SOCl_3]^-$ might form.
(c) The reaction expected is an oxidation of the graphite layer :

$$C_n \rightarrow C_n^+ + e$$
$$e + 3HNO_3 \rightarrow NO_2(g) + H_3O^+ + 2NO_3^-$$

The nitrate ion would be expected to form strong H bonds to additional nitric acid molecules.

6.17 Problem: CuZn has the CsCl structure. (a) Draw a unit cell for CuZn. (b) To what Bravais lattice does this structure belong? (c) How do you expect CuZn to differ from CsCl?

6.17 Solution: (a)

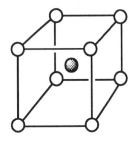

(b) The lattice is simple cubic. The environment is *not* the same for a Cu as for a Zn, so this is *not a body-centered cubic lattice!* Above 727 K the atom sites are randomly occupied by Cu or Zn, and the lattice may then properly be described as body-centered cubic (*bcc*).
(c) Cs and Cl have very different electronegativities, and CsCl is best described as an ionic compound. Cu and Zn are neighboring metals, and CuZn may be described as an intermetallic compound, one of several brasses (the high temperature form being known as β-brass). It also belongs to a class of compounds known as Hume-Rothery phases, which are characterized by the ratio of valence electrons to atoms. In this scheme Cu is considered to have one valence electron, Zn two, Ga three, etc., but Fe, Co, and Ni contribute no valence electrons to such compounds.

Thus intermetallic compounds with such diverse stoichiometries as Cu_5Sn, Cu_3Al, and FeAl all show β-brass phases with (at high temperatures) atoms randomly occupying *bcc* lattice sites. These may be viewed as the solid state equivalent of isoelectronic molecules (those with the same number of valence electrons and atoms) and may be rationalized as a stoichiometry where the Fermi surface touches the Brillouin zone boundary for the particular structure.

6.18 Problem: A number of binary superconductors belong to the cubic system with A atoms at 0 0 0 and ½ ½ ½, and B atoms at ½ 0 ¼) and ½ 0 ¾) where) indicates cyclic permutation of *x y z* (thus ¼ ½ 0 and 0 ¼ ½, etc.). **(a)** Draw a unit cell (for simplicity the ½ ½ ½ point and hidden planes may be omitted). **(b)** What is the point symmetry at the A sites? How are the B sites related to each other? Why is the unit cell described as simple cubic? **(c)** What is the formula when A is Ge and B is Nb? **(d)** If T_C of the Nb-Ge compound is 23.2 K, what is the ratio of the number of superconducting electrons at 20.0 K compared to 0 K? How does this ratio compare to that of a 93 K superconductor at 20 K?

6.18 Solution: (a)

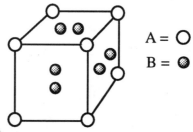

(b) All A sites have T_h point symmetry. The B sites form a set of chains parallel to the crystallographic axes *a*, *b*, and *c*. Planes may be found which contain any single set of chains and corner atoms *or* the body-centered atoms *but not both*. Thus the unit cell is a primitive cubic cell not a body-centered cubic cell. (The body-centered position is related to the corner positions by a glide plane operation.)

(c) # of Ge atoms/cell = $\frac{1}{8}$ (8) + 1 = 2

# of Nb atoms/cell = $\frac{1}{2}$ (6 x 2) = 6

Formula is Nb_3Ge

(d) From the 2-fluid model we have $n_n/n = (T/T_C)^4$, where n_n is the fraction of "normal" electrons and n is the sum of the "normal" and "superconducting" electrons.

At 20 K, $n_n/n = (20/23.2)^4 = 0.55$

so the superconducting fraction would be (1 − 0.55) or 0.45. At 0 K all of the electrons would be superconducting based on this model, so 0.45 is also the ratio of superconducting electrons at 20 K compared to those at 0 K. The same type of calculation shows that this ratio is 0.998 at 20 K for a superconductor having a T_C of 93 K. Safe practice is to use superconducting temperatures below 75% of T_C.

6.19 Problem: A buckminsterfullerine cage is shown below. The solid consists of a cubic closest packed array of such cages and has a completely filled h_u band and a completely empty t_{1u} band at an energy of approximately 1.7 eV above the h_u band. The solid may readily be intercalated with alkali metals.

(a) Explain the normal state conduction properties expected for C_{60}, K_3C_{60}, and K_6C_{60}.

(b) K_3C_{60} is a superconductor with a T_C of 19 K, a London penetration depth, λ, falling in the range 2400 Å to 4800 Å, and a coherence length, ξ, of about 26 Å. Is K_3C_{60} a type I or type II superconductor?

(c) The penetration depth decreases as the density of states (DOS) at the Fermi surface increases, while the T_C increases with an increase in DOS at the Fermi surface. Rb_3C_{60} is also a superconductor. Compare the penetration depth and T_C expected for this compound with the values for K_3C_{60}. Is there a limit to changes expected as the radius of M in M_3C_{60} increases?

(d) If the upper critical field of K_3C_{60} is 50 tesla, what value is expected for the lower critical field?

(e) How many IR-active vibrations would be expected for a C_{60} molecule?

(f) Why may the claim be made that C_{60} is the purest of C? Do you feel that it is a valid claim? Explain.

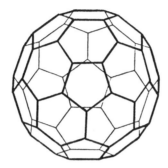

C_{60}
Buckminsterfullerene

6.19 Solution: (a) Solid C_{60} should be an insulator since there is a 1.7 eV band gap between filled and empty states. K_3C_{60} should display metallic conductivity since the $3e/C_{60}$ would half-fill the t_{1u} band. K_6C_{60} should be an insulator since the t_{1u} band would be completely filled. This asumes the next higher empty band is sufficiently removed in energy so as to not show thermal population. Experiment shows K_6C_{60} to be an insulator.

(b) The ratio λ/ξ is around 100. When $(\lambda/\xi) > \sqrt{2}$, the superconductor is a type II superconductor.

(c) Increasing the size of the alkali metal intercalate ion will push the C_{60}^{3-} ions further apart, decreasing the overlap between the t_{1u} orbitals on separate C_{60}^{3-} ions and thus decreasing the dispersion of the half-filled t_{1u} band. Therefore, the DOS at the Fermi surface is expected to increase as the radius of the alkali metal increases. T_C should increase and λ decrease as the radius of M^+ increases. At some point the overlap becomes so small the electrons are localized on individual C_{60} cages and the M_3C_{60} will cease to be a conductor (or a superconductor) and becomes an insulator.

(d) Using the values of λ and ξ from part (b), and the H_{C_2} value given, we may calculate H_{C_1} from the relationship

$$\frac{\lambda}{\xi} = \frac{1}{\sqrt{2}} \left(\frac{H_{C_2}}{H_{C_1}} \right)^{\frac{1}{2}}$$

and obtain a range of 7 to 28 gauss (1 T = 10,000 gauss). Normally H_{C_1} is experimentally more accessible than H_{C_2} and one would not obtain H_{C_1} in this manner.

(e) We first assign the point group, here I_h from the noncoincident C_5 axes and i. The steps in finding the infrared vibrations are as follows:

(1) Find the number of unshifted atoms under an operation of each class in the point group (line 1 of the table below).

(2) Find the symmetry species for x, y, and z and sum the corresponding characters (in this case x, y, and z transform under I_h as t_{1u} and the character string is given on line 2).

(3) Take the direct product of the $\Gamma_{\text{unshifted atoms}}$ and Γ_{xyz} (line 3).

(4) Reduce the reducible represntation from line 3 (we show on line 4 only the number of t_{1u} species contained).

(5) Subtract symmety species corresponding to rotations (R_x, R_y, and R_z) (here t_{1g}) and to translation (T_x, T_y, T_z, here simply one t_{1u} species). The remaining symmetry species correspond to the real molecular vibrations. The IR active vibrations have the same symmetry species designation as those for the x, y, and z coordinates (corresponding to the symmetry species of the dipole moment operator). In the C_{60} case, x, y, and z transform as a degenerate set, i.e., t_{1u}, hence the reason we could ignore the other symmetry species. The others will either transform as quadratic functions and be Raman active, or belong to neither set and be silent (not seen by either type of spectroscopy).

I_h	E	$12C_5$	$12C_5^2$	$20C_3$	$15C_2$	i	$12S_{10}$	$12S_{10}^3$	$20S_6$	15σ	
$\Gamma_{\text{unshifted atoms}}$	60	0	0	0	0	0	0	0	0	4	line 1
Γ_{xyz}	3	p	q	0	−1	−3	−q	−p	0	1	line 2
$\Gamma_{\text{Reducible}}$	180	0	0	0	0	0	0	0	0	4	line 3

of $t_{1u} = \frac{1}{120}[1\cdot 3\cdot 180 + 0 + 0 \quad + 0 \quad + 0 \quad + 0 + 0 \quad + 0 \quad + 0 \quad + 15\cdot 1\cdot 4] = 5$

$$5t_{1u} - 1t_{1u} \text{ (translation)} = 4 \text{ IR active} \qquad \left(p = \frac{1+\sqrt{5}}{2} \qquad q = \frac{1-\sqrt{5}}{2} \right)$$

The four expected IR bands are indeed found for C_{60} at 527, 576, 1183, and 1428 cm^{-1} and provide a method for checking the purity of C_{60} samples.

(f) Both graphite and diamond would have unsaturated valences at their surfaces—these are usually saturated by H or O. C_{60} has no unsaturated valences, so, in theory, it should be the most pure form. In practice the soot from which C_{60} is isolated contains many other fullerines and other forms of carbon. The C_{60} is extracted from these using toluene and further purified by chromatography. The large interstices in solid C_{60} make final solvent separation difficult. Thus in practice, for macroscopic quantities, diamond remains the most pure form of carbon.

II. CHEMICAL REACTIONS
7
Acids and Bases

7.1 Problem: Which has broader applications in terms of definitions of acids and bases, the Arrhenius/Ostwald or the Brønsted/Lowry definitions?

7.1 Solution: In both theories an acid is a proton donor. In the Arrhenius/Ostwald theory a base provides OH^-. The Brønsted/Lowry theory emphasizes the role of solvent. Protons are transferred to and from the solvent and to the conjugate base. Other species besides OH^- are recognized as bases or proton acceptors.

7.2 Problem: What is the great departure in terms of acid-base definitions in the Lewis theory compared to those of Arrhenius/Ostwald and Brønsted/Lowry?

7.2 Solution: The Arrhenius/Ostwald and Brønsted/Lowry theories focus on proton transfer. The Lewis theory includes proton transfer, but defines acid-base processes in terms of sharing an electron-pair between a base (the electron pair donor) and an acid (the electron pair acceptor).

7.3 Problem: (a) Indicate the conjugate bases of the following: NH_3, NH_2^-, NH^{2-}, H_2O, HI.
(b) Indicate the conjugate acids of the above species.
(c) What relationship exists between the strength of a conjugate acid and a conjugate base for a neutral substance, such as NH_3 in parts (a) and (b) above?

7.3 Solution: An acid becomes a "conjugate" base when it loses a proton; a base becomes a "conjugate" acid when it gains a proton.

(a) | acid | → | conjugate base | + H^+ |
|---|---|---|---|
| NH_3 | → | NH_2^- | + H^+ |
| NH_2^- | → | NH^{2-} | + H^+ |
| NH^{2-} | → | N^{3-} | + H^+ |
| H_2O | → | OH^- | + H^+ |
| HI | → | I^- | + H^+ |

109

(b)

base	+	H⁺	→	conjugate acid
NH_3	+	H^+	→	NH_4^+
NH_2^-	+	H^+	→	NH_3
NH^{2-}	+	H^+	→	NH_2^-
H_2O	+	H^+	→	H_3O^+
HI	+	H^+	→	H_2I^+

(c) Let us consider behavior in aqueous solution. We know that HI(aq) contains mainly I⁻(aq) and no detectable H_2I^+(aq), H_2O contains equal, but small, amounts of H_3O^+(aq) and OH^-(aq), while NH_3(aq) contains some NH_4^+(aq) but no detectable NH_2^-. From these observations we might conclude that substances with strong conjugate acids have weak conjugate bases and *vice versa*. In liquid NH_3 the autoprotolysis is

$$NH_3(l) \rightleftharpoons NH_4^+ + NH_2^-$$

$$K_s = \frac{[NH_4^+][NH_2^-]}{[NH_3]^2}$$

In water the reactions are:

$$NH_3 + H_2O \rightarrow NH_2^- + H_3O^+ \quad\quad K_a(NH_3)$$
$$NH_3 + H_2O \rightarrow NH_4^+ + OH^- \quad\quad K_b(NH_3)$$

from which

$$K_a K_b = \frac{[NH_4^+][NH_2^-]}{[NH_3]^2} K_w = K_s K_w$$

For gross approximations $[NH_4^+][NH_2^-]/[NH_3]^2$ may be replaced with K_s for $NH_3(l)$, i.e., 10^{-30}. Using $K_b = 10^{-9.5}$ for NH_3 and our approximation $K_a K_b = K_s K_w$ gives a value of 10^{-35} for K_a for NH_3 in aqueous solutions.

7.4 Problem: The heat of neutralization of H^+(aq) and OH^-(aq) is considerably higher in concentrated salt solution ($\Delta H_{neut.} = -85.4$ kJ/mol at $\mu = 16$) than in dilute aqueous solution ($\Delta H_{neut.} = -56.5$ kJ/mol at $\mu \rightarrow 0$). Explain what might cause the increase in the heat of neutralization when a high concentration of NaCl is present.

7.4 Solution: The water activity in concentrated salt solutions is decreased because of the large amount of water tied up in ion hydration. This will cause a decrease in the average number of water molecules associated with the H^+, thereby increasing the availability of the proton, that is, less energy is lost by the breakup of $(H_2O)_n H^+$ when the proton is transferred. The same reasoning applies to the solvation of the OH^- ion. Lowering the water activity of strong acid or base solutions thus increases the effective strength of the acid or base.

7.5 Problem: Write a thermochemical cycle from which you can obtain the hydride affinity of a positive ion (E^+) from the bond dissociation energy of EH and other suitable data.

7.5 Solution:

$$E^+ + H^- \xrightarrow{1} EH \qquad \text{Hydride affinity } (H^-A) = -\Delta H_1$$

$$4 \uparrow \qquad \qquad \downarrow 2 \qquad \qquad \Delta H_2 = D_{E-H}$$

$$E^+ + H + e \xleftarrow{3} E + H \qquad \Delta H_3 = IE(E)$$

$$\qquad \qquad \qquad \qquad \qquad \qquad \Delta H_4 = -EA(H)$$

$$\Delta H_1 + \Delta H_2 + \Delta H_3 + \Delta H_4 = 0$$
$$H^-A(E^+) = D_{E-H} + IE_E - EA_H$$

7.6 Problem: Why is there a smaller difference in the proton affinities (PA) of PH_3 and PF_3 compared to those of NH_3 and NF_3?

7.6 Solution: There is a great inductive effect on comparing NH_3 and NF_3. The highly electronegative F withdraws electron density from N decreasing the PA by 259 kJ/mol. There is a smaller effect in going from PH_3 to PF_3 even though the electronegativity of P is lower than that of N. There can only be single bonding for NH_3, NF_3, and PH_3. For PF_3 double bonding (F→P π donation) reduces the inductive effect.

7.7 Problem: Why is $H^+(aq)$ the strongest acid and $OH^-(aq)$ the strongest base possible in water?

7.7 Solution: A very strong acid in water transfers the proton to give $H^+(aq)$ and a very strong base accepts a proton from H_2O to form $OH^-(aq)$.

7.8 Problem: Are the solution acidity scales compressed more for solvents that stabilize lyonium and lyate ions greatly or for those that stabilize them minimally?

7.8 Solution: The maximum acidity range occurs in the gas phase. The greater the solvation of lyonium and lyate ions the greater compression of the acidity scale.

7.9 Problem: What would be a good solvent for studying
(a) a series of very weak bases?
(b) a series of very weak acids?
(c) a wide range of strengths of acids and bases?

7.9 Solution: (a) An acidic solvent will enhance ionization of weak bases. H_2SO_4 or another acidic solvent would be suitable.
(b) A basic solvent will enhance ionization of weak acids. $NH_3(l)$ or an amine would be suitable.
(c) Studying a wide range of strengths of acids and bases would require low solvation. An alcohol would be suitable.

7.10 Problem: In what media might one obtain protonated species of very weakly basic substances such as alkanes?

7.10 Solution: An extremely weak base such as an alkane requires the strongest acids possible. The "super acids" such as SbF_5 in HSO_3F protonate alkanes.

7.11 Problem: Give the approximate pK_a values for the following acids.
(a) H_3PO_3 $pK_1 =$ $pK_2 =$
(b) HNO_3 $pK_1 =$
(c) $HClO_4$ $pK_1 =$
(d) H_5IO_6 $pK_1 =$ $pK_2 =$

7.11 Solution: Pauling's rules for estimating the strength of acids of the general formula $XO_n(OH)_m$ are

$$pK_1 = 7 - 5n$$

and successive ionization steps give pK_2, etc., which are successively 5 units greater. In the following we have rewritten the formulas of acids to apply more readily Pauling's rules. The values of n and pK_a are thus:

(a) H_3PO_3 $HPO(OH)_2$ $n = 1$ $pK_1 = 2$ $pK_2 = 7$
(b) HNO_3 $(HO)NO_2$ $n = 2$ $pK_1 = -3$
(c) $HClO_4$ $(HO)ClO_3$ $n = 3$ $pK_1 = -8$
(d) H_5IO_6 $(HO)_5IO$ $n = 1$ $pK_1 = 2$ $pK_2 = 7$

7.12 Problem: Select the best answer and give the basis for your selection.
(a) Thermally most stable: PH_4Cl PH_4Br PH_4I
(b) Strongest acid: H_2O H_2S H_2Se H_2Te
(c) Acidic oxide: Ag_2O V_2O_5 CO Ce_2O_3
(d) Strongest acid: MgF_2 $MgCl_2$ $MgBr_2$
(e) Stronger base (toward a proton): PH_2^- NH_2^-

7.12 Solution:
(a) $PH_4X \xrightarrow{\Delta} PH_3 + HX$

HI is the strongest acid (or I⁻ has the lowest proton affinity) in the series so PH_4I is most stable.
    (Lattice energy effects, which would be in the opposite direction, *do not* override the primary acid strength effects here.)
(b) Strongest acid: H_2Te is the strongest protonic acid in this sequence. The major factor (as in the case of the hydrogen halides) is the decreasing bond X—H energy as we descend the group. (This changes much more than the EAs of SH, SeH, etc.)
(c) Acidic oxide: V_2O_5 is the most acidic oxide here, the charge on the metal ion being the major consideration.
(d) Strongest acid: $MgBr_2$ is the strongest acid here. The lattice energy of MgF_2 makes acidic behavior difficult to detect. The factors are probably the same as with the BX_3 series

although far less experimental work has been done here.
(e) NH_2^- is the stronger base. Important factors are similar to those in part (b) of this problem.

7.13 Problem: Give equations to explain why adding ammonium acetate to either zinc amide(s) in liquid ammonia or zinc acetate(s) in acetic acid causes the solid to dissolve.

7.13 Solution: Zinc salts tend to show amphoteric behavior. In liquid ammonia, $NH_4CH_3CO_2$ is an acid since it increases the positive ion involved in the autoionization of the solvent. In glacial acetic acid, $NH_4CH_3CO_2$ is a base since it increases the concentration of the negative ion involved in the autoionization reaction. Consequently, the zinc salt dissolves in the acidic or basic medium produced. Equations are:

$$Zn(NH_2)_2(s) + 2NH_4^+ \xrightarrow{NH_3(l)} Zn^{2+}(\text{solvated}) + 4NH_3$$
$$\xrightarrow{NH_3(l)} [Zn(NH_3)_4]^{2+}$$
$$Zn(OAc)_2(s) + 2OAc^- \xrightarrow{HOAc} [Zn(OAc)_4]^{2-}$$

7.14 Problem: What are the acid-base properties expected for
(a) Mn^{2+} (b) Mn^{IV} (c) Mn^{VII}

7.14 Solution: (a) Mn^{2+} has low charge and is basic. $Mn(OH)_2$ is a weak base.

(b) Mn^{IV} has little acidic or basic character. MnO_2 dissolves in acid only by reduction and it does not dissolve in bases.

(c) Mn^{VII} is acidic. Mn_2O_7 (Explosive!) dissolves in water to give $HMnO_4$, a strong acid.

7.15 Problem: Transition metal ions usually are encountered in igneous rocks in low oxidation states. Why are many of their ores found in sedimentary rocks formed by the weathering of igneous rocks?

7.15 Solution: Transition metal ions in low oxidation states tend to be oxidized by weathering in the oxygen atmosphere. The higher oxidation states are more acidic and hydrolyze to separate as oxides. The oxidation of Mn^{2+} in igneous rocks and deposition of MnO_2 in sedimentary rocks is an excellent example. Weathering has concentrated some elements in sedimentary rocks.

7.16 Problem: Would the following *increase, decrease,* or have *no effect* on acidity of the solution?
(a) Addition of Li_3N to liquid NH_3
(b) Addition of HgO to an aqueous KI solution
(c) Addition of SiO_2 to molten Fe + FeO
(d) Addition of $CuSO_4$ to aqueous $(NH_4)_2SO_4$
(e) Addition of $Al(OH)_3$ to NaOH
(f) Addition of $KHSO_4$ to H_2SO_4
(g) Addition of CH_3CO_2K to liq. NH_3

7.16 Solution: See Problem 7.13 for guide.
(a) $N^{3-} + 2NH_3 \rightarrow 3NH_2^-$

Li_3N increases the NH_2^- concentration, behaves as base, decreases acidity.

(b) $HgO + 4KI + H_2O \rightarrow [HgI_4]^{2-} + 4K^+ + 2OH^-$

HgO increases the OH^- ion concentration, behaves as base, decreases acidity.

(c) $FeO + SiO_2 \rightarrow FeSiO_3$

SiO_2 is an oxide ion acceptor, behaves as a Lux–Flood acid, increases acidity

(d) $SO_4^{2-} + H_2O \rightarrow HSO_4^- + OH^-$

$CuSO_4$ forms a complex with NH_3, but Cu^{2+} does not displace H^+ from NH_4^+. Hydrolysis is more important.

(e) $Al(OH)_3(s) + OH^-(aq) \rightarrow [Al(OH)_4]^-(aq)$

$Al(OH)_3$ combines with OH^-, therefore acidity increases.

(f) $KHSO_4$ behaves as a base in H_2SO_4 since it increases the HSO_4^- ion concentration. Therefore the acidity of the solution decreases.

(g) CH_3CO_2K does not solvolyze since acetic acid is a strong acid in liquid NH_3. Since neither K^+ nor OAc^- are ions common to the autoionization of liquid ammonia this salt will be neutral in $NH_3(l)$.

7.17 Problem: Some transition metals such as vanadium and molybdenum in their highest oxidation state form aggregated poly acids. Are the poly acids formed to the greater extent in
(a) strongly acidic solution   **(b)** strongly basic solution or   **(c)** intermediate pH values?

7.17 Solution: Polyvanadates and polymolybdenates are encountered in neutral or slightly acidic solution. In strongly acidic solution cations such as VO_2^+ or oxides V_2O_5 or MoO_3 precipitate. In strongly basic solution simple anions, e.g., VO_4^{3-} or MoO_4^{2-}, are formed.

7.18 Problem: In general, anhydrous nitrites of a given metal ion have lower thermal stability than the nitrates, and similarly for the sulfites and sulfates. How may this be explained?

7.18 Solution: This problem is similar to Problem 7.12a, Nitrites and nitrates may be considered as oxide complexes of N_2O_3 and N_2O_5; sulfites and sulfates may be considered to be oxide complexes of SO_2 and SO_3. The nonmetals in the higher oxidation state is the better oxide ion acceptor (or stronger acid) and hence the nitrates and sulfates will show greater thermal stability. It should be noticed that whereas the sulfites and sulfates will generally decompose on strong heating to give SO_2 and SO_3, respectively, decomposition of the nitrites and nitrates is more complex.

7.19 Problem: Select the best response within each horizontal group and indicate the major factor governing your choice.
Strongest protonic acid

(a) SnH_4	SbH_3	H_2Te
(b) NH_3	PH_3	SbH_3
(c) H_5IO_6	H_6TeO_6	HIO
(d) $Fe(H_2O)_6^{3+}$	$Fe(H_2O)_6^{2+}$	H_2O
(e) $Na(H_2O)_x^+$	$K(H_2O)_x^+$	

Strongest Lewis acid
- **(f)** BF$_3$ BCl$_3$ BI$_3$
- **(g)** BeCl$_2$ BCl$_3$
- **(h)** B(*n*-Bu)$_3$ B(*t*-Bu)$_3$

More basic toward BMe$_3$
- **(i)** Me$_3$N   Et$_3$N
- **(j)** 2-MePy   4-MePy   Py (= pyridine)
- **(k)** 2-MeC$_6$H$_4$CN   C$_6$H$_5$CN

7.19 Solution: For **(a)** and **(b)** you may give an answer based on your knowledge of periodic trends in acidity of binary hydrides. A more complete understanding can be obtained by analyzing an appropriate thermodynamic cycle similar to the type written for Problem 7.5.

(a) H$_2$Te		Changes in EA of central atom
(b) SbH$_3$		Changes in X—H bond energy
(c) H$_5$IO$_6$		Based on Pauling's rules (see Problem 7.11)
(d) Fe(H$_2$O)$_6^{3+}$		Higher charge on central atom repels protons most strongly
(e) Na(H$_2$O)$_x^+$		Smaller ionic radius of Na$^+$ leads to greater electrostatic repulsion of proton
(f) BI$_3$		Least π-conjugation
(g) BCl$_3$		More highly developed positive charge on central atom
(h) B(*n*-Bu)$_3$		Less strain in adducts
(i) Me$_3$N		Less strain in adducts
(j) 4-MePy		Combination of *e*-releasing inductive effect of Me group and least steric effect when group is in the *para* -position
(k) 2-MeC$_6$H$_4$CN		The steric effect of the 2-Me is not appreciable here since the reactive center (N) is farther removed than part **(j)** above.

7.20 Problem: Tetrahydrofuran (thf) has an ionization energy (IE) of 11.1 eV and diethyl ether has an IE of 9.6 eV. How can we account for this difference? What difference is expected in their PA's? Which is higher? thf is much more basic toward MgCl$_2$ and Grignard reagents than Et$_2$O. How might this observation be rationalized?

7.20 Solution: The large difference in IE of thf and Et$_2$O cannot be rationalized readily on the basis of electronic effects of a ring vs. an open chain. This suggests that rehybridization takes place on ionization—going from essentially *sp*^{3} for the neutral to *sp*^{2} for the ion (this is known to occur with NH$_3$). The thf would suffer from ring strain in the ion, which would be reflected in its higher IE. The higher IE of thf is reflected in a lower proton affinity for this compound as compared to Et$_2$O. The greater basicity of thf, compared to Et$_2$O, toward Lewis acids may be explained by the lower steric requirements of thf.

7.21 Problem: Geochemical classifications distinguish metals as **lithophile** if they tend to occur as oxides and **chalcophile** if they tend to occur as sulfides. What properties of the metals determine this distinction? Give two examples of each classification.

7.21 Solution: Lithophile metals form hard ions such as Na^+, K^+, Ca^{2+}, Al^{3+}, etc. Chalcophile metals, preferring the softer S^{2-}, form soft metal ions such as Cu^+, Ag^+, Hg^{2+}, Pb^{2+}, etc.

7.22 Problem: Indicate which of the following Lewis bases (ligands) should show ambidentate character, and indicate for such bases the hard and soft ends: NO_3^-, SO_3^{2-}, $S_2O_3^{2-}$, $S_2O_7^{2-}$, NO_2^-, Me_2SO, CN^-, $SeCN^-$.

7.22 Solution: NO_3^- and $S_2O_7^{2-}$ are O donors. Ambidentate ligands must have more than one atom capable of donating an e pair in bond formation. This rules out the nitrate ion, since here only the O atoms have unshared pairs. Chelating agents are not classified as ambidentate since these ligands generally do not show preferential ligation through a single atom. In the following ligands the soft end is the one of lower electronegativity.

SO_3^{2-}	S	soft	NO_2^-	N	soft
$S_2O_3^{2-}$	O	hard		O	hard
CN^-	C	soft	$SeCN^-$	Se	soft
	N	hard		N	hard

7.23 Problem: Given the following C and E parameters for the Drago–Wayland Equation, select the stronger base in each pair.
(a) Acetone or dimethylsulfoxide (dmso) **(b)** Dimethylsulfide or dimethylsulfoxide
Comment on your conclusions regarding possible ambiguity.

	E_B	C_B
Acetone	1.74	1.26
Dimethylsulfoxide	2.40	1.47
Dimethylsulfide	0.25	3.75

Select from among the following acids to support your conclusions.

	E_A	C_A
I_2	0.50	2.00
$HCCl_3$	1.49	0.46
SO_2	0.56	1.52
C_6H_5OH	2.27	1.07

7.23 Solution: The enthalpy of interaction of an acid and a base is given by

$$-\Delta H = E_A E_B + C_A C_B \quad \text{(kcal/mol)}$$

(a) Since both E_B and C_B are greater for dmso than for acetone, dmso is the stronger base in all cases.

(b) The situation here is ambiguous. Thus Me$_2$S is more basic than dmso towards acids with $C_A \geq E_A$ but is a weaker base for acids where $E_A \gg C_A$. Thus with iodine, Me$_2$S releases

$$Q = -\Delta H = (0.50)(0.25) + (2.00)(3.75) = 7.6 \text{ kcal}$$

Whereas with I$_2$, dmso releases only

$$Q = -\Delta H = (0.50)(2.40) + (2.00)(1.47) = 4.1 \text{ kcal}$$

On the other hand, with C$_6$H$_5$OH, Me$_2$S is expected to release

$$Q = -\Delta H = (2.27)(0.25) + (1.07)(3.75) = 4.1 \text{ kcal}$$

whereas with C$_6$H$_5$OH, dmso releases

$$Q = -\Delta H = (2.27)(2.40) + (1.07)(1.47) = 7.0 \text{ kcal.}$$

7.24 Problem: Use values for the E and C parameters given below to rank the basicities of NH$_3$, (CH$_3$)$_2$NH, (CH$_3$)$_3$P, and (CH$_3$)$_2$S toward H$_2$O and B(C$_2$H$_5$)$_3$. See the Drago–Wayland Equation in Problem 7.23.

Base	E_B	C_B	Acid	E_A	C_A
NH$_3$	2.31	2.04	H$_2$O	1.54	0.13
(CH$_3$)$_2$NH	1.80	4.21	B(C$_2$H$_5$)$_3$	1.70	2.71
(CH$_3$)$_3$P	1.46	3.44			
(CH$_3$)$_2$S	0.25	3.75			

7.24 Solution: The Drago-Wayland equation allows you to predict heats of acid-base reactions:

$$-\Delta H = E_A E_B + C_A C_B \quad \text{(kcal/mol)}$$

For any given acid, the stronger the base the more negative ΔH. The following values of ΔH are in kcal/mol:

Base	H$_2$O	BEt$_3$
NH$_3$	–3.83	–9.46
Me$_2$NH	–3.32	–14.47
Me$_3$P	–2.70	–11.80
Me$_2$S	–0.89	–10.59

Hence the rankings are:

toward H$_2$O NH$_3$ > Me$_2$NH > Me$_3$P $\gg$ Me$_2$S

toward BEt$_3$ Me$_2$NH > Me$_3$P > Me$_2$S > NH$_3$

7.25 Problem: Neglecting the metals with partially filled p orbitals, in what regions of the periodic table are the class (a) metal acceptors (those that prefer N, O, or F donors) and the class (b) metal acceptors (those that prefer ligand atoms from the third period or a later period)? How does this classification relate to size, charge, electronegativity, and the hard–soft acid–base classification?

7.25 Solution: The class (a) metal acceptors are the large metal atoms with low electronegativity found on the left side of the periodic table. The class (b) metals are near Au in the periodic table—these are the highly electronegative and highly polarizing (also highly polarizable) metals, particularly in low oxidation states. The class (b) metals are soft acids, preferring soft bases.

7.26 Problem: Indicate the order of increasing values of the proton affinity expected for the following species: NH_2^-, NH_3, NH^{2-}, C_5H_5N, and CH_3CN.

7.26 Solution: In Problem 2.35 we found the order of the electronegativities of the conjugate acids of the above compounds to be $NH_2^- < NH_3 < NH_4^+ <, C_5H_5NH^+ < CH_3CNH^+$.

The attraction for protons by atoms of the same element would be expected to be just the opposite of the attraction for electrons, so the order of PA's would be expected to be
$$NH^{2-} > NH_2^- > NH_3 > C_5H_5N > CH_3CN.$$
Experimentally, the PA of C_5H_5N is greater than that of NH_3 but the other compounds fall in the predicted order. The inversion of the NH_3 and C_5H_5N order has been rationalized on the basis of the greater polarizability of pyridine than that of ammonia. In aqueous solution this "self-solvation" becomes much less important and NH_3 is more basic then C_5H_5N.

7.27 Problem: Transmission of signals by optical waveguides is now practical. One of the factors used in calculating the wavelength for minimum signal loss is the "average electronic excitation gap". Rationalize the following data on the band gap (in eV) of various materials (from K. Nassau, *The Bell System Technical Journal* **1981**, *60* , 327).

LiF	16.5	LiCl	11.0	LiBr	9.5	BeO	13.7		
KF	14.7	KCl	10.5	KBr	9.2	MgO	11.4		
ZnF_2	13	$ZnCl_2$	9	$ZnBr_2$	7.5	CaO	9.9		
HgF_2	9	—		—		ZnO	6.1	ZnTe	4.4
						B_2O_3	12.4	BN	10.6

7.27 Solution: The above data may be rationalized on the basis of absolute hardnesses from HOMO–LUMO gaps. The band gap is greatest for hard–hard interactions and least for soft–soft interactions.

III. CHEMICAL REACTIONS
8
Oxidation-Reduction Reactions

8.1 Problem: (a) Should the value of the heat of reaction (ΔH) calculated from the variation of K_p with temperature be affected by the units in which K_p is expressed?
(b) Under what circumstances would ΔH be unaffected and under what circumstances affected by the selection of units for the K_{eq}?

8.1 Solution: (a) The units in which K_p is expressed will affect the ΔG^0 (since the numerical value of K_p is affected) but will not affect ΔH^0 *if ideal gas behavior is observed.* (The *slope* of $\log K_p$ vs. 1/T will remain constant.) Note that ΔS^0 will be affected and the change in ΔG^0 will be the same as the $-T \Delta S^0$ term.
(b) ΔH^0 will be affected if the change in the units of K involves a change in standard states which cannot be converted from one to another by a simple multiplicative factor, that is, K_n, K_c, K_p, etc.

8.2 Problem: Write balanced ionic half-reactions for
(a) HgO $\xrightarrow{base}$ Hg   (d) $S_2O_3^{2-}$ $\xrightarrow{acid}$ S
(b) VO^+ $\xrightarrow{acid}$ $HV_2O_5^-$ (e) F_2O $\xrightarrow{acid}$ HF
(c) $HGeO_3^-$ $\xrightarrow{base}$ Ge   (f) $C_{12}H_{22}O_{11}$ $\xrightarrow{acid}$ CO_2

8.2 Solution: The sequence of steps used to balance a half-reaction is:
1. Select those species which undergo a chemical change and balance with respect to atoms other than H and O. Thus in part (e) we write $F_2O \rightarrow 2\,HF$; in part (f) $C_{12}H_{22}O_{11} \rightarrow 12\,CO_2$. Where two or more species are produced from a single species, two half-reactions will usually be written.
2. Balance the oxygen atoms by adding H_2O to the appropriate side of the equation. In part (f) we obtain $13\,H_2O + C_{12}H_{22}O_{11} \rightarrow 12\,CO_2 + ...$

3. Balance the hydrogen atoms by adding H⁺ to the appropriate side. In part **(f)** we obtain
$$13\,H_2O + C_{12}H_{22}O_{11} \rightarrow 12\,CO_2 + 48\,H^+ + ...$$
4. Balance the equation with respect to charge by adding electrons to the appropriate side. Again, to balance part **(f)** completely we have $13\,H_2O + C_{12}H_{22}O_{11} \rightarrow 12\,CO_2 + 48\,H^+ + 48\,e$.

5. For reactions occurring in base, we add a number of OH⁻ equal to that of the H⁺ in the half-reaction to each side. The n (H⁺ + OH⁻) are treated as n H₂O and water appearing on both sides of the equation, and are cancelled.

The half-reactions are thus

(a) $2e\ +\ HgO\ +\ 2H^+ \rightarrow Hg\ +\ H_2O$

$$\underline{+\ 2OH^- +\ 2OH^-}$$
$2e\ +\ HgO\ +\ H_2O \rightarrow Hg\ +\ 2OH^-$

(b) $2\,VO^+\ +\ 3H_2O\ \rightarrow\ HV_2O_5^-\ +\ 5H^+\ +\ 2e$

(c) $4e\ +\ 5H^+\ +\ HGeO_3^- \rightarrow Ge\ +\ 3H_2O$

$$\underline{+\ 5OH^- +\ 5OH^-}$$
$4e\ +\ 2\,H_2O\ +\ HGeO_3^- \rightarrow Ge\ +\ 5OH^-$

(d) $4e\ +\ 6H^+\ +\ S_2O_3^{2-} \rightarrow 2\,S\ +\ 3H_2O$

(e) $4e\ +\ 4\,H^+\ +\ F_2O \rightarrow 2\,HF\ +\ H_2O$

(f) $13\,H_2O + C_{12}H_{22}O_{11} \rightarrow 12\,CO_2 + 48\,H^+ + 48\,e$

8.3 Problem: The following are somewhat more challenging equations to balance.
(a) $H_2O + P_2I_4 + P_4 \rightarrow PH_4I + H_3PO_2$
(b) $ReCl_5 + H_2O \rightarrow Re_2Cl_9^{2-} + ReO_4^- + Cl^- + H^+$
(c) $B_{10}H_{12}CNH_3 + NiCl_2 + NaOH \rightarrow Na_4[B_{10}H_{10}CNH_2]_2Ni + NaCl + H_2O$
(d) $ICl + H_2S_2O_7 \rightarrow I_2^+ + I(HSO_4)_3 + HS_3O_{10}^- + HSO_3Cl + H_2SO_4$

8.3 Solution:
(a) $(32e\ +\ 32\,H^+\ +\ 2\,P_2I_4 + P_4 \rightarrow 8\,PH_4I)$
$(8\,H_2O + P_4 \rightarrow 4\,H_3PO_2 + 4\,H^+ + 4e\,)8$

$$\overline{64\,H_2O + 2\,P_2I_4 + 9\,P_4 \rightarrow 8\,PH_4I + 32\,H_3PO_2}$$

(b) $(3e^- + 2\,ReCl_5 \rightarrow Re_2Cl_9^{2-} + Cl^-)2$

$(ReCl_5 + 4\,H_2O \rightarrow ReO_4^- + 5\,Cl^- + 8\,H^+ + 2e^-)3$

$7\,ReCl_5 + 12\,H_2O \rightarrow 2\,Re_2Cl_9^{2-} + 3\,ReO_4^- + 17\,Cl^- + 24\,H^+$

(c) $(B_{10}H_{12}CNH_3 \rightarrow B_{10}H_{10}CNH_2^{2-} + 3\,H^+ + e^-)2$

$\quad 2e^- + NiCl_2 \rightarrow Ni^0 + 2\,Cl^-$

$\quad 6\,NaOH \rightarrow 6\,Na^+ + 6\,OH^-$

$2\,B_{10}H_{12}CNH_3 + NiCl_2 + 6\,NaOH \rightarrow Na_4[B_{10}H_{10}CNH_2]_2Ni + 2\,NaCl + 6\,H_2O$

(d) $(e^- + 2\,ICl \rightarrow I_2^+ + 2\,Cl^-)2$

$(ICl + 3\,HSO_4^- \rightarrow I(HSO_4)_3 + Cl^- + 2e^-)$

$5\,ICl + 3\,HSO_4^- \rightarrow 2\,I_2^+ + I(HSO_4)_3 + 5\,Cl^-$

$(H_2S_2O_7 + Cl^- \rightarrow HSO_3Cl + HSO_4^-)5$

---

(1) $5\,ICl + 5\,H_2S_2O_7 \rightarrow 2\,I_2^+ + I(HSO_4)_3 + 5\,HSO_3Cl + 2\,HSO_4^-$

(2) $2\,H_2S_2O_7 \rightarrow H_2S_3O_{10} + H_2SO_4$

(3) $2\,H_2S_3O_{10} + 2\,HSO_4^- \rightarrow 2\,HS_3O_{10}^- + 2\,H_2SO_4$

===

(1) + 2 x (2) + (3) =

$5\,ICl + 9\,H_2S_2O_7 \rightarrow 2\,I_2^+ + I(HSO_4)_3 + 2\,HS_3O_{10}^- + 5\,HSO_3Cl + 4\,H_2SO_4$

8.4 Problem: Calculate E^0 values for the following cells and write the reactions for which these apply.

(a) Pt, H_2 | HCl || KCl | Hg_2Cl_2(s), Hg. (E^0 Hg_2Cl_2(s) $\xrightarrow{0.268}$ Hg + Cl$^-$)

(b) Cu | Cu^{2+} || I$^-$ | CuI(s), Cu. (E^0 CuI(s) $\xrightarrow{-0.185}$ Cu + I$^-$)

(c) Pt, CuI(s) | Cu^{2+} || I$^-$ | CuI(s), Cu.

From parts (b) and (c), comment on the necessity of knowing the half-cells involved in making predictions of (1) the spontaneity of a reaction and (2) the free-energy change of a reaction.

8.4 Solution:

(a) Pt, H_2 | HCl || KCl | Hg_2Cl_2(s), Hg

$H_2 \rightarrow 2H^+ + 2e$	$E^0 = 0.00$
$2e + Hg_2Cl_2 \rightarrow 2Hg + 2Cl^-$	$E^0 = 0.268$
$H_2 + Hg_2Cl_2 \rightarrow 2Hg + 2Cl^- + 2H^+$	$E^0 = 0.268$ V

(b) Cu | Cu^{2+} || I^- | CuI(s), Cu

$Cu \rightarrow Cu^{2+}$	$E^0 = -0.337$
$2(e + CuI \rightarrow Cu + I^-)$	$E^0 = -0.185$
$2CuI \rightarrow Cu^{2+} + Cu + 2I^-$	$E^0 = -0.523$ V

(c) Pt, CuI(s) | Cu^{2+} || I^- | CuI(s), Cu

$CuI \rightarrow Cu^{2+} + I^- + e$	$E^0 = -0.86$
$e + CuI \rightarrow Cu + I^-$	$E^0 = -0.185$
$2CuI \rightarrow Cu^{2+} + Cu + 2I^-$	$E^0 = -1.045$ V

The spontaneity depends on the *sign* of E^0 which is independent of the reaction path. ΔG^0 is a function of nFE where n depends on the reaction path but ΔG^0 does not.

8.5 Problem: Calculate the emf values for the following cells. Which of these reactions might be too slow to observe?
- **(a)** Pt, H_2 | $H^+(a = 0.1)$ || $H^+(a = 10^{-7})$ | H_2, Pt.
- **(b)** Zn | $Zn^{2+}(a = 1)$ || $Cu^{2+}(a = 10^{-4})$ | Cu.
- **(c)** Fe, $Fe(OH)_2$ | $OH^-(a = 0.1)$ || $Fe^{2+}(a = 1)$ | Fe.

8.5 Solution:

These reactions involve ions, with no breaking or forming covalent bonds, and none is expected to be too slow to observe.

(a)

$H_2 \rightarrow H^+(0.1) + 2e$

$E = E^0 - \dfrac{0.059}{n} \log [H^+]^2$

$E = 0.00 - \dfrac{0.059}{2} \log (0.1)^2 = 0.06$

$H^+(10^{-7}) \rightarrow H_2$

$E = 0.00 - \dfrac{0.059}{2} \log \dfrac{1}{(10^{-7})^2} = -0.42$

$2H^+(10^{-7}) \rightarrow H^+(0.1)$ $\qquad E = -0.36$ V

(b) Zn → Zn²⁺ + 2e E^0 = 0.76

2e + Cu²⁺(10⁻⁴) → Cu $E = 0.337 - \dfrac{0.059}{2} \log \dfrac{1}{10^{-4}}$ = 0.22

_____

Zn + Cu²⁺(10⁻⁴) → Cu + Zn²⁺ E = 0.98 V

(c) Fe²⁺ + 2e → Fe E^0 = −0.44

Fe + 2 OH⁻(0.1) → Fe(OH)₂ + 2e $E = 0.887 - \dfrac{0.059}{2} \log \dfrac{1}{0.1}$ = 0.22

_____

Fe²⁺ + 2 OH⁻(0.1) → Fe(OH)₂(s) E = 0.42

8.6 Problem: Describe the conditions of acidity most appropriate for the following processes.
(a) $Mn^{2+} \rightarrow MnO_4^-$ (d) $ClO_4^- \rightarrow ClO_3^-$ (g) $H_2O_2 \rightarrow O_2$
(b) $CrO_4^{2-} \rightarrow Cr_2O_7^{2-}$ (e) $C_2O_4^{2-} \rightarrow 2\,CO_2$
(c) $Fe^{3+} \rightarrow FeO_4^{2-}$ (f) $H_2O_2 \rightarrow H_2O$

8.6 Solution:
(a) $Mn^{2+} + 4H_2O \rightarrow MnO_4^- + 8H^+ + 5e$ Reaction favored in base
(b) $2\,CrO_4^{2-} + 2H^+ \rightarrow Cr_2O_7^{2-} + H_2O$ Reaction favored in acid
(c) $Fe^{3+} + 4H_2O \rightarrow FeO_4^{2-} + 8H^+ + 3e$ Reaction favored in base
(d) $ClO_4^- + 2H^+ + 2e \rightarrow ClO_3^- + H_2O$ Reaction favored in acid
(e) $C_2O_4^{2-} \rightarrow 2\,CO_2 + 2e$ Independent of acidity except for the removal of oxalate ion by protonation
(f) $H_2O_2 + 2H^+ + 2e \rightarrow 2H_2O$ Reaction favored in acid
(g) $H_2O_2 \rightarrow O_2 + 2H^+ + 2e$ Reaction favored in base

Whether a total reaction involving these half-cells will be favored in acid or base depends, of course, on the behavior of the other half-reaction involved at the pH in question.

8.7 Problem Given the following emf diagram in acidic solution:

E_A^0 $Sb_2O_5(s) \xrightarrow{0.48} Sb_2O_4(s) \xrightarrow{0.68} Sb^{III} \xrightarrow{0.20} Sb \xrightarrow{-0.51} SbH_3$

$O_2 \xrightarrow{1.23} H_2O$

(a) At a pH of 4 the $Sb^{III} \rightarrow Sb$ emf is = 0.04 V. From these data determine whether the predominate species for Sb^{III} in the pH range of 0 to 4 is Sb_2O_3, SbO_2^-, SbO^+, or Sb^{3+}.
(b) Which of the oxidation states of Sb are unstable in acidic solution? Write balanced equations for all reactions of the unstable species occurring spontaneously in acid.
(c) What is the emf for $Sb_2O_5 \rightarrow SbH_3$ in acidic solution?

8.7 Solution (a) At pH = 4 $E = 0.04$ for $Sb^{III} \to Sb$.

From the Nernst equation for $Sb^{III}O_x + 2x\,H^+ + 3e \to Sb + x\,H_2O$

$$E = 0.04 = 0.20 - \frac{0.059}{3} \log\left(\frac{1}{[H^+]^{2x}}\right)$$

$$-\frac{0.48}{0.059} = 2x(-4) \qquad x = 1$$

therefore, $SbO^+ + 2H^+ + 3e \to Sb + H_2O$ and SbO^+ is the predominant species.

(b) Sb_2O_4 disproportionates

$$2Sb_2O_4 + 2H^+ \to 2SbO^+ + Sb_2O_5 + H_2O \qquad E^0 = 0.20\text{ V}$$

SbH_3 liberates H_2 in water

$$2\,SbH_3 \to 2\,Sb + 3\,H_2 \qquad E^0 = 0.51\text{ V}$$

(c) $Sb_2O_5 \to SbH_3 \qquad E^0 = 0.03\text{ V}$

$$\frac{(0.48 \times 1) + (0.68 \times 1) + (0.20 \times 3) - (0.51 \times 3)}{8} = 0.03$$

8.8 Problem: The emf is more positive for the Fe^{III}/Fe^{II} couple in the presence of o-phenanthroline and more negative in the presence of F^- as compared to the value for the hydrated ions. What do these values tell us about the tendency of these ligands to stabilize Fe^{2+} or Fe^{3+}?

8.8 Solution: o-phenanthroline stabilizes Fe^{II} since the reduction is more favorable. F^- stabilizes Fe^{III} since reduction is less favorable.

8.9 Problem: K_{sp} for $Cu(OH)_2$ is 1.6×10^{-19}.
(a) Calculate the emf for $Cu(OH)_2 \to Cu$ in $1\,M$ base.
(b) What information other than the emf values for Fe^{III}-Fe^{II} in acid and in base is needed to calculate the K_{sp} values for both $Fe(OH)_2$ and $Fe(OH)_3$?

8.9 Solution: (a) $K_{sp} = [Cu^{2+}][OH^-]^2 = 1.6 \times 10^{-19}$

for $[OH^-] = 1$, $[Cu^{2+}] = K_{sp}$

$$E_A^0 = 0.340 \qquad E_B^0 = 0.340 - \frac{0.059}{2}\log\left(\frac{1}{1.6 \times 10^{-19}}\right)$$

$$= 0.340 - \frac{0.059(18.80)}{2}$$

$$= 0.340 - 0.554 = -0.21\text{ V}$$

(b) To calculate K_{sp} for $Fe(OH)_2$ and $Fe(OH)_3$ from E_A^0 and E_B^0 for $Fe^{III} \to Fe^{II}$, we need E_A^0 and E_B^0 for $Fe^{II} \to Fe$ or $Fe^{III} \to Fe$.

8.10 Problem: By extrapolation from known data for Cr, W, and Mo, Seaborg estimated the following emf values for the couples involving possible species for element 106.

$$E_A^0 \quad MO_3 \xrightarrow{-0.5} M_2O_5 \xrightarrow{-0.2} MO_2 \xrightarrow{-0.7} M^{3+} \xrightarrow{0.0} M$$

Predict the results of the following.
(a) M is placed in 1 M HCl.
(b) MCl_3 is placed in an acidic solution containing $FeSO_4$.
(c) MO_2 and MO_3 are in contact in acidic solution.
(d) M^{3+} and MO_3 are in contact in acidic solution.
(e) M is treated with excess concentrated HNO_3.

8.10 Solution:
(a) No reaction occurs. $E^0 = 0.0$ for the evolution of H_2 with the formation of M^{3+}, therefore, overvoltage effects would be sufficient to prevent this reaction.

(b) $2 M^{3+} + 3 H_2O + SO_4^{2-} \rightarrow 2 MO_2 + H_2SO_3 + 4 H^+$ $\quad E^0 = 0.87$ V

(c) No reaction. ($E^0 = -0.3$ for $MO_2 + MO_3 \rightarrow M_2O_5$)

(d) $2 M^{3+} + MO_3 + 3 H_2O \rightarrow 3 MO_2 + 6 H^+$ $\quad E^0 = 0.35$ V

(e) $M + 2 HNO_3 \rightarrow MO_3 + 2 NO + H_2O$ $\quad E^0 = 0.72$ V

8.11 Problem: Given the following half-cell emf diagram for osmium.

$$E_A^0 \quad OsO_4(s) \xrightarrow{1.0} OsCl_6^{2-} \xrightarrow{0.85} OsCl_6^{3-} \xrightarrow{0.4} Os^{2+} \xrightarrow{0.85} Os$$

(a) Which of the above species, if any, would be unstable in 1M HCl? Give balanced equations for any reactions that occur.
(b) Which couple(s) would remain unchanged in their emf value on altering the pH?
(c) Which couple(s) would remain unchanged in their emf value on altering the chloride ion concentration?
(d) Calculate the value of E_A^0 for $OsO_4(s) \rightarrow Os$.
(e) Predict the results of mixing excess osmium with solid OsO_4 in contact with 1M HCl.

8.11 Solution: This problem should be attacked in the same fashion as problem 8.10.
(a) Only Os^{2+} disproportionates.
$$3 Os^{2+} + 12 Cl^- \rightarrow Os + 2 OsCl_6^{3-} \quad E^0 = 0.45$$

(b) To be unaffected by the hydrogen ion concentration, the half-reaction must not involve H^+ ion. Thus all couples except those involving OsO_4 would be unaffected by a change in pH.

(c) To be unaffected by the chloride ion concentration, the half-reaction must not involve free Cl^- ion. Such couples would be Os/Os^{2+}; Os/OsO_4; Os^{2+}/OsO_4; and $OsCl_6^{3-}/OsCl_6^{2-}$.

(d) $OsO_4(s) \xrightarrow{1.0} OsCl_6^{2-} \xrightarrow{0.85} OsCl_6^{3-} \xrightarrow{0.4} Os^{2+} \xrightarrow{0.85} Os$

$$\frac{4(1.0) + 1(0.85) + 1(0.4) + 2(0.85)}{8} = 0.87$$

(e) $4 Os + 3 OsO_4 + 42 Cl^- + 24 H^+ \rightarrow 4 OsCl_6^{3-} + 3 OsCl_6^{2-} + 12 H_2O$ $\quad E^0 = 0.3$ V

8.12 Problem: (a) What E is required to reduce the concentration of V^{3+} to 10^{-6} m at pH 1 in the presence of 1 m VO^{2+}?
(b) What E is required to reduce the VO^{2+} concentration to 10^{-6} m at pH 1 in the presence of 1 m V^{3+}?

$$VO^{2+} + 2H^+ + e \rightarrow V^{3+} + H_2O \quad E^0 = 0.34 \text{ V}$$

8.12 Solution: (a) $E = 0.34 - 0.059 \log \dfrac{[V^{3+}]}{[VO^{2+}][H^+]^2}$

$= 0.34 - 0.059 \log 10^{-4} = 0.58$ V

(b) $E = 0.34 - 0.059 \log \dfrac{1}{(10^{-6})(10^{-1})} = 0.34 - 0.059 \log 10^8$

$= -0.13$ V

8.13 Problem: (a) Reconstruct the emf diagrams for Cl_2 in acidic and basic solutions, showing only species stable with respect to disproportionation.

$E_A^0 \quad ClO_4^- \xrightarrow{1.20} ClO_3^- \xrightarrow{1.18} HClO_2 \xrightarrow{1.70} HClO \xrightarrow{1.63} Cl_2 \xrightarrow{1.36} Cl^-$

$E_B^0 \quad ClO_4^- \xrightarrow{0.37} ClO_3^- \xrightarrow{0.30} ClO_2^- \xrightarrow{0.68} ClO^- \xrightarrow{0.42} Cl_2 \xrightarrow{1.36} Cl^-$

(b) Construct a Pourbaix diagram for Cl with unit activity for predominant species.

8.13 Solution: (a) A species will disporportionate if the sum of its oxidation half-cell emf and reduction half-cell emf is positive. In basic solution we find that Cl_2 will spontaneously yield Cl^- and ClO^- ($E^0 = 0.94$). Calculating a new E_B^0 for $ClO^- \rightarrow Cl^-$ as 0.89, we find that ClO^- is also unstable with respect to disproportionation. (The emf for the intermediate step, $ClO^- \xrightarrow{0.42} Cl_2$, however, suggests an activation energy which may cause ClO^- to undergo such slow autooxidation that it may be metastable, that is, the reaction is slow although the species is thermodynamically unstable. Such is the actual case here.) The following diagrams result from this procedure.

$E_A^0 \quad ClO_4^- \xrightarrow{1.40} Cl_2 \xrightarrow{1.36} Cl^-$

$E_B^0 \quad ClO_4^- \xrightarrow{0.56} Cl^-$

(b) A Pourbaix diagram shows the E vs. pH dependence of stable (or sometimes metastable) species. The Cl_2/Cl^- couple has no pH dependence so a plot of E vs. pH for unit activities of Cl_2 and Cl^- is a horizontal line with $E^0 = 1.36$ V. The ClO_4^-/Cl_2 couple shows the expected pH dependence given by the Nernst equation for

$$ClO_4^- + 8H^+ + 7e \rightarrow \tfrac{1}{2} Cl_2 + 4H_2O$$

$$E = E^0 - \dfrac{RT}{n} \ln Q = 1.40 - \dfrac{0.059}{7} \log \left(\dfrac{1}{[H^+]^8} \right)$$

$$E = 1.40 - 0.067 \, pH$$

In a similar fashion, for ClO_4^-/Cl^- we have

$$ClO_4^- + 8H^+ + 8e \rightarrow Cl^- + 4H_2O$$

$$E = E^0 - 0.059 \log\left(\frac{1}{[H^+]^8}\right) = 1.39 - 0.059 \, pH$$

The Pourbaix diagram is thus:

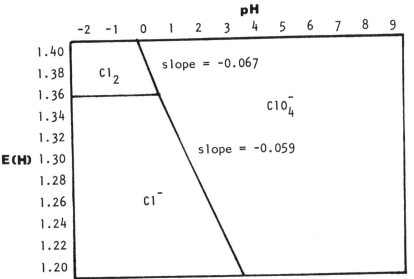

Pourbaix diagram constructed for Cl species in the low pH range

8.14 Problem: The ionization energies and sublimation energies of Li and Be are regular for their families. The emf for reduction of Li$^+$ is very much out of line for the family, but Be^{2+} is not. Explain.

8.14 Solution: The trends for ionization, sublimation, and hydration energies are the same for Groups 1 and 2. The high hydration of Li$^+$ makes its emf out of line for the family. The *very* high overall ionization energy of Be overrides the very hydration of Be^{2+} and the emf for Be→Be^{2+} follows the family trend. The emf is determined by ΔG^0 for $M(s) \rightarrow M(H_2O)_x^{n+}$.

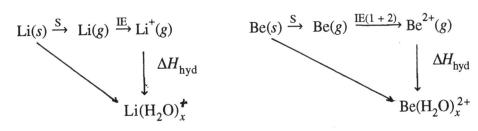

8.15 Problem: The oxidation of $C_2O_4^{2-}$ by MnO_4^- in acidic solution is usually very slow initially, but it can be speeded up by the addition of Mn^{2+}. Explain how the Mn^{2+} might speed the process.

8.15 Solution: Mn^{2+} catalyzes the oxidation of $C_2O_4^{2-}$ by the formation of Mn^{3+} stabilized by forming a complex with oxalate ion. The Mn^{3+} is reduced to Mn^{2+} by electron transfer from the oxalate ion.

8.16 Problem: From the following potential diagrams for Np:

Acidic Solution

$$NpO_2^{2+} \xrightarrow{1.24} NpO_2^+ \xrightarrow{0.64} Np^{4+} \xrightarrow{0.15} Np^{3+} \xrightarrow{-1.79} Np$$

Basic Solution

$$NpO_2(OH)_2 \xrightarrow{0.6} NpO_2OH \xrightarrow{0.3} Np(OH)_4 \xrightarrow{-2.1} Np(OH)_3 \xrightarrow{-2.23} Np$$

(a) Write equations for the reactions which would occur for unstable oxidation states in acidic solution and in basic solution.
(b) Give a suitable preparation for each stable oxidation starting with Np.

8.16 Solution:
(a) $2\,Np + 6\,H^+ \rightarrow 2\,Np^{3+} + 3\,H_2$ $\quad\quad Np + 4\,H_2O \rightarrow Np(OH)_4 + 2\,H_2$

$$2\,Np(OH)_3 + 2\,H_2O \rightarrow 2\,Np(OH)_4 + H_2$$

No species disproportionate.
The E^0 is favorable for $NpO_2(OH)_2$ to oxidize OH^- to O_2 in basic solution, but the potential is too low to overcome the overvoltage.
(b) Cl_2 or a comparable oxidizing agent will produce Np^{VI} in acidic or basic solution.
Br_2 will oxidize lower oxidation states to NpO_2^+ in acidic solution.

$$2\,Np + 6\,HCl \rightarrow 2\,Np^{3+} + 3\,H_2$$
$$Np + 2\,I_2 \rightarrow Np^{4+} + 4\,I^-$$
$$Np + 2\,H_2O + 3\,Cl_2 \rightarrow NpO_2^{2+} + 6\,Cl^- + 4\,H^+$$
$$Np + 4\,H_2O \rightarrow Np(OH)_4 + 2\,H_2$$
$$2\,Np(OH)_4 + 2\,OH^- + I_2 \rightarrow 2\,NpO_2OH + 4\,H_2O + 2\,I^-$$
$$Np + 6\,OH^- + 3\,Cl_2 \rightarrow NpO_2(OH)_2 + 2\,H_2O + 6\,Cl^-$$

8.17 Problem: Give the products expected for the following reactions:
(a) Cl_2(excess) + HI (1 M)   **(d)** HClO(small amt.) + NaI(excess)
(b) $HClO_4$ (1 M) + NaI (1 M)   **(e)** Which of these reactions might be too slow to observe?
(c) HClO(excess) + NaI(small amt.)

8.17 Solution:
(a) Cl_2(excess) + HI (1 M) → Cl^- + I_2
(b) $HClO_4$ (1 M) + NaI (1 M) → IO_3^- + Cl^- (from emf's, but it is too slow to observe)
(c) HClO(excess) + NaI(small amount) → Cl_2 + IO_3^-
(d) HClO(small amount) + NaI(excess) → Cl^- + I_2

(e) Simple electron transfer such as $Cl_2 + I^-$ are expected to be fast. Atom transfers that involve bond breaking and bond formation are often slow. Reactions (**b, c, and d**) involving oxoanions or oxoacids are expected to be slow. The reaction of $HClO_4$ with I^- is too slow to observe at room temperature.

8.18 Problem: Calculate E_{cell} at pH = 0, 14, and 3.5 for
(a) Cl_2 → HOCl + Cl^- (b) Br_2 → HOBr + Br^-

8.18 Solution: $X_2 + H_2O$ → X^- + HOX + H^+

At pH = 0   for Cl_2 E_A^0 = 1.36 – 1.63 = –0.27 V Unfavorable
            for Br_2 E_A^0 = 1.06 – 1.60 = –0.54 V Unfavorable
At pH = 14 for Cl_2 E_B^0 = 1.36 – 0.42 = +0.94 V Favorable
            for Br_2 E_B^0 = 1.06 – 0.46 = +0.60 V Favorable

At pH = 3.5 $[H^+]$ = 3.2 x 10^{-4} m
For Cl_2 E = –0.27 – $\frac{0.0591}{1}$ log $[Cl^-][HOCl][H^+]$

K_a(HOCl) = 4.0 x 10^{-8} = $\frac{[H^+][OCl^-]}{[HOCl]}$ = $\frac{(3.2 \times 10^{-4})[OCl^-]}{1 - [OCl^-]}$

3.2 x $10^{-4}$$[OCl^-]$ ≅ 4.0 x 10^{-8}
$[OCl^-]$ = 1.25 x 10^{-4}, verifying that [HOCl] = 1.0 – 1.25 x 10^{-4} ≅ 1.0 m
E = –0.27 – 0.0591 log (1)(10)(3.2 x 10^{-4})
= –0.06 V for disproportionation of Cl_2 at pH 3.5

For Br_2 E = –0.54 – $\frac{0.0591}{1}$ log $[Br^-][HOBr][H^+]$

pK_a = 8.7, K_a = 2.0 x 10^{-9}

Since HOBr is a weaker acid than HOCl, [HOBr] ≅ 1.0
E = –0.54 – 0.0591 log (1)(1)(3.2 x 10^{-4})
= –0.33 V for disproportionation of Br_2 at pH 3.5

8.19 Problem: A Pt electrode maintained at a potential of 0.525 V relative to the standard hydrogen electrode is immersed in an aqueous solution of Fe^{2+} and Fe^{3+} salts. What is the value of $[Fe^{3+}]/[Fe^{2+}]$?

$Fe^{3+} \xrightarrow{0.771} Fe^{2+}$

8.19 Solution: The Nernst equation for the Fe^{2+}/Fe^{3+} couple is

$$E = E^0 - \frac{0.059}{1} \log \frac{[Fe^{2+}]}{[Fe^{3+}]}$$

If E is arbitrarily fixed at 0.525 V:

$$0.525 = 0.771 - 0.059 \log [Fe^{2+}]/[Fe^{3+}]$$

From which we obtain $[Fe^{2+}]/[Fe^{3+}] = 1.4 \times 10^4$.

8.20 Problem: Given the following estimated half-cell emf values (A. Stiller, Ph.D. Dissertation, University of Cincinnati, 1973).

$$E_A^0 \quad SeO_2 \xrightarrow{0.72} Se_4^{2+} \xrightarrow{1.0} Se_8^{2+} \xrightarrow{0.8} Se_8$$
$$\text{colorless} \quad \text{yellow} \quad \text{green} \quad \text{red}$$

Write equations to describe the following reactions:
(a) Se_8 dissolves in $H_2S_2O_7$ to give a yellow solution.
(b) On diluting the solution from part (a) to about 98% H_2SO_4, the solution turns green.
(c) On long standing in H_2SO_4 having a concentration above 80%, Se_8 dissolves to give green solutions from which red Se_8 can be recovered almost quantitatively by dilution with water.

8.20 Solution: It should be noted that the emf values are those appropriate for a hydrogen activity of one; in fuming sulfuric acid the hydrogen ion activity would be much greater than one. Accordingly, the $SeO_2 \rightarrow Se_4^{2+}$ half-reaction would show a more positive value than the E_A^0 value, but the other couples above would not be affected since hydrogen ion is not involved in the remaining half-reactions. The reactions are thus written

(a) $Se_8 + 8H_2S_2O_7 \rightarrow 2Se_4^{2+} + 4HS_2O_7^- + 6H_2SO_4 + 2SO_2$

It may be noted that SO_3 becomes a strong oxidizing agent at high acidities. The $HS_2O_7^-$ and H_2SO_4 products arise from H balance and from the reaction of $H_2S_2O_7$ with water produced simultaneously with SO_2.

(b) $\quad 8H_2O + Se_4^{2+} \rightarrow 4SeO_2 + 16H^+ + 14e$

$\quad\quad 7(2e + 2Se_4^{2+} \rightarrow Se_8^{2+})$

$\quad\quad 8H_2O + 15Se_4^{2+} \rightarrow 7Se_8^{2+} + 4SeO_2 + 16H^+$

Reducing the hydrogen ion activity allows this disproportionation reaction to occur. The hydrogen ion would be taken up by the anions present.

(c) $3H_2SO_4 + Se_8 \rightarrow Se_8^{2+} + 2HSO_4^- + 2H_2O + SO_2$

At lower acid concentration, the H_2SO_4 oxidizes the Se_8 only to Se_8^{2+}. Again, reducing the hydrogen ion activity allows a disproportionation reaction to occur.

$$16H_2O + Se_8^{2+} \rightarrow 8SeO_2 + 32H^+ + 30e$$
$$15(2e + Se_8^{2+} \rightarrow Se_8)$$
_____

$$16H_2O + 16Se_8^{2+} \rightarrow 15Se_8 + 8SeO_2 + 32H^+$$

The disproportionation reaction yields 15/16th of the Se_8 initially dissolved.

8.21 Problem: Calculate ΔG^0 for the following reactions:
(a) $2Al + Fe_2O_3 \rightarrow Al_2O_3 + 2Fe$
(b) $3Cs_2O + 2Al \rightarrow Al_2O_3 + 6Cs(g)$

ΔG_f^0 Al_2O_3, 1580 kJ/mol; Fe_2O_3, 740 kJ/mol;
Cs_2O, 280 kJ/mol; $Cs(g)$ 51 kJ/mol

8.21 Solution: The ΔG^0 for a reaction is the sum of the sum of the ΔG_f^0 of the products minus the sum of the ΔG_f^0 of the reactants. Elements in their standard states have a ΔG_f^0 value of zero. For the above reactions we obtain:

(a) $\Delta G^0 = -1580 - (-740) = -840$ kJ at 25°C. This reaction, known as the thermite reaction, is thermodynamically favored even at room temperature. However, its reaction rate is not perceptible at room temperature, so it is initiated at high temperature (usually by an Mg fuse), after which the reaction sustains itself.

(b) $\Delta G^0 = 6(51) + (-1580) - 3(-280) = -434$ kJ at 25°C. This reaction is also carried out at high temperature, the Cs vapor being removed, the positive entropy change making the free energy change even more negative than the above calculation for 25°C.

IV. COORDINATION CHEMISTRY
9
Models and Stereochemistry of Coordination Compounds

9.1 Problem: Name the following compounds according to the IUPAC rules:

K_2FeO_4 $K[Co(edta)]$

$[Cr(NH_3)_6]Cl_3$ $[Cr(OH)(H_2O)_3(NH_3)_2](NO_3)_2$

$[Cr(Cl)_2(NH_3)_4]Cl$

$K[Pt(Cl)_3(C_2H_4)]$ $[(NH_3)_4Co\overset{OH}{\underset{NH_2}{\diamondsuit}}Co(en)_2]Cl_4$

$K_3[Al(C_2O_4)_3]$

$K_2[Co(N_3)_4]$

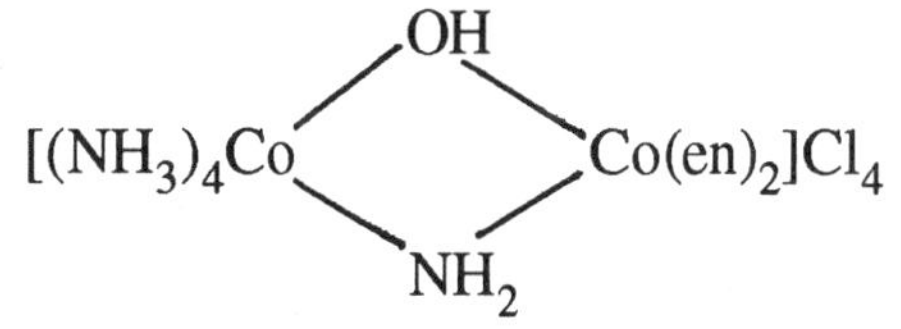

9.1 Solution: (See DMA, 3rd ed., Appendix B)
Potassium tetraoxoferrate(VI)
Hexaamminechromium(III) chloride
Tetraamminedichlorochromium(III) chloride   (alphabetical order)
Potassium trichloro(ethylene)platinate(II)
Potassium tris(oxalato)aluminate
Potassium tetraazidocobaltate(II)   (N_3^- is azide ion)
Potassium ethylenediaminetetraacetatocobaltate(III)
Diamminetriaquahydroxochromium(III) nitrate   (alphabetical order)
Tetraamminecobalt(III)-µ-amido-µ-hydroxo-bis(ethylenediamine)cobalt(III) chloride

9.2 Problem: The values for formation constants for each step in the formation of $[Ni(en)_3]^{2+}$ are $\log K_1 = 7.52$, $\log K_2 = 6.28$, and $\log K_3 = 4.26$ at 30° C in 1.0 M KCl.

(a) What is $\log \beta_3$ for the overall formation from Ni^{2+} and 3 en?

(b) Why do values of K decrease with $K_1 > K_2 > K_3$?

9.2 Solution: (a) $\beta_3 = K_1 K_2 K_3$ $\quad \log \beta_3 = \log K_1 + \log K_2 + \log K_3$

$$\log \beta_3 = 7.52 + 6.28 + 4.26 = 18.06$$

(b) The observed trend is the common one for stepwise formation constants in the absence of changes in geometry, spin state, or other "hidden factors". At one time such trends were attributed to "saturation" or a decreasing bond strength as the electron density near the metal increased as more basic ligands replaced water. With charged ligands this argument has merit. However, enthalpy data for ethylenediamine complexation do not support this rationalization. In this case the ΔH (kJ/mol) values are −37.7, −38.4, and −40.6, respectively, which, although the differences are small, would lead to the opposite trend to that observed. The explanation must therefore lie in the entropy changes. For a gas phase case (1st appoximation), we could obtain the entropy change for the reaction from the change in symmetry numbers of the species involved. The symmetry number, σ, is the number of different equivalent orientations available by rotation alone, which is the order of the rotation group for finite groups. Thus

O_h	C_{2v}	C_{2v}	C_{2v}
σ 24	2	2	2

$$\Delta S^0_\sigma = -3R \ln 2 + R \ln 24 + R \ln 2 = R \ln \frac{24 \times 2}{2 \times 2 \times 2} = 15.1 \text{ J/mol K}$$

At 298 K $\quad T\Delta S = +4.50$ kJ/mol (contribution to K_1)

C_{2v}	C_{2v}	D_{2h}	C_{2v}
σ 2	2	4	2

$$\Delta S^0 = -R \ln \frac{4 \times 2 \times 2}{2 \times 2} = -11.6 \text{ J/mol K}$$

At 298 K $\quad T\Delta S = -3.45$ kJ/mol (contribution to K_2)

D_{2h}	C_{2v}	D_3	C_{2v}
σ 4	2	6	2

$$\Delta S^0 = -R \ln \frac{6 \times 2 \times 2}{4 \times 2} = -R \ln 3 = -9.13 \text{ J/mol K}$$

At 298 K $\quad T\Delta S = -2.73$ kJ/mol (contribution to K_3)

If entropy changes depended only on σ, and if the solution ΔS were the same as the gas phase ΔS, then we would expect phen and bipy to show the same K_1:K_2:K_3 ratios as en. This is not the case and suggests that changes in solution may depend on the primary vs. tertiary nature of the amine and resultant changes both in bulk water structure as the hydrophobic-hydrophilic nature of the complex is changed and in the character of the donor ability as the charge dispersal through hydrogen bonding is altered.

9.3 Problem: Considering the charge densities of the cations, what is the expected order of decreasing stabilities of F^- complexes of Be^{2+}, Mg^{2+}, and Ca^{2+}?

9.3 Solution: The cations are hard acids and F^- is a hard base so the order of stabilities is that expected from electrostatic interactions.
$$Be^{2+} > Mg^{2+} > Ca^{2+}$$

9.4 Problem: The K_1 for HgI^+ is greater than that for MgI^+ even though Mg^{2+} has a considerably higher charge density. How can this be explained?

9.4 Solution: Mg^{2+} is a hard acid. Hg^{2+} is a soft acid and I^- is a soft base. Hg^{2+} has a stronger affinity for soft I^- because of the mutual polarization.

9.5 Problem: K_1 = 10.7 for $[Cu(en)]^{2+}$ and β_2 = 7.8 for $[Cu(NH_3)_2]^{2+}$. K_1 = 4.7 for $[Ag(en)]^+$ and β_2 = 7.2 for $[Ag(NH_3)_2]^+$.
(a) Why is the formation constant for adding one en to Cu^{2+} greater than for adding two NH_3?
(b) Why is the order the reverse for Ag^+?

9.5 Solution: (a) Cu^{2+} shows the normal chelate effect forming a more stable complex with a didentate ligand compared to a monodentate with about the basicity. The chelate effect is more important than the greater basicity of en compared to NH_3.

(b) The chelate effect is not operative for Ag^+ since it tends to form linear complexes. The en cannot span the 180° angle and serves as a monodentate ligand with Ag^+.

9.6 Problem: For complexes of Fe^{3+} and Ag^+ with SCN^-, would you expect coordination of SCN^- through S or N to these cations? (*Hint:* Consider hardness/softness.)

9.6 Solution: The hard acid Fe^{3+} should attach through the harder base N and the soft acid Ag^+ should attach through the softer base S.

9.7 Problem: Which of the following complexes obey the 18-electron rule (EAN rule)?
(a) $[Cu(NH_3)_4]^{2+}$, $[Cu(en)_3]^{2+}$, $[Cu(CN)_4]^{3-}$ (d) $[Fe(CN)_6]^{3-}$, $[Fe(CN)_6]^{4-}$, $[Fe(CO)_5]$
(b) $[Ni(NH_3)_6]^{2+}$, $[Ni(CN)_4]^{2-}$, $[Ni(CO)_4]$ (e) $[Cr(NH_3)_6]^{3+}$, $[Cr(CO)_6]$
(c) $[Co(NH_3)_6]^{3+}$, $[CoCl_4]^{2-}$

9.7 Solution: To obey the rule of 18 (EAN), the sum of the number of electrons in the metal valence shell [$(n-1)dnsnp$] and those donated by the ligands should be 18, filling the low-energy metal orbitals.

(a) $[Cu(NH_3)_4]^{2+}$ $9 + 8 = 17e$ $[Cu(CN)_4]^{3-}$ $10 + 8 = \underline{18e}$

    $[Cu(en)_3]^{2+}$ $9 + 12 = 21e$

(b) $[Ni(NH_3)_6]^{2+}$ $8 + 12 = 20e$ $[Ni(CO)_4]$ $10 + 8 = \underline{18e}$

    $[Ni(CN)_4]^{2-}$ $8 + 8 = 16e$

(c) $[Co(NH_3)_6]^{3+}$ $6 + 12 = \underline{18e}$ $[CoCl_4]^{2-}$ $7 + 8 = 15e$

(d) $[Fe(CN)_6]^{3-}$ $5 + 12 = 17e$ $[Fe(CO)_5]$ $8 + 10 = \underline{18e}$

    $[Fe(CN)_6]^{4-}$ $6 + 12 = \underline{18e}$

(e) $[Cr(NH_3)_6]^{3+}$ $3 + 12 = 15e$ $[Cr(CO)_6]$ $6 + 12 = \underline{18e}$

9.8 Problem: Calculate the magnetic moment for the spin-only contribution for:

(a) $[Ni(CN)_4]^{2-}$ (planar)

(b) $[Ni(NH_3)_6]^{2+}$

(c) $[Cu(NH_3)_4]^{2+}$ (planar)

(d) $[Ag(NH_3)_2]^+$ (linear)

(e) $[Co(NH_3)_6]^{3+}$

(f) $[CoCl_4]^{2-}$ (tetrahedral)

(g) $[Cr(NH_3)_6]^{3+}$

(h) $[Ni(CO)_4]$ (tetrahedral)

9.8 Solution: (a) d^8, planar, diamagnetic, $n = 0$, $\beta_M = 0$.

(b) d^8, $n = 2$, $\beta_M = \sqrt{2(2+2)} = 2.83$ Bohr magnetons.

(c) d^9, $n = 1$, $\beta_M = 1.73$ B.m.

(d) d^{10}, $n = 0$, $\beta_M = 0$.

(e) low-spin d^6, $n = 0$, $\beta_M = 0$.

(f) d^7, $n = 3$, $\beta_M = 3.87$ B.m.

(g) d^3, $n = 3$, $\beta_M = 3.87$ B.m.

(h) d^{10}, $n = 0$, $\beta_M = 0$.

9.9 Problem: Determine the number of unpaired electrons and the LFSE for each of the following.

(a) $[Fe(CN)_6]^{4-}$

(b) $[Fe(H_2O)_6]^{3+}$

(c) $[Co(NH_3)_6]^{3+}$

(d) $[Cr(NH_3)_6]^{3+}$

(e) $[Ru(NH_3)_6]^{3+}$

(f) $[PtCl_6]^{2-}$

(g) $[CoCl_4]^{2-}$ (tetrahedral)

9.9 Solution:

		Unpaired e	LFSE			Unpaired e	LFSE
(a)	t_{2g}^6	0	$-24Dq$	(e)	t_{2g}^5	1	$-20Dq$
(b)	$t_{2g}^3 e_g^2$	5	$0Dq$	(f)	t_{2g}^6	0	$-24Dq$
(c)	t_{2g}^6	0	$-24Dq$	(g)	$e^4 t_2^3$	3	$-12Dq$
(d)	t_{2g}^3	3	$-12Dq$				

9.10 Problem: Discuss briefly the factors working for and against the maximum spin state of d electrons in transition metal complexes.

9.10 Solution: Maximum (high) spin is favored by maximum charge delocalization and exchange energy. The low-spin case is favored by maximum LFSE.

9.11 Problem: Explain why square-planar complexes of transition metals are limited (other than those of planar ligands such as porphyrins) to those of **(a)** d^7, d^8, and d^9 ions and **(b)** d^8 configurations for very strong field ligands which can serve as π acceptors.

9.11 Solution: (a) These are the cases for maximum LFSE for the square-planar case. The most important feature of the energy-level diagram shown is one orbital of *very* high energy, so that d^8 complexes are diamagnetic.
(b) Only very strong field ligands, particularly π acceptors, give sufficiently strong fields to cause spin pairing for d^7 and d^8.

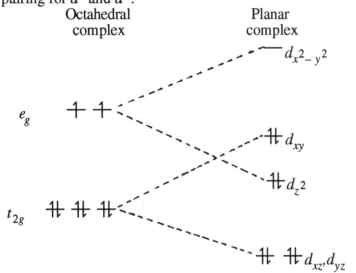

Energy levels for square-planar Ni^{II} complexes

9.12 Problem: Construct a table of ligand field stabilization energies (LFSE) for all d^n configurations in tetrahedral complexes with weak fields. Which d^n configurations would be diamagnetic for a strong field?

9.12 Solution:

	d^0	d^1	d^2	d^3	d^4	d^5	d^6	d^7	d^8	d^9	d^{10}	
t_2	0	0	0	1	2	3	3	3	4	5	6	electrons
e	0	1	2	2	2	2	3	4	4	4	4	electrons
LFSE	0	–6	–12	–8	–4	0	–6	–12	–8	–4	0	Dq

Diamagnetic configurations for a strong field are d^0, d^{10}, and d^4 (low spin).

9.13 Problem: For which d^n configurations would no Jahn-Teller splitting be expected for the tetrahedral case (ignore possible low-spin cases)?

9.13 Solution: Without spin pairing, no gain in energy will result for d^2 since the d_{z^2} and $d_{x^2-y^2}$ orbitals will have one electron each. No gain in energy results for high-spin d^5 or for d^{10} since each orbital is singly occupied (d^5) or filled (d^{10}).

9.14 Problem: Give the orbital occupancy (identify the orbitals) for the Jahn-Teller splitting expected for tetrahedral complexes with high-spin d^3 and d^4 configurations. Indicate the nature of distortions expected.

9.14 Solution: For d^3 the tetrahedron should be elongated along z. This causes the orbitals in the xy plane derived from e and t_2 to have lower energy (the ligands are farther from the xy plane). One t_2 orbital (d_{xy}) has lower energy. For d^4 the tetrahedron should be flattened along z. This raises the energy of the orbitals in the xy plane and lowers the others. Two electrons are accommodated in the two lower energy orbitals from t_2.

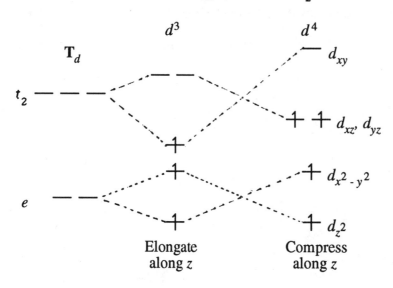

9.15 Problem:
(a). Why are low-spin complexes usually not encountered for tetrahedral coordination?
(b) Octahedral splitting is expressed as $10Dq$. What would be the spliting for ML_8 with cubic coordination? Assume the same ligands at the same distance as for the octahedral and tetrahedral cases.

9.15 Solution: (a) The LFSE is small for the tetrahedral case since there is small splitting. There are four ligands instead of six for the octahedron and the ligands do not approach along the directions of the d orbital lobes. The LFSE is usually smaller than the pairing energy.
(b) Tetrahedral splitting (4L) is $(4/9)(10Dq)_{oct}$. Cubic splitting (8L) is twice as great or $(8/9)(10Dq)$. The cube consists of two interpenetrating tetrahedra (four alternating apices of a cube define a tetrahedron).

9.16 Problem: Sketch the possible geometrical isomers for the following complexes and indicate which of these would exhibit optical activity.
(a) $[CoBrCl(en)(NH_3)_2]^+$.
(b) $[CoCl(NH_2CH_2CO_2)_2(NH_3)]$.
(c) $[PtBrCl(NO_2)(NH_3)]^-$
(d) $[CoCl_2(trien)]^+$ (consider the different ways of linking trien to Co).
(e) $[(gly)_2Co\underset{OH}{\overset{OH}{\diagup\diagdown}}Co(gly)_2]$.

9.16 Solution:

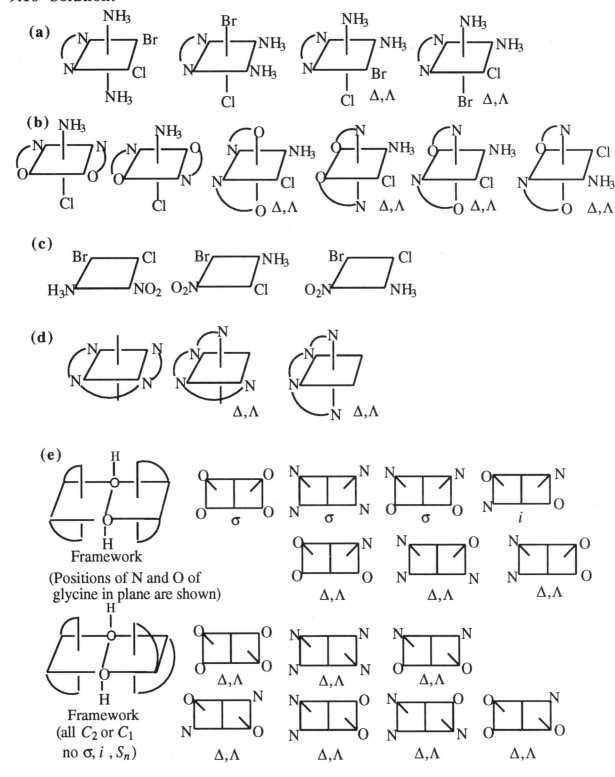

9.17 Problem: Draw *all* possible isomers for M(a)₂bcd assuming the complex forms a square pyramid.

9.17 Solution:

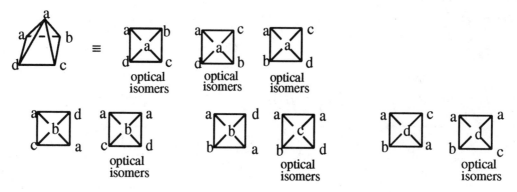

9.18 Problem: How might one distinguish between the following isomers?
(a) [CoBr(NH₃)₅]SO₄ and [Co(SO₄)(NH₃)₅]Br.
(b) [Co(NO₂)₃(NH₃)₃] and [Co(NH₃)₆][Co(NO₂)₆].
(c) *cis-* and *trans-* [CoCl₂(en)₂]Cl.
(d) NH₄⁺*cis-* and *trans-*[Co(NO₂)₄(NH₃)₂]⁻.
(e) *cis-* and *trans-* [Pt(gly)₂].

9.18 Solution:
(a) Add Ba^{2+} or Ag^+ to detect $BaSO_4$ or AgBr precipitates. The uncoordinated counterion should precipitate immediately.
(b) Molecular weight determination in a solvent of low dielectric constant or conductance in a solvent of high dielectric constant should distinguish between the isomers since their molecular weights differ by a factor of two.
(c) The *cis* isomer can be resolved into optical isomers, the *trans* isomer is inactive. Absorption spectra can be used also, the *cis* isomer is violet and the *trans* isomer is green. The striking colors make it unnecessary to take spectra.
(d) One possibility is the reaction with $C_2O_4^{2-}$. The *trans* isomer should give only one product, *trans*-[Co(C₂O₄)₂(NH₃)₂]⁻. The *cis* isomer should give several products, including resolvable isomers. There would be some differences in the absorption spectra of the original nitro isomers also. Splitting of the lower energy absorption band is usually greater for the *trans* isomer. The colors of these isomers are similar.
(e) The isomers should differ in absorption spectra. If a suitable solvent can be found, the *trans* isomer would be found to have zero dipole moment.

9.19 Problem: Give examples of the following types of isomerism:
(a) Hydrate isomerism.   (d) Ionization isomerism.
(b) Coordination isomerism.   (e) Geometrical isomerism.
(c) Linkage isomerism.

9.19 Solution:
(a) Hydrate isomerism— [Cr(H₂O)₆]Cl₃ and [Cr(H₂O)₅Cl]Cl₂·H₂O

(b) Coordination isomerism— $[Co(NH_3)_6][Cr(CN)_6]$ and $[Cr(NH_3)_6][Co(CN)_6]$
$[Pt(NH_3)_4][PdCl_4]$ and $[Pd(NH_3)_4][PtCl_4]$

(c) Linkage isomerism— $[Ir(NCS)(NH_3)_5]^{2+}$ and $[Ir(SCN)(NH_3)_5]^{2+}$
$[Co(NO_2)(NH_3)_5]Cl_2$ (N bonded) and $[Co(ONO)(NH_3)_5]Cl_2$ (O bonded)

(d) Ionization isomerism— $[Pt(NH_3)_4Cl_2]Br_2$ and $[Pt(NH_3)_4Br_2]Cl_2$

(e) Geometrical isomerism— *cis*-$[CrCl_2(NH_3)_4]^+$ and *trans*-$[CrCl_2(NH_3)_4]^+$
cis-$[PtCl_2(NH_3)_2]$ and *trans*-$[PtCl_2(NH_3)_2]$

9.20 Problem: Draw all isomers possible for octahedral M(*abcdef*). (*Hint*: Calculate the number of stereoisomers for [M(*abcdef*)]. Write unique pairings *trans* to *a* and calculate the number of possible isomers for each of these. Then consider unique *trans* pairings for the remaining groups.

9.20 Solution: For n different ligands the total number of stereoisomers possible is $n!/\sigma$, where σ is the order of the rotation group except for linear molecules. For M(abcdef) the rotation group is **O**, of order 24. The number of stereoisomers is $6!/24 = 30$. For each *trans*-M(xy) isomer (C_4 is thr rotation group of order 4 for the framework) there are $4!/4 = 6$ stereoisomers. The geometrical isomers can be drawn by examining all *trans* pairings:

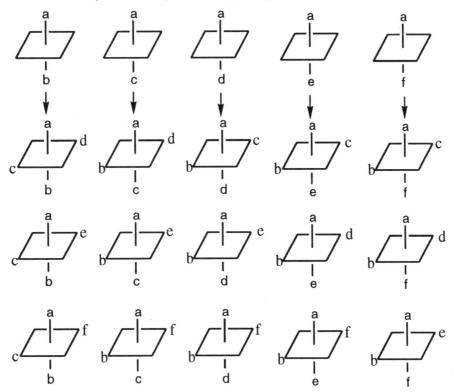

Each of the 15 geometrical isomers is one of a pair of enantiomers. The other enantiomer can be sketched as the mirror image of the first or by interchanging the positions of the last pair of ligands.

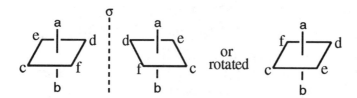

9.21 Problem: Draw all possible isomers for
(a) M($abcdef$) ($\widehat{ef}$ is didentate).
(b) M($\widehat{aa}\widehat{bc}d_2$) (*Hint*: leave $\widehat{bc}$ in a fixed position in all drawings except for enantiomers; $\widehat{aa}$ is a symmetrical didentate ligand and $\widehat{bc}$ is an unsymmetrical didentate ligand.)

9.21 Solution: (a) The framework belongs to the rotation group C_1 with order 1. The number of isomers is $4!/1 = 24$.

For each of these there are 2 geometrical isomers, for example:

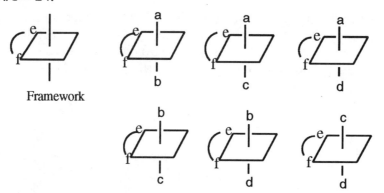

and each is optically active, giving 24 stereoisomers. We can get the mirror image either by reflecting through a mirror plane or by interchanging the axial pair.

(b) The didentate ligand $\widehat{aa}$ can be either in the same plane as $\widehat{cb}$, or not. There are 3 geometrical isomers and 5 stereoisomers.

9.22 Problem: Sketch the isomers possible for a trigonal prismatic complex M(*aa*)$_3$, where *aa* is a planar didentate ligand. Could any of these be optically active? Assign the point groups.

9.22 Solution:

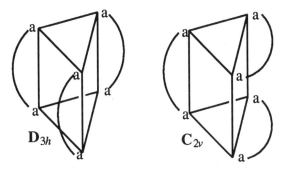

Neither is optically active since both have planes of symmetry. Didentate ligands do not span the diagonal positions of the tetragonal faces.

9.23 Problem: Sketch octahedral and trigonal prismatic [M(NO$_3$)$_3$] complexes to show that the N atoms are in a trigonal planar arrangement in each case. Assign the point groups.

9.23 Solution:

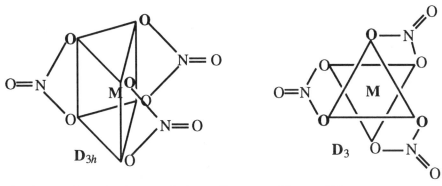

The N atoms are in the σ_h plane for D_{3h} and along C_2 axes of D_3 and D_{3h}.

9.24 Problem: Determine the number of Λ and Δ pairs of chelate rings for the isomer of [Co(edta)]$^-$ shown. The pairs of chelate rings attached to a common point and those spanning parallel edges are omitted.

9.24 Solution: The IUPAC rules for designating chirality of octahedral complexes (See DMA, 3rd ed., p. 419) are based on the helices defined by non-parallel and non-intersecting octahedral edges spanned by chelate rings. A pair of edges viewed as shown (the solid edge B—B is closer to the viewer) defines a right helix (Δ) if B—B tips down to the right and a left helix (Λ) if B—B tips down to the left. The same system applies to the conformation of chelate rings. With the coordinated N atoms of $NH_2C_2H_4NH_2$ spanning an octahedral edge oriented horizontally (A—A), the conformation is δ (right-handed helix) if the C—C bond (replacing B—B) tips down to the right and λ (left handed) if it tips down to the left. The drawings of $[Co(edta)]^-$ shown below are all oriented in the same way. They need to be reorienated with the edge farther away from you held horizontally. A model of an octahedron (see Problem 3.24) is helpful.

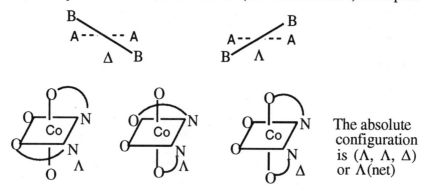

The absolute configuration is (Λ, Λ, Δ) or Λ(net)

9.25 Problem: Which chelate ring conformation is favored for coordinated S-(+)-propanediamine, and why?

9.25 Solution: The δ conformation is favored in metal complexes of pn because of crowding of the axial methyl group by other ligands if the pn has the λ conformation.

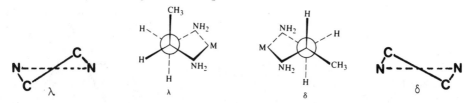

Enantiomeric conformations of a *gauche* chelate ring of S-(+)-1,2-propanediamine placing the methyl group in axial (λ) and equatorial (δ) positions. Viewed along the C—C bond.

9.26 Problem: For *trans*-$[CoCl_2(trien)]^+$ (trien = $NH_2C_2H_4NHC_2H_4NHC_2H_4NH_2$) there are three isomers possible, depending on the chelate ring conformations. Two are optically active and one is meso with the central diamine ring in an eclipsed "envelope" conformation. Sketch the two optically active *trans* isomers (all rings *gauche*) and assign the absolute configurations (δ or λ) to the chelate rings. (See D. A. Buckingham, P. A. Marzilli, and A. M. Sargeson, *Inorg. Chem.* **1967**, *6*, 1032.)

9.26 Solution: The three isomers of *trans*-$[CoX_2(trien)]^+$ differ in the conformations of the chelate rings. For the *meso* isomer the central chelate ring is achiral and the terminal rings have enantiomeric conformations. See Problems 9.24 and 9.25 for descriptions of ring conformations.

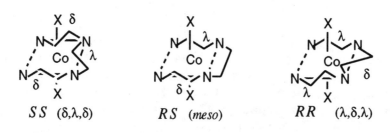

SS (δ,λ,δ)   RS (meso)   RR (λ,δ,λ)

9.27 Problem: What is (are) the number(s) of d electrons for metals usually encountered with the following stereochemistries?
(a) Linear.
(b) Square planar.
(c) Trigonal prismatic.
(d) Dodecahedral

9.27 Solution: (a) Linear complexes are encountered for Ag^+, Cu^+, and Hg^{2+}, all d^{10} ions. An $sp_z d_{z^2}$ hybrid provides strong bonding and a nonbonding orbital for the electron pair from d_{z^2}.

(b) Square planar complexes are found for the following cases: d^7, Co^{2+}; d^8, Ni^{2+}, Pd^{2+}, Pt^{2+}, Ag^{3+}, Rh^+; d^9, Cu^{2+}, Ag^{2+}. Square planar coordination involves one very high energy orbital ($d_{x^2-y^2}$). The most favorable situation is with this orbital empty (d^7 or d^8) or half filled (d^9).

(c) For C.N. 6, the expected geometries are octahedral and trigonal prismatic. The trigonal prism is not expected for d configurations providing high octahedral stabilization energy $\{d^3, d^6$ (low-spin), and $d^8\}$. For $\mathbf{D}_{3h}$ symmetry, one d orbital (d_{z^2}) is lowest in energy. Trigonal prismatic coordination is found for d^0, Mo^{VI}; d^5, Mn^{II}; d^{10}, Zn^{II}; d^1, Re^{VI}; d^7 (high-spin), Co^{II}.

(d) For the dodecahedron there is one very low energy d orbital, so maximum stabilization is achieved by occupancy of this orbital only: d^1, $[Mo(CN)_8]^{3-}$ and d^2, $[Mo(CN)_8]^{4-}$.

9.28 Problem: ^{63}Cu has $I = \frac{1}{2}$. When CuI is dissolved in $P(OMe)_3$, the ^{63}Cu NMR spectrum shows a five-line pattern with relative intensities 1:4:6:4:1. What inference can be made about the environment of Cu in this solution? (See J. K. M. Saunders and B. K. Hunter, *Modern NMR Spectroscopy, A Guide for Chemists*, Oxford University Press, Oxford, 1987, p. 48.)

9.28 Solution: The quintet pattern indicates that the Cu signal is split by four chemically equivalent nuclei having $I = \frac{1}{2}$; obviously four ^{31}P are coordinated to give $[Cu\{P(OMe)_3\}_4]^+$. Either a square-planar or tetrahedral geometry is consistent with the NMR spectrum; however, since Cu^I is d^{10}, no LFSE is possible so that minimum ligand repulsion dictates a tetrahedral complex.

9.29 Problem: The Figure shows the low- and high-temperature limiting ^1H NMR spectra in the hydride region of $(H)_2Fe[P(Ph)(OPh)_2]_2$ in toluene. Account for the appearance of the spectra. (*Hint:* In this case, coupling between two H's and between H and P in the same plane in the *cis*-isomer is too small to produce noticeable splittings.)

Figure reprinted with permission from J. P. Jesson and E. L. Muetterties, in *Dynamic Nuclear Magnetic Resonance Spectroscopy*, L. M. Jackman and F. A. Cotton, Eds., Academic Press, New York, 1975, p. 289.

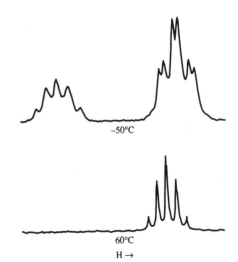

−50°C

60°C
H →

9.29 Solution: A complex with six ligands is expected to be octahedral. The low-temperature spectrum shows the presence of both *trans*- and *cis*- $(H)_2Fe[Fe(Ph)(OPh)_2]_4$. The low-field quintet indicates a chemically equivalent pair of H's split by the four chemically equivalent ^{31}P in the plane of the *trans* isomer. The other signal can be rationalized as a pair of *cis*- H split by the *trans*- P in the plane into a doublet. Each component of the doublet is split into a triplet by coupling to the pair of axial P. Axial P–H coupling is larger than the in-plane coupling; otherwise, a doublet of triplets would be seen. In the high-temperature spectrum some process averages all P ligands so that both H's are in an equivalent environment and are coupled to the four chemically equivalent P's. (This is likely to be some motion of the H's rather than the bulkier P ligands.) That the high-temperature spectrum is not just due to the presence of the *trans*- isomer is seen by the difference in the chemical shift of the *trans*- isomer at low temperature and the high-temperature quintet.

9.30 Problem: Draw all of the isomers expected for **(a)** $[B(gly)_2]^+$ and **(b)** $[Pt(gly)_2]$. What factors account for the difference in your expectations?

9.30 Solution: The boron complex is expected to be tetrahedral since it has no d electrons and would have no LFSE and tetrahedral geometry minimizes ligand-ligand repulsion. Alternatively, tetrahedral geometry is expected for sp^3 hybridization. The Pt complex would be expected to be square planar since the LFSE would be high for a d^8 configuration of a heavy metal; or, the geometry would be expected from dsp^2 hybridization. The boron complex should thus give optical isomers whereas the Pt complex would give geometric isomers, as shown below.

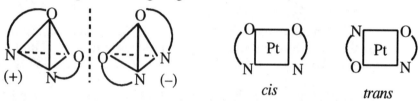

9.31 Problem: Sketch all possible isomers of [Pt(pn*)BrCl] where pn* is an optically active ligand 1,2-propanediamine (propylenediamine) MeC*HCH$_2$NH$_2$.
NH$_2$

9.31 Solution: For each geometric isomer a pair of optical isomers exists having the R or S configuration at C*.

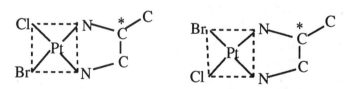

9.32 Problem: Since a square-planar complex with four different ligands, M(abcd), is optically inactive, sketch a square-planar complex which is optically active even though the free ligands are not.

9.32 Solution: See the solution to Problem 9.33 for the general conditions for optical activity. If we design a complex lacking a center or plane of symmetry, it is likely to be optically active. The presence of an S_n ($n > 2$) axis without σ or i is unusual. In order to eliminate the planes of symmetry usually found for a square planar complex, we can use two diamines substituted differently. The use of two difference eliminates i and σ between the ligands. In the example shown the diamine on the right has two identical substituents on one carbon only, eliminating the plane of symmetry cutting through both ligands and perpendicular to the PtN$_4$ plane. Having the same substituent on each carbon of the other diamine eliminates σ in the PtN$_4$ plane. The ligand on the left has a plane of symmetry itself, so it is an optically inactive *meso* form. This complex would have a plane of symmetry (and be optically inactive) if it were tetrahedral.

Possible configurations of a Pt^{II} complex. (H atoms on N are omitted.)

9.33 Problem: The complex Zr(OC$_6$H$_4$CH=NC$_2$H$_5$)$_4$ has **S$_4$** symmetry. It cannot be optically active even though it lacks a center of symmetry and a plane of symmetry. Explain.

9.33 Solution: A compound is optically active (its mirror images are nonsuperimposable) only if it lacks *any* S_n axis. The center of symmetry is equivalent to S_2 and the plane of symmetry is equivalent to S_1. (See DMA, 3rd ed., p. 430 or D. C. Bradley, M. B. Hursthouse, and L. F. Randall, *Chem. Commun.* **1970,** 368 for a drawing of the complex.)

IV. COORDINATION CHEMISTRY
10
Spectra And Bonding of Coordination Compounds

10.1 Problem: Identify the ground-state with the spin multiplicity for the following cases: **(a)** octahedral complexes and **(b)** tetrahderal complexes.

Cu^{2+}, V^{3+}, Cr^{3+}, Mn^{2+}, Fe^{2+}, and Ni^{2+}.

10.1 Solution: Spectral terms can be derived for any electron configuration (See DMA, 3rd ed., p. 32.). These are tabulated in many places (See DMA, 3rd ed., p. 38). The ground state is D for d^1, d^4, d^6, and d^9. It is F for d^2, d^3, d^7, and d^8. These are identified in Orgel diagrams and Tanabe-Sugano diagrams here and in Problem 10.5. The splitting of spectral terms in octahedral and tetrahedral fields are shown in these diagrams also (See DMA, 3rd ed., Table 10.1, p. 443, and Appendix D). A D term gives E_g and T_{2g} in octahedral and tetrahedral (omit g subscripts) fields. An F term gives T_{1g}, T_{2g}, and A_{2g} in an octahedral field (omit g for a tetrahedral field).

The ground state for d^5 has 5 unpaired electrons (maximum spin multiplicity), or a half-filled orbital set. Any empty, half-filled, or filled s, p, d, or f orbital set gives an S term. The high-spin d^5 gives ^6S.

For d^5 and d^6 there are low-spin configurations in addition to the ground states. Low-spin T_d cases are unlikely for d^5 and d^6. The low-spin configuration for d^6 (O_h) is t_{2g}^6. The filled t_{2g} orbitals are totally symmetric, $^1A_{1g}$. For low-spin $d^5(O_h)$ the configuration is t_{2g}^5. There is one vacancy (1 hole) in t_{2g} and the energy state is $^2T_{2g}$ for the ground state. Cu^{2+} (d^9) has the ground state term ^2D, giving ^2E_g for O_h and $^2T_2$ for T_d as ground states.

	Cr³⁺	Ni²⁺	Mn²⁺		Fe²⁺	
Number of *d* electrons	3	8	5		6	
Free ion ground state	^4F	^3F	^6S		^5D	
			High spin	Low spin	High spin	Low spin
O_h	$^4A_{2g}$	$^3A_{2g}$	$^6A_{1g}$	$^2T_{2g}(t_{2g}^5)$	$^5T_{2g}$	$^1A_{1g}(t_{2g}^6)$
T_d	$^4T_1$	$^3T_1$	$^6A_1$		^5E	

$d^3 \ B = \begin{matrix} 0.75 \text{ kK 22 for V}^{II} \\ 0.92 \text{ kK for Cr}^{III} \end{matrix}$

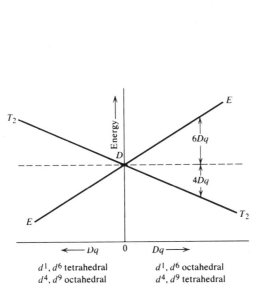

Orgel diagram   Tanabe-Sugano diagram for d^3

10.2 Problem: For $[V(H_2O)_6]^{2+}$ absorption bands are observed at 12,300, 18,500, and 27,900 cm⁻¹. Assign the bands and calculate or estimate Dq, B', and the bending. Why would transitions to $(^2E_g, ^2T_{1g})$ and $^2T_{2g}$ not be expected to be observed?

10.2 Solution: $[V(H_2O)_6]^{2+}$ has the d^3 configuration.

$^4A_{2g} \rightarrow {}^4T_{2g}$ $\nu_1 = 10Dq$
$^4A_{2g} \rightarrow {}^4T_{1g}(F)$ $\nu_2 = 18Dq - c$
$^4A_{2g} \rightarrow {}^4T_{1g}(P)$ $\nu_3 = 12Dq + 15B' + c$

From ν_1 $Dq = 1,230$ cm⁻¹ without bending.

From ν_2 and Dq $18,500 = 18(1230) - c$ and $c = 3600$ (bending).

From ν_3, $27,900 = 14,760 + 15 B' + 3600$ and $B' = 640$ cm⁻¹ (2 sig. figures). $B = 860$ for V²⁺, so this is reasonable.

Using the T-S diagram for d^3 (see above) we see that at $Dq/B' = 1.9$, the energy of $(^2E, ^2T_1)$ is close to that of $^4T_{2g}$ and $^2T_{2g}$ is close to that of $^4T_{1g}(F)$, so their weak peaks would probably be masked by the spin-allowed bands.

10.3 Problem For $[V(H_2O)_6]^{3+}$ (in a hydrated crystal) absorption bands are observed at 17,800 and 25,700 cm^{-1}. Assign the bands. Calculate or estimate the values of Dq, B', and bending. What difficulty is encountered in calculating the three parameters in this case?

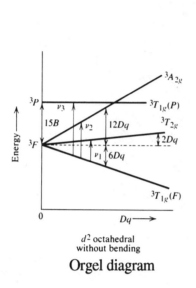

Orgel diagram

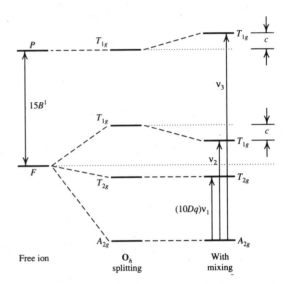

Splitting and mixing of states for d^3 and d^8 ions in an octahedral field.

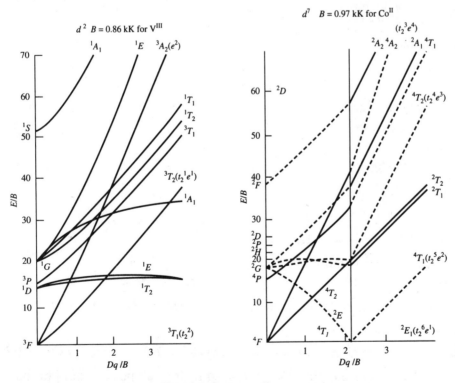

Tanabe-Sugano diagrams

10.3 Solution: $[V(H_2O)_6]^{3+}$ has the d^2 configuration.

$$^3T_{1g} \to {}^3T_{2g} \qquad v_1 = 8Dq + c$$
$$\to {}^3A_{2g} \qquad v_2 = 18Dq + c$$
$$\to {}^3T_{1g}(P) \qquad v_3 = 15B' + 6Dq + 2c$$

We have only two bands so two equations and cannot calculate 3 parameters.
From $v_1 = 17,800 = 8Dq$ + bending

Neglecting bending, $Dq \simeq 2200$ cm^{-1}, but it must be lower because of bending. If $Dq \simeq 2,000$ cm^{-1}, $v_2 = 36,000 +$ bending, or above 36,000 cm^{-1}, so the second band must be v_3 with v_2 too high in energy to be observed (and likely of low intensity as a 2-electron transition).
If we use $Dq \simeq 2000$ cm^{-1}, $v_1 = 17,800 = 16,000 + c$ and $c = 1,800$ cm^{-1}

For V^{3+} $B = 860$ cm^{-1} from the top of the T-S diagram. If we use the estimate $B' = 0.8 \times 860 \simeq 700$ cm^{-1}. Calculating the energy of v_3 from the estimates of parameters

$$v_3 = 15(700) + 12,000 + 2(1800) = 26,100 \text{ cm}^{-1}$$

Considering the rough estimates made, this is reasonable agreement. Using these parameters the energy of v_2 would be 37,800 cm^{-1}.

10.4 Problem: Common glass used for windows and many bottles is colored because of Fe^{2+} present. It is decolorized by adding MnO_2 to form Fe^{3+} and Mn^{2+}. Why is the glass decolorized? Would you expect broad, very weak absorption peaks for Fe^{3+} and Mn^{2+} or many weak peaks?

10.4 Solution: Fe^{3+} and Mn^{2+} (high-spin d^5) have only spin-forbidden transitions of low intensity. There are many spin-forbidden transitions. There is only one way to place 5 unpaired electrons in the d orbitals, but many ways for spin pairing, giving many excited energy states.

10.5 Problem: Identify the lowest-energy spin-allowed transitions for high-spin $[CoHi_6]^{2+}$ and low-spin $[CoLo_6]^{2+}$.

10.5 Solution: From the Tanabe-Sugano diagram for d^7 shown above,
 For high-spin $[CoHi_6]^{2+}$ $\quad {}^4T_{1g} \to {}^4T_{2g}$
 For low-spin $[CoLo_6]^{2+}$ $\quad {}^2E_g \to {}^2T_{1g}$ and ${}^2E_g \to {}^2T_{2g}$ at slightly higher energy.

10.6 Problem: For $[Cr(NH_3)_6]^{3+}$ there are two absorption bands observed at 21,500 cm^{-1} and 28,500 cm^{-1} and a very weak peak at 15,300 cm^{-1}. Assign the bands and account for any missing spin-allowed bands. Calculate Dq for NH_3 using the Orgel diagram. Account for any discrepancy between the observed position of any of the spin-allowed bands and that expected from the Orgel diagram.

10.6 Solution: (Cr^{3+} is d^3)
O_h $\quad v_1 \quad$ 21,500 cm^{-1} $\qquad {}^4A_{2g} \to {}^4T_{2g}$ (lower energy)
$\quad\quad v_2 \quad$ 28,500 cm^{-1} $\qquad {}^4A_{2g} \to {}^4T_{1g}(F)$

$^4A_{2g} \rightarrow {}^4T_{1g}(P)$ can be assumed to be too high in energy to be observed. The weak peak is spin-forbidden. v_1 is $10Dq$ or $Dq = 2{,}150$ cm^{-1}. v_2 is $18Dq$, but it is lowered because of bending (configuration interaction). $18Dq = 38{,}700$ cm^{-1}, but 28,500 is observed for v_2. The weak low-energy band must be $^4A_{2g} \rightarrow {}^2E_g, {}^2T_{1g}$ (unresolved). The $^4A_{2g} \rightarrow {}^2T_{2g}$ peak would be between the spin-allowed bands and does not appear.

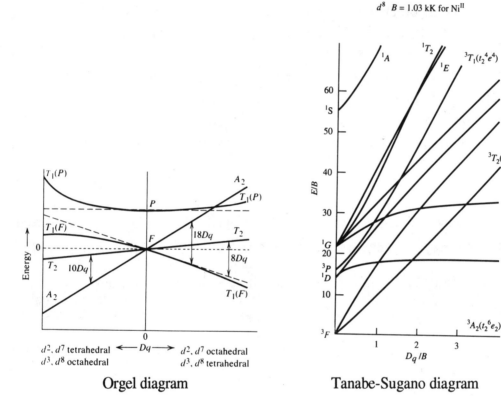

Orgel diagram   Tanabe-Sugano diagram

10.7 Problem: The following data are available for the complexes identified:

[Ni(H$_2$O)$_6$]$^{2+}$	[Ni(NH$_3$)$_6$]$^{2+}$	
8,600 cm^{-1}	10,700 cm^{-1}	
13,500	17,500	
25,300	28,300	
15,400 }	15,400 }	Very weak peaks for both complexes.
18,400 }	18,400 }	

Assign the bands. Calculate $10Dq$ and the expected positions of the spin-allowed bands. Account for any discrepancy between the experimental and calculated energies of the bands. Account for the relative positions of corresponding bands for the two complexes.

10.7 Solution: (See the splitting pattern and the Tanabe-Sugano diagram for d^8 above.)

d^8		$[Ni(H_2O)_6]^{2+}$	$[Ni(NH_3)_6]^{2+}$	
$10Dq =$	v_1	8,600 cm^{-1}	10,700 cm^{-1}	$^3A_{2g} \rightarrow {}^3T_{2g}$
	v_2	13,500	17,500	$\rightarrow {}^3T_{1g}(F)$
	v_3	25,300	28,300	$\rightarrow {}^3T_{1g}(P)$
		15,400	15,400	$\rightarrow {}^1E_g$
		18,400	18,400	$\rightarrow {}^1A_{1g}$
	Dq	860	1,070	
	B'	ca. 800	ca. 800	(about 0.8 of the value 1.03 kK from the Tanabe-Sugano diagram for d^8.)
v_2 ($18Dq$)		15,500	19,300	(bending neglected)
v_3 ($12Dq + 15B'$)		22,300	24,800	(bending neglected)

The discrepancies from observed values result from bending (configuration interaction). The bands for the NH$_3$ complex are at higher energy because of the higher field strength. The energies of the spin-forbidden transitions are independent of Dq because these involve pairing of the electrons within the e_g orbitals, not promotion from one set to another.

10.8 Problem: Interpret the following comparisons of intensities of absorption bands for transition metal complexes.

(a) Two isomers of a Co^{III} complex believed to be *cis-* and *trans*-isomers give the following spectral features:
  Both give two absorption bands in the visible region. Complex A has two symmetrical bands with $\varepsilon = 60\text{-}80$. The lower energy band for complex B is broad with a shoulder and has lower intensity. Assign the isomers. Explain.

(b) An octahedral complex of Co^{III}, with an amine and Cl$^-$ coordinated, gives two bands with $\varepsilon = 60\text{-}80$, one very weak peak with $\varepsilon = 2$, and a high energy band with $\varepsilon = 20,000$. What is the presumed nature of these transitions? Explain.

(c) Two complexes of Ni^{II} are believed to be octahedral and tetrahedral. Each has three absorption bands, but complex C has $\varepsilon \simeq 10$ and D has $\varepsilon \simeq 150$. Which probably is the tetrahedral complex? Explain. Measurement of what physical property would exclude the possibility of either complex being square planar?

10.8 Solution: (a) A is the *cis* isomer because there is smaller splitting (see Problem 10.11) and higher intensity (intensity is usually higher for lower symmetry). B is the *trans* isomer—larger splitting and lower intensity.

(b) We expect 2 spin-allowed bands with $\varepsilon < 100$. The very weak peak is expected to be spin-forbidden and the very intense band is allowed, presumably a charge-transfer band involving a Cl$\rightarrow$Co transition.

(c) D is expected to be the tetrahedral complex—higher intensity for the noncentrosymmetric complex. Measurement of magnetic susceptibility would distinguish diamagnetic complexes of Ni^{2+} from paramagnetic octahedral or tetrahedral complexes.

10.9 Problem: Calculate the relative energies of the d orbitals for an ML_6 complex with trigonal prismatic coordination (D_{3h}), assuming that the ligands are at the same angle relative to the xy plane as for a regular tetrahedron. [*Hint:* Start with the tetrahedral case, but allow for three ligands up and down, instead of two. The degeneracy of the d orbitals is the same as for other D_{3h} complexes, such as trigonal (ML_3) and trigonal bipyramidal (ML_5) cases.]

Relative d orbital energies for three primary geometric configurations

Configuration	Relative energies in units of Dq				
	d_{z^2}	$d_{x^2-y^2}$	d_{xy}	d_{xz}	d_{yz}
M–L (along z axis)	5.14	–3.14	–3.14	0.57	0.57
ML_2 (two ligands at 90°, along x and y)	–2.14	6.14	1.14	–2.57	–2.57
ML_4 (regular tetrahedron)	–2.67	–2.67	1.78	1.78	1.78

See R. Krishnamurthy and W. B. Schaap, *J. Chem. Educ.* **1969**, *46*, 799.

10.9 Solution: We start with the tetrahedral case, 2 ligands above the xy plane and 2 below and multiply by $\frac{3}{2}$ ($\frac{6}{4}$) to give the effect of 3 ligands above the 3 below the xy plane. The relative energies of the d orbitals are given by:

	d_{z^2}	$d_{x^2-y^2}$	d_{xy}	d_{xz}	d_{yz}	
ML_4 tetrahedron	–2.67	–2.67	1.78	1.78	1.78	Dq
For 6L ($ML_4 \times \frac{3}{2}$)	–4.00	–4.00	2.67	2.67	2.67	Dq
		average				
ML_6 D_{3h}	–4.00	–0.67	–0.67	2.67	2.67	Dq

Since for D_{3h} symmetry the $d_{x^2-y^2}$ and d_{xy} orbitals are degenerate (E' representation) we average the energies for these orbitals. The d_{xz} and d_{yz} orbitals are degenerate as required (E'' representation).

10.10 Problem: Calculate the relative energies of the d orbitals for the following complexes, assuming Dq (X) = 1.40Dq (Y) and that where X and Y are along the axis (*trans* to one another), the field strength is the same as for two equivalent ligands with the average field strength of X and Y. Use z as the unique axis.

(a) MX_5Y
(b) *trans*-[MX_4Y_2]
(c) *cis*-[MX_4Y_2] (both Y's in the xy plane)
(d) *facial*-[MX_3Y_3]

10.10 Solution: In each case we start with an octahedral MX_6 complex, subtract the effects of the missing ligands and add the effects of the Y ligands, or just add the components of the particular case individually. $Dq(X) = 1.40 Dq(Y)$ or $Dq(Y) = 0.714 Dq(X)$

(a) MX_5Y Oct. $MX_6 - (X)_z + (Y)_z$

	d_{z^2}	$d_{x^2-y^2}$	d_{xy}	d_{xz}	d_{yz}	
Oct. MX_6	6.00	6.00	–4.00	–4.00	–4.00	Dq
– (X)$_z$	–5.14	3.14	3.14	–0.57	–0.57	Dq
+ (Y)$_z$	3.67	–2.24	–2.24	0.41	0.41	Dq
MX_5Y	4.53	6.90	–3.10	–4.16	–4.16	Dq (X)

(b) *trans*-[MX$_4$Y$_2$]   Oct. MX$_6$ − 2(X)$_z$ + 2(Y)$_z$ or Square planar MX$_4$ + 2(Y)$_z$

	d_{z^2}	$d_{x^2-y^2}$	d_{xy}	d_{xz}	d_{yz}	
Sq. MX$_4$, 2 × ML$_2$	−4.28	12.28	2.28	−5.14	−5.14	
+ 2(Y)$_z$	7.35	−4.49	−4.49	0.82	0.82	
trans-[MX$_4$Y$_2$]	3.07	7.79	−2.21	−4.32	−4.32	*Dq* (X)

(c) *cis*-[MX$_4$Y$_2$]   Oct. MX$_6$ − 2(X)$_{xy}$ + 2(Y)$_{xy}$ or Linear MX$_2$ + 2(X)$_{xy}$ + 2(Y)$_{xy}$

	d_{z^2}	$d_{x^2-y^2}$	d_{xy}	d_{xz}	d_{yz}	
Linear MX$_2$	10.28	−6.28	−6.28	1.14	1.14	
+2(X)$_{xy}$	−2.14	6.14	1.14	−2.57	−2.57	
+ 2(Y)$_{xy}$	−1.53	4.38	0.81	−1.83	−1.83	
cis-[MX$_4$Y$_2$]	6.61	4.24	−4.33	−3.26	−3.26	*Dq* (X)

(d) *facial*-[MX$_3$Y$_3$]   (X)$_z$ +(Y)$_z$ + 2(X)$_{xy}$ + 2(Y)$_{xy}$

		d_{z^2}	$d_{x^2-y^2}$	d_{xy}	d_{xz}	d_{yz}	
MX(z)		5.14	−3.14	−3.14	0.57	0.57	
MY(z)	0.714 x	(5.14	−3.14	−3.14	0.57	0.57)	
MX$_2$ (xy)		−2.14	6.14	1.14	−2.57	−2.57	
MY$_2$ (xy)	0.714 x	(−2.14	6.14	1.14	−2.57	−2.57)	
facial-[MX$_3$Y$_3$]		5.14	5.14	−3.43	−3.43	−3.43	*Dq* (X)

The X–Y ligands are *trans* along each axis and the effective field is the average $\frac{Dq(X) + Dq(Y)}{2} =$ 0.86 *Dq*(X), so the splitting pattern is the same as for the **O**$_h$ case, hence, this is the "cubic" case.

10.11 Problem: Compare the splitting of the e_g (**O**$_h$) orbitals (difference in energies of d_{z^2} and $d_{x^2-y^2}$) for the cases in Problem 10.10 with the spectral results for *cis*- and *trans*-[Co(F)$_2$(en)$_2$]$^+$.

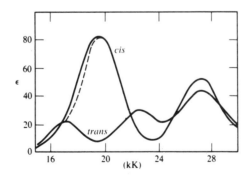

The absorption spectra of *cis*- and *trans*-[Co(F)$_2$(en)$_2$]NO$_3$.
The dashed line outlines the main Gaussian band.
(From F. Basolo, C. J. Ballhausen, and J. Bjerrum,
Acta Chem. Scand. **1955**, *9*, 810.)

10.11 Solution: In Problem 10.10 you calculated energies for d orbitals in some geometries of symmetry lower than O_h. The splitting of the degeneracy of the e_g ($d_{x^2-y^2}$ and d_{z^2}) orbitals in lower symmetries can lead to an increase in the number of $d \rightarrow d$ bands as compared to the octahedral case. Replacement of one, two, or three ligands in MX_6 lowers the symmetry from O_h (cubic). The effective symmetry can be regarded as "tetragonal" when the symmetry is lowered so that the average field along one axis (z) differs from that along the x and y axes. Tetragonal complexes include MX_5Y, trans-$[MX_4Y_2]$ and cis-$[MX_4Y_2]$ (X—X along z and X–Y along x and y). In these terms facial-$[MX_3Y_3]$ is effectively "cubic" having X–Y along all three axes.

	Energy difference ($d_{x^2-y^2} - d_{z^2}$)
(a) MX_5Y	$2.37\ Dq$
(b) trans-$[MX_4Y_2]$	$4.72\ Dq$, approximately 2 times **(a)**
(c) cis-$[MX_4Y_2]$	$-2.37\ Dq$, approximately – **(a)** or ½ **(b)**
(d) facial-$[MX_3Y_3]$	0, no splitting, the "cubic" case

The usual observation is that splitting of the lower energy band is about twice as great for trans-$[MX_4Y_2]$ compared to $[MX_5Y]$ or cis-$[MX_4Y_2]$, and no splitting is observed for facial-$[MX_3Y_3]$.

10.12 Problem: Calculate the relative energies of the d orbitals (See DMA, 3rd ed., Section 10.3.1) for ML_5 in the TBP (D_{3h}) and SP (C_{4v}) geometries. For high-spin d^6 to d^9 metal ions, for which cases is there a significant preference for one of the two geometries?

10.12 Solution: The energies of the d orbitals for the planar trigonal case are 3/2 times the energies for the ML_2 case with two ligands at 90°, but allowing for the degeneracy of the $d_{x^2-y^2}$ and d_{xy} orbitals and the d_{xz} and d_{yz} orbitals (as required for D_{3h} symmetry):

	d_{z^2}	$d_{x^2-y^2}$	d_{xy}	d_{xz}	d_{yz}	
ML_3 (3/2 × ML_2)	–3.21	9.21	1.71	–3.85	–3.85	Dq
		average		already degenerate		
Planar ML_3, D_{3h}	–3.21	5.46	5.46	–3.85	–3.85	
Linear ML_2	10.28	–6.28	–6.28	1.14	1.14	
TBP, ML_5, D_{3h}	7.07	–0.82	–0.82	–2.71	–2.71	Dq

The square pyramidal case is the sum of the energies for a square planar arrangement (2 × ML_2 at 90°) plus the contribution for one ligand along z:

	d_{z^2}	$d_{x^2-y^2}$	d_{xy}	d_{xz}	d_{yz}	
M–L (z)	5.14	–3.14	–3.14	0.57	0.57	
ML_4 (planar) (2 × ML_2, 90°)	–4.28	12.28	2.28	–5.14	–5.14	
ML_5 (SP, C_{4v})	0.86	9.14	–0.86	–4.57	–4.57	Dq

	LFSE (TBP)	LFSE (SP)
d^6	$-2.70\,Dq$	$-4.57\,Dq$
d^7	-5.41	-9.14
d^8	-6.23	-10.00
d^9	-7.05	-9.14

For d^6
TBP $-3(2.71) - 2(0.82) + 7.07 = -2.70$
SP $\;\;\;-3(4.57) - 0.86 + 0.86 + 9.14 = -4.57$

For all cases the LFSE is greater for the SP, but particularly for d^7 and d^8. The ligand-ligand repulsion is greater for the SP.

10.13 Problem: Explain why the ligand field (d–d) bands are shifted only slightly for $[CoX(NH_3)_5]^{2+}$ ions ($X^- = F^-$, Cl^-, Br^-, or I^-), but charge-transfer bands are shifted greatly for the series.

10.13 Solution: The LF bands shift slightly because the field strengths of the halides differ little. The energy of the charge-transfer band depends on the ease of loss of an electron from X^- (oxidation), and there are great differences in ease of oxidation from F^- to I^-.

10.14 Problem: Selection rules for electric dipole transitions (those usually observed for transition metals) require that the symmetry of the transition (the direct product of the representations for the ground and excited states) be the same as the representation for one of the electric dipole moment operators, and these correspond to the representations for x, y, and z. Selection rules for magnetic dipole transitions (observed for lanthanide metals) require that the symmetry of the transition be the same as the representation of one of the magnetic dipole moment operators, and these correspond to rotations about x, y, and z (R_x, R_y, and R_z in the character tables). For CD and ORD, the transitions must obey the selection rules for electric *and* magnetic dipole transitions. For $[Cr(NH_3)_6]^{3+}$ (O_h) the $d \rightarrow d$ transitions are $^4A_{2g} \rightarrow {}^4T_{2g}$ and $^4A_{2g} \rightarrow {}^4T_{1g}$. For $[Cr(en)_3]^{3+}$ (D_3) the transitions are $^4A_2 \rightarrow {}^4E$ and $^4A_2 \rightarrow {}^4A_1$ (both derived from $^4A_{2g} \rightarrow {}^4T_{2g}$ by lowering symmetry). Which of these are electric-dipole-allowed? Which are *both* electric- and magnetic-dipole-allowed?

10.14 Solution:
$$\int \psi_{ground}(\text{operator})\psi_{excited}\,d\tau$$

For the integral describing an electronic transition to be nonzero (an allowed transition) the symmetry of the transition (product of the representations for the wavefunctions for the ground and excited states) must be the same as that of the operator. For electric dipole transitions the operator belongs to the same representations as x, y, and z. For magnetic dipole transitions the operator belongs to the same representations as R_x, R_y, and R_z.

For $[Cr(NH_3)_6]^{3+}$ ($\mathbf{O}_h$)

Transition	Symmetry of transition	Symmetry of operators	
$A_{2g} \rightarrow T_{2g}$	$A_{2g} \times T_{2g} = T_{1g}$	x, y, z	T_{1u}
$A_{2g} \rightarrow T_{1g}$	$A_{2g} \times T_{1g} = T_{2g}$	R_x, R_y, R_z	T_{1g}

Both transitions are electric dipole forbidden (neither is T_{1u}). They appear as low intensity absorption bands because of vibronic coupling. The T_{1g} ($A_{2g} \rightarrow T_{2g}$) transition is magnetic dipole allowed, but $A_{2g} \rightarrow T_{1g}$ is magnetic dipole forbidden. There is no ORD or CD (circular dichroism) for $[Cr(NH_3)_6]^{3+}$ since optical activity requires that *both* integrals be finite. We know that $[Cr(NH_3)_6]^{3+}$ is not optically active because the $\mathbf{O}_h$ symmetry includes i and σ.

For $[Cr(en)_3]^{3+}$ ($\mathbf{D}_3$)

Transition	Symmetry of transition	Symmetry of operators	
$A_2 \rightarrow E$	$A_2 \times E = E$	z, A_2	$(x, y), E$
$A_2 \rightarrow A_1$	$A_2 \times A_1 = A_2$	R_z, A_2	$(R_x, R_y), E$
$A_2 \rightarrow A_2$	$A_2 \times A_2 = A_1$		

The transitions of E and A_2 symmetry obey both selection rules and should appear in the CD (or ORD) spectrum. The A_1 ($A_2 \rightarrow A_2$) transition is electric and magnetic dipole forbidden (See DMA, 3rd ed., pp. 446, 467).

10.15 Problem (a) For a linear ML_2 complex ($\mathbf{D}_{\infty h}$) with the ligands lying along the z axis, indicate the metal orbitals that may participate in σ bonds. Sketch the ligand group orbitals that could enter into σ MOs. (b) Identify the metal orbitals for π bonds.

10.15 Solution (a) For a linear complex with bonding along z, the metal orbitals for σ bonding are those along this direction:

L—M—L----z   M s, d_{z^2} a_{1g} and p_z a_{1u}

L_2 s L_1 L_2 d_{z^2} L_1 L_2 p_z L_1

(b) M π orbitals (p_x, p_y), e_{1u} and (d_{xz}, d_{yz}), e_{1g}

10.16 Problem: (a) Identify the metal orbitals for σ bonding for a trigonal planar complex, ML_3 ($\mathbf{D}_{3h}$). (b) Sketch the σ MOs involving s and p orbitals.

10.16 Solution: (a) The M σ orbitals in planar ML_3 are those contained in the xy plane. M σ orbitals s, a_1'; (p_x, p_y), e'; and $(d_{x^2-y^2}, d_{xy})$, e'

(b) Using the atomic orbitals as templates:

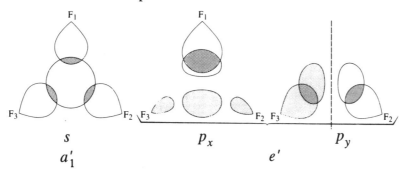

10.17 Problem: Use the group theoretical approach to obtain the representations and LGO for σ bonding in trigonal planar complexes, ML$_3$.

10.17 Solution: Examining the effects of one symmetry operation of each class, the σ orbitals left unmoved are:

D_{3h}	E	C_3	C_2	σ_h	S_3	σ_v
Γ_σ	3	0	1	3	0	1

$= a_1' + e'$

$s, a_1'; e', (p_x, p_y);$ and $e', (d_{xz}, d_{yz})$.

Using the projection operator, we operate on σ_1, using all operations:

D_{3h}	E	C_3	C_3'	C_2	C_2'	C_2''	σ_h	S_3	S_3'	σ_v	σ_v'	σ_v''
σ_1	σ_1	σ_2	σ_3	σ_1	σ_3	σ_2	σ_1	σ_2	σ_3	σ_1	σ_3	σ_2

Multipying these by the characters for a_{1g} and e', we get

a_{1g} $\sigma_1 + \sigma_2 + \sigma_3$ or $a_{1g} = 1/\sqrt{3}(\sigma_1 + \sigma_2 + \sigma_3)$

$e'(1)$ $2\sigma_1 - \sigma_2 - \sigma_3$ or $e'(1) = 1/\sqrt{6}(2\sigma_1 - \sigma_2 - \sigma_3)$

the other e' must have a node through σ_1 giving a coefficient zero for σ_1 and opposite signs for σ_2 and σ_3.

$e''(2)$ $\sigma_2 - \sigma_3$ or $e'(2) = 1/\sqrt{2}(\sigma_2 - \sigma_3)$

10.18 Problem: Negative ions might be expected to create stronger ligand fields than neutral molecules on the basis of the pure crystal-field model. Explain why OH$^-$ has a weaker ligand field ligand than does H$_2$O. Why does CO have such a strong ligand field?

10.18 Solution: OH$^-$ is a stronger π donor than H$_2$O and π donation (L→M) decreases $10Dq$ because the metal electrons must be in the antibonding π* orbitals (see Figure). CO is a π acceptor, the metal t_{2g} orbitals are lowered in energy since they are bonding and electrons are delocalized onto the ligands, increasing $10Dq$. See the effect of L→M bonding in Problem 10.19.

10.19 Problem: For the MO diagram for [Ni(NH$_3$)$_6$]$^{2+}$ (below). **(a)** Identify the portion of the diagram considered by crystal field theory. **(b)** Designate the portion of the diagram considered by valence bond (VB) theory. Why does the d^8 case present a problem for VB theory for an octahedral complex? **(c)** Sketch the effect on the e_g^* levels in $[NiL_6]^{2+}$ with a π-donor ligand.

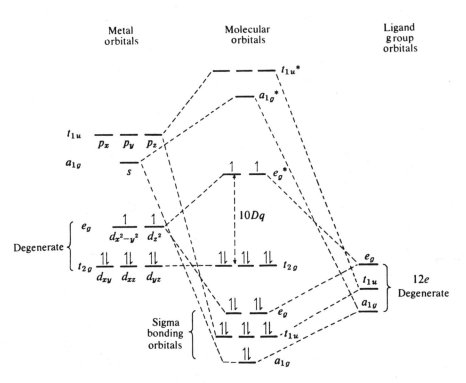

Qualitative diagram for the MO's of an octahedral d^8 complex such as $[Ni(NH_3)_6]^{2+}$ (σ only).

10.19 Solution: (a) The crystal-field portion of the MO diagram is the central portion including t_{2g} and e_g^* separated by $10Dq$.
(b) The VB theory portion includes the bonding d^2sp^3 orbitals, a_{1g}, t_{1u}, and e_g. VB theory does not deal with antibonding e_g^* orbitals so there is not a place for the two metal d electrons.
(c)

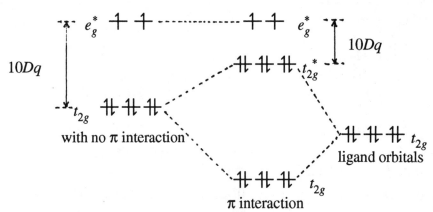

10.20 Problem: For a square-planar complex with the ligands lying along the x and y axes, indicate all of the metal orbitals that may participate in σ bonds. Sketch one ligand group orbital that could enter into a σ_g MO and one that could enter into a σ_u MO. Repeat the above for π bonds.

10.20 Solution: The σ orbitals of a metal in a square-planar complex ($\mathbf{D}_{4h}$) are those directed along the x and y axes, including the spherical s and d_{z^2} with the donut in the xy plane: p_x, p_y, s, $d_{x^2-y^2}$ and d_{z^2}. The ligand group orbitals are those with symmetry corresponding to these orbitals:

σ Group Orbitals

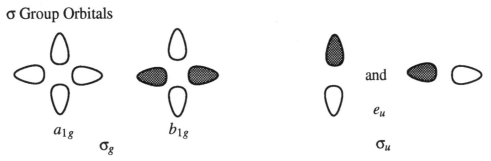

The π orbitals of a metal in a square planar complex are those with opposite signs for the lobes in planes perpendicular to the σ bonds: p_z, d_{xy}, d_{xz}, d_{yz}, p_x, and p_y.

π Group Orbitals (π vertical)

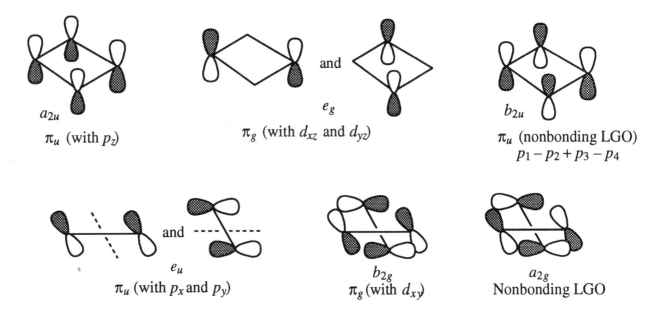

10.21 Problem: Give a pictorial approach to obtain the bonding MO's for σ and π bonding in square-planar complexes.

10.21 Solution: Sigma bonding for square-planar complexes. The metal orbitals that overlap with the ligand orbitals for σ bonding are s, p_x, p_y, and $d_{x^2-y^2}$—the dsp^2 hybrids. For bonding the LGO's match the signs of the metal orbitals.

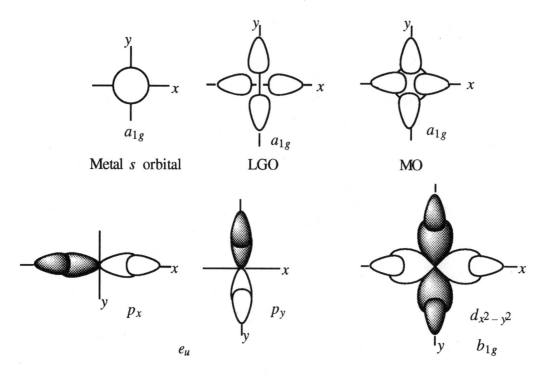

Pi bonding. The M orbitals with π symmetry for overlap with the LGO's parallel to the C_4 axis are p_z and d_{xz}, d_{yz}. Using these orbitals as templates, we can sketch the corresponding LGO's.

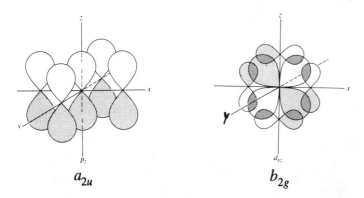

In the xy plane the M orbitals that can form π bonds are p_x, p_y, and d_{xy}. The p_x and p_y orbitals give better overlap for σ bonding and are expected to be involved primarily in σ bonding.

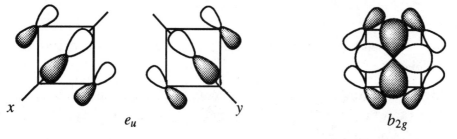

10.22 Problem: Use the group theoretical approach to obtain the representations and LGO's for σ bonding in square-planar complexes.

10.22 Solution: The vectors left unchanged by the operations of the D_{4h} group are given below. The reducible representation obtained reduces to give $A_{1g} + B_{1g} + E_u$.

D_{4h}	E	C_4	C_2	C_2'	C_2''	i	S_4	σ_h	σ_v	σ_d			
Γ_σ	4	0	0	2	0	0	0	4	2	0	$= A_{1g}$	$+ B_{1g}$	$+ E_u$
								M orbitals			s	$d_{x^2-y^2}$	p_x, p_y

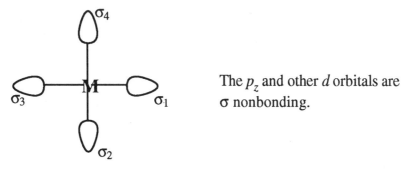

The p_z and other d orbitals are σ nonbonding.

Applying the projection operator:

D_{4h}	E	C_4	C_4'	C_2	$(C_2')_a$	$(C_2')_b$	$(C_2'')_a$	$(C_2'')_b$	i	S_4	S_4'	σ_h	σ_v	σ_d	σ_d'
σ_1	σ_1	σ_2	σ_4	σ_3	σ_1	σ_3	σ_2	σ_4	σ_3	σ_2	σ_4	σ_1	σ_3	σ_2	σ_4

Multiplying by the characters for the representations,

a_{1g} $4\sigma_1 + 4\sigma_2 + 4\sigma_3 + 4\sigma_4$ or $a_{1g} = \frac{1}{2}(\sigma_1 + \sigma_2 + \sigma_3 + \sigma_4)$

b_{1g} $4\sigma_1 - 4\sigma_2 + 4\sigma_3 - 4\sigma_4$ or $b_{1g} = \frac{1}{2}(\sigma_1 - \sigma_2 + \sigma_3 - \sigma_4)$

e_u $4\sigma_1 - 4\sigma_3$ or $e_u(1) = 1/\sqrt{2}(\sigma_1 - \sigma_3)$

rotate $e_u(1)$ by C_4 to get $e_u(2) = 1/\sqrt{2}(\sigma_2 - \sigma_4)$

10.23 Problem: Using Pauling's electroneutrality principle (see Problem 2.14) and considering the degree of covalent character expected from ligand electronegativities, would you consider CN^- as a ligand to favor higher or lower oxidation states in transition metal complexes compared to the ligand NH_3? Does the metal–ligand π bonding of CN^- favor high or low oxidation states? Which oxidation states are favored?

10.23 Solution: Because of the lower electronegativity of the donor C of CN^- compared to N of NH_3, the M—CN bond is more covalent than the M—NH_3 bond and more electron density is transferred to the metal for CN^- as a σ ligand. This favors a high oxidation state to maintain electroneutrality. The metal-to-ligand π donation that greatly stabilizes cyanide complexes uses filled t_{2g} metal orbitals, favoring low oxidation states. The back bonding lowers the accumulation of negative charge on the metal atom and low oxidation states are favored in cyanide complexes.

10.24 Problem: The d–d electronic transitions, forbidden by the symmetry (LaPorte) selection rule for centrosymmetric point groups, can become partially allowed through mixing between d and p orbitals. However, d–p mixing is prohibited for some noncentrosymmetric point groups. Identify two of these groups.

10.24 Solution: For centrosymmetric point groups d (g) and p (u) orbitals necessarily belong to different representations and cannot mix. The d–p mixing is prohibited for any point group where there is no representation to which at least one of the p orbitals and one of the d orbitals belongs:

$\mathbf{C}_{5h}$, $\mathbf{D}_{5h}$, $\mathbf{D}_{4d}$, $\mathbf{D}_{6d}$, $\mathbf{S}_8$, $\mathbf{O}$, and $\mathbf{I}$.

10.25 Problem: It is possible for symmetry fobidden d–d transitions to gain intensity from d–f mixing for most noncentrosymmetric groups. For which noncentrosymmetric groups is d–f mixing prohibited? What other consideration makes d–f mixing less important than d–p mixing, at least for the first transition series metals?

10.25 Solution: There are no representations in common for d and f orbitals for $\mathbf{D}_{7h}$, $\mathbf{D}_{6d}$, $\mathbf{D}_{8d}$, and $\mathbf{I}$. The greater energy separation between $3d$ and $4f$ orbitals (compared to the separation between $3d$ and $4p$ orbitals) makes mixing of little importance.

10.26 Problem: The spin-allowed transitions of $[Cr(NH_3)_6]^{3+}$ and $[Cr(en)_3]^{3+}$ have similar intensities. Is this expected from the selection rules for the full symmetry of the ions? Why is it true?

10.26 Solution: We might expect greater intensities of ligand-field bands for the noncentrosymmetric $[Cr(en)_3]^{3+}$ because of possible d–p mixing. In fact, the intensities are very similar for the visible spectra for $[Cr(NH_3)_6]^{3+}$ and $[Cr(en)_3]^{3+}$. This observation indicates that the effective symmetry is $\mathbf{O}_h$ for the CrN_6 chromophore in each case.

10.27 Problem: Show the splitting of the energy levels resulting from compressing an octahedral field along the z axis (the equivalent of having four weak-field ligands in the xy plane and two strong-field ligands along z). The resulting symmetry is $\mathbf{D}_{4h}$. Give the appropriate labels for the d orbitals and indicate how you know which orbitals are lower in energy.

10.27 Solution: The correlations can be obtained from the character tables for O_h and D_{4h} (identifying the representations for the d orbitals) or from a correlation table for O_h:

O_h		D_{4h}			
$(d_{z^2}, d_{x^2-y^2})$	e_g	(d_{z^2})	a_{1g},	$(d_{x^2-y^2})$	b_{1g}
(d_{xy}, d_{xz}, d_{yz})	t_{2g}	(d_{xy})	b_{2g},	(d_{xz}, d_{yz})	e_g

Since the resulting field is stronger along the z axis, d_{z^2} will be higher in energy than $d_{x^2-y^2}$ and (d_{xz}, d_{yz}) will be higher in energy than d_{xy}. The d_{z^2} orbital is raised in energy by the same amount by which $d_{x^2-y^2}$ is lowered. The splitting of t_{2g} is smaller since the orbitals are not directed toward the ligands. The single b_{2g} orbital is lowered by twice as much as the two e_g orbitals are raised (the net stabilization must be zero for three or six t_{2g} electrons).

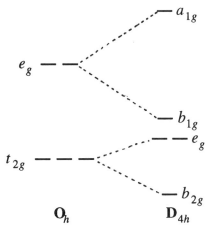

10.28 Problem: Given the relative energies of the d orbitals in Dq units in a ligand field of octahedral symmetry and of square-planar symmetry, determine the splitting pattern in a field of pentagonal-pyramidal symmetry.

	d_{z^2}	$d_{x^2-y^2}$	d_{xy}	d_{xz}	d_{yz}
Octahedral	6.00	6.00	−4.00	−4.00	−4.00
Square planar	−4.28	12.28	2.28	−5.14	−5.14

At what d-orbital populations would a pentagonal pyramidal geometry give greater ligand-field-stabilization energy (LFSE) *in a weak field* than octahedral geometry?

10.28 Solution: Using the procedures of Problems 10.9 and 10.10, we obtain the relative Dq values for a linear MX_2 complex by substracting the square-planar configuration values from those fot the octahedral configuration; half of these values gives the values for a linear MX configuration. To the linear MX Dq values we add 5/4 of the values for the square-planar configuration (since there are 5 ligands in the base of the pentagonal pyramid), and average the values for orbitals that are degenerate in C_{5v} geometry. The resulting Dq values for the pentagonal pyramidal MX_6 complex are:

d_{z^2}	$d_{x^2-y^2}$	d_{xy}	d_{xz}	d_{yz}
−0.21	5.96	5.96	−5.86	−5.86 Dq

From these values, and those of the octahedral configuration, we may calculate the following LFSE for weak-field complexes:

	LFSE	
	Pentagonal pyramid	Octahedron
d^1, d^6	−5.86	−4.0
d^2, d^7	−11.72	−8.0
d^3, d^8	−11.93	−12.0
d^4, d^9	−5.97	−6.0
d^5, d^{10}	0	0

Thus, we find that the LFSE is greater for the pentagonal pyramid than for the octahedron for d^1, d^2, d^6, and d^7. The pentagonal pyramid is not favored, however, since ligand–ligand repulsion is much greater than in the octahedron.

10.29 Problem: Using the appropriate Tanabe-Sugano diagram for Cr^{III} (see Problem 10.6) in an octahedral complex having a value of Dq/B of approximately 3, compare the following transitions with regard to:
 (1) frequency or wavenumber
 (2) breadth of band expected
 (3) intensity of band expected

(a) $^4A_{2g} \rightarrow {}^2T_{1g}$
(b) $^4A_{2g} \rightarrow {}^4T_{2g}$

Explain the fact that the energy of the $^4T_{2g}$ arising from the ^4F term is linear with ligand field strength, while $^4T_{1g}$ arising from ^4F is not.

10.29 Solution: Using a Tanabe-Sugano diagram for a d^3 configuration, we find the following.
 Transition (a) is approximately 2/3 the energy of (b) and will thus occur at approximately 2/3 of the frequency. It will be a sharp band since the energy of the transition is relatively independent of field strength and will not be affected by bond stretching. In contrast band (b) should be broad. Transition (a) is both spin and symmetry (Laporte) forbidden, and so will be very weak; (a) is spin allowed, but symmetry forbidden ($g \rightarrow g$) so it will be more intense, although still weak.
 The $^4T_{1g}$ state from the ^4F term interacts with the $^4T_{1g}$ from the ^4P, so they diverge with increasing Dq. The $^4T_{2g}$ is the only state of this symmetry designation that appears and is thus not influenced by other states.

10.30 Problem: A general knowledge of the features of $d \to d$ and charge transfer spectra of octahedral complexes should enable you to match each lettered spectrum with the correct complex.

Complex	Maxima (in kK)			
A	22.4(weak)	25.9(weak)	36.8(strong)	41.7(strong)
B	18.1(weak)	22.2(weak)	30.1(strong)	33.9(strong)
C	23.9(weak)	30.1(weak)		
D	32.7(weak)	39.1(weak)		
E	19.3(weak)	24.3(weak)	39.2(strong)	
F	16.6(weak)	24.9(weak)		

(Weak means $\varepsilon \simeq 100$, strong means $\varepsilon \simeq 10{,}000$)

Complexes: $[IrBr_6]^{3-}$, $[Co(H_2O)_6]^{3+}$, $[RhBr_6]^{3-}$, $[Rh(NH_3)_6]^{3+}$, $[RhCl_6]^{3-}$, $[Rh(H_2O)_6]^{3+}$

10.30 Solution: All the complexes are expected to be low-spin since most complexes of these metals are low-spin for the d^6 configuration. Low-spin d^6 complexes give two weak (spin-allowed, symmetry-forbidden) bands: $^1A_{1g} \to {}^1T_{2g}$ and $^1A_{1g} \to {}^1T_{1g}$. Intense (allowed) charge-transfer bands of the L $\to$ M type might be seen also for the halide complexes. The expected order or increasing energy for the $d \to d$ bands is:

$Br^- < Cl^- < H_2O < NH_3$ (Spectrochemical series)
$Co^{III} < Rh^{III} < Ir^{III}$

The complexes with intense bands (A, B, and E) are expected to be the halide complexes. The complexes without charge-transfer bands, listed in the order of increasing energy of the first band, are expected to be:

F	C	D
16.6	23.9	32.7 kK
$[Co(H_2O)_6]^{3+}$	$[Rh(H_2O)_6]^{3+}$	$[Rh(NH_3)_6]^{3+}$

Of the remaining complexes $[RhBr_6]^{3-}$ is expected to correspond to the lowest energy first band (B). We expect a small increase in the energy of this band for $[RhCl_6]^{3-}$, so E (19.3) is more likely than A (22.4). Also we expect the Cl $\to$ Rh charge-transfer band to be at much higher energy than Br $\to$ Rh, as is the case of E. The spectrum for complex A agrees with that expected for $[IrBr_6]^{3-}$, the $d \to d$ bands are at higher energy and the same pattern of charge-transfer bands is seen as for $[RhBr_6]^{3-}$.

Problem 10.31: To which point groups must optically active molecules belong?

Solution 10.31: Optically active molecules must lack σ planes (S_1), i (a center of symmetry, which is the same as S_2), or *any* S_n axis. These are the pure rotation groups: $\mathbf{C}_n$, $\mathbf{D}_n$, $\mathbf{T}$, $\mathbf{O}$, and $\mathbf{I}$.

Problem 10.32: Which of the spin-allowed transitions of [Ni(en)$_3$]$^{2+}$ (**D$_3$**) might be expected to appear in the circular dichroism spectrum based on selection rules? (Actually the complex is too labile to be resolved.)

Solution 10.32: For [Ni(en)$_3$]$^{2+}$ (d^8):

O$_h$ transitions	**D$_3$** transitions	Transition Symmetry	μ_e	μ_m	**D$_3$** representations for μ_e and μ_m
$^3A_{2g} \rightarrow {}^3T_{2g}$	$^3A_2 \rightarrow {}^3A_1 + {}^3E$	$A_2 \times A_1 = A_2$	z	R_z	A_2
$\rightarrow {}^3T_{1g}(F)$	$\rightarrow {}^3A_2 + {}^3E$	$A_2 \times A_2 = A_1$			
$\rightarrow {}^3T_{1g}(P)$	$\rightarrow {}^3A_2 + {}^3E$	$A_2 \times E = E$	(x,y)	(R_x, R_y)	E

Transitions are allowed if the transition symmetry belongs to the same representation as that of one of the operators. In the case, the representations are the same for the corresponding operators for electric dipole (μ_e) and magnetic dipole (μ_m) transitions.

$^3A_2 \rightarrow {}^3A_1$ and $^3A_2 \rightarrow {}^3E$ are electric- and magnetic-dipole-allowed and should appear in CD spectra. It was noted above that this cannot be verified experimentally since the complex racemizes too rapidly.

COORDINATION CHEMISTRY
11
Reaction Mechanisms of Coordination Compounds

11.1 Problem: The following are approximate rates of exchange of solvent water with bound water for some metal aqua complexes:

Metal	Na^+	Sr^{2+}	Mg^{2+}	Ni^{2+}	Be^{2+}	Ga^{3+}	Al^{3+}	Ru^{2+}	Cr^{3+}
$k\ (s^{-1})$	$\sim 10^9$	$\sim 10^9$	$\sim 10^6$	$\sim 10^5$	$\sim 10^3$	$\sim 10^3$	~ 10	$\sim 10^{-2}$	$\sim 10^{-6}$

Explain how these data are compatible with a dissociative **d** mechanism for water exchange.

11.1 Solution: A **d** mechanism is one in which the breaking of the bond between a metal and the leaving ligand exerts a dominant influence on the rate. The rates given tend to be slower for smaller, highly charged cations which should form strongest M—O bonds, thus supporting a **d** mechanism. This trend is somewhat complicated by effects of d-electron configuration in the transition series. (Compare Ni^{2+} and Ru^{2+} or Al^{3+} and Cr^{3+}.)

11.2 Problem: Account for the difference in rate constants for the following two reactions.
$[Fe(H_2O)_6]^{2+} + Cl^- \rightarrow [FeCl(H_2O)_5]^+ + H_2O \quad k = 10^6\ M^{-1}s^{-1}$
$[Ru(H_2O)_6]^{2+} + Cl^- \rightarrow [RuCl(H_2O)_5]^+ + H_2O \quad k = 10^{-2}\ M^{-1}s^{-1}$

11.2 Solution: The Fe^{2+} complex involving a first-row transition metal is high-spin d^6 and labile. Ru^{2+}, like other second- and third-row transition metal ions, forms low-spin complexes. Low-spin d^6 species are kinetically inert.

11.3 Problem: Octahedral complexes in which the loss of LFSE is large on going to the activated complex were said to substitute slowly. A model of the transition state for **D** substitution is a five-coordinate, square pyramid. Using the approach of Krishnamurthy and Schaap (DMA, 3rd ed., p.459), calculate the LFSE for strong-field d^n configurations. The difference LFSE(oct.) − LFSE(sq. py.) is the ligand field contribution to the activation energy for substitution, the so-called Ligand Field Activation Energy (LFAE). Compute this quantity for each configuration.

11.3 Solution: The energies of the d orbitals for a five-coordinate square-pyramidal complex are:

	d_{z^2}	$d_{x^2-y^2}$	d_{xy}	d_{xz}	d_{yz}
	0.86	9.14	−0.86	−4.57	−4.57 Dq

	LFSE(oct.)	LFSE(sq. py.)	LFAE
d^1	4 Dq	4.57 Dq	−0.57 Dq
d^2	8	9.14	−1.14
d^3	12	13.71	−1.71
d^4	16	18.28	−2.28
d^5	20	19.14	0.86
d^6	24	20.00	4.00
d^7	18	19.14	−1.14
d^8	12	18.28	−6.28
d^9	6	9.14	−3.14
d^{10}	0	0	0

The negative values are artifacts since, if distortion lowers the LFSE, such distortion would have occurred in the ground state. The most postive values of LFAE occur for strong-field d^5 and d^6 which are substitution-inert. The inertness of the d^3 configuration is not reflected in this calculation. (However, ligand field effects account for only $\simeq$ 10% of the bond energy.) The negative value of LFAE for d^8 reflects the stabilization which occurs on lowering the coordination number; stong-field d^8 complexes generally are square planar. A similar calculation for the weak-field case gives LFSAE = +2.00 Dq for d^8. (LFAE for d^3 in a weak field is +2.00 Dq. For a seven-coordinate intermediate, see N. S. Hush, *Aust. J. Chem.* **1962**, *15*, 378.)

11.4 Problem: Distinguish between the intimate and stoichiometric mechanism of a reaction.

11.4 Solution: The intimate mechanism deals with the factors which contribute primarily to the activation energy—whether bond breaking by the leaving group in substitution reactions or bond making by the entering group. The stoichiometric mechanism is the sequence of *detectable* elementary steps and describes the occurrence or non-occurrence of reaction intermediates. (We use lower case letters in referring to intimate mechanisms and upper case for stoichiometric mechanisms.)

11.5 Problem: What is the significance of the following facts taken together for the mechanism of substitution at Co^{III} in aqueous solution?
(a) The rates of aquation are always given by the expression rate = k_{aq}[CoX(NH$_3$)$_5$$^{2+}$].
(b) No direct replacement of X$^-$ by Y$^-$ is ever observed. Instead, water enters first and is subsequently replaced by Y$^-$.

11.5 Solution: (a) Since water is the solvent, its concentration cannot be varied. Hence, it never appears as an observable in the rate law and its concentration is incorporated into the value of k_{obs}.
(b) The observation implies that the energetic significance of bond making by the entering group is small. Hence, water is most often the entering ligand for statistical reasons. Ultimately, the thermodynamically stable product is formed with Y$^-$ replacing water.

11.6 Problem: The following data have been reported for aquation at 298.1K of complexes *trans* -[Co(N$_4$)LCl]$^{n+}$ where N$_4$ is a tetradentate chelate.

L	$k \times 10^4$ (s^{-1}) (N$_4$) = cyclam	(N$_4$) = tet-b
Cl$^-$	0.011	9.3
NCS$^-$	0.000011	0.0070
NO$_2^-$	0.43	410
N$_3^-$	0.036	210
CN$^-$	0.0048	3.4
NH$_3$	0.00073	---

What do these data suggest about the intimate mechanism for aquation? (See. W.-K. Chau, W.-K. Lee and C. K. Poon, *J. Chem. Soc., Dalton Trans.* **1974**, 2419.)

11.6 Solution: Tet-b is a more sterically crowded ligand than cyclam. (See Figure 11.3, DMA, 3rd ed.) The fact that aquation of tet-b complexes proceeds 10^2–10^3 times as fast as the comparable cyclam complex is in line with expectations for a **d** mechanism. Further of interest is the fact that π-donor L does not lead to a rate increase as compared to π-acceptor L. This suggests that a trigonal bipyramid is not the transition state or intermediate. The rigid macrocyclic ligands probably prevent rearrangement and impose a square-pyramidal geometry.

11.7 Problem: The reactions [Cr(NCS)$_6$]$^{3-}$ + solv → [Cr(NCS)$_5$(solv)]$^{2-}$ + NCS$^-$ have been investigated and found to have the following rate constants near 70°C.

Solvent	k (s^{-1})
dimethylacetamide	9.5 x 10^{-5}
dimethylformamide	12.4 x 10^{-5}
dimethylsulfoxide	6.2 x 10^{-5}

What do these values suggest about the intimate mechanism of these reactions? (See S. T. D. Lo and D. W. Watts, *Aust.. J. Chem.* **1975**, *28*, 1907.)

11.7 Solution: The different groups all enter at about the same rate. Hence a **d** mechanism is indicated.

11.8 Problem: The figure shows plots of k_{obs} vs [X$^-$] for the anation reactions in dmso solvent.
[Co(NO$_2$)(en)$_2$(dmso)]$^{2+}$ + X$^-$ → [Co(NO$_2$)(en)$_2$X]$^+$ + dmso
All these reactions are presumed to have the same mechanism. The dashed line is the rate of dmso exchange.
(a) What is the significance of the shapes of the curves?
(b) What is the significance of the fact that the first-order limiting rate constants are smaller than that for dmso exchange?
(c) If the mechanism were **D**, to what would the limiting rate constants correspond?

(d) If the mechanism were I_d, to what would the limiting rate constants correspond?
(e) The limiting rate constants are $5.0 \times 10^{-5}\ s^{-1}$ and 1.2×10^{-4} for Cl^- and NO_2^-, respectively. For NCS^-, the limiting rate constant can be estimated as $1 \times 10^{-5} s^{-1}$. Do these values constitute evidence for a **D** or an I_d mechanism?
(See W. R. Muir and C. H. Langford, *Inorg. Chem.* **1968**, *7*, 1032.)

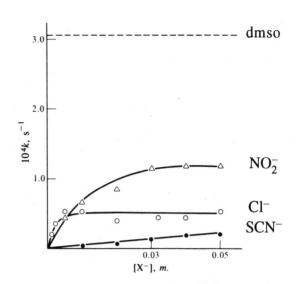

11.8 Solution:
(a) From the lack of linearity, we conclude that for $X^- = NO_2^-$ and Cl^-, the data show mixed first- and second-order kinetics. Presumably NCS^- would also behave similarly at higher concentrations. This makes the reactions candidates for I_a, **D**, or I_d mechanisms.
(b) Unless some other entering group could be found which entered faster than dmso, the anations should be regarded as having a **d** intimate mechanism.
(c) k_1 (the rate constant for Co–O bond breaking). (See Table 11.7 in DMA, 3rd ed.)
(d) kK (the product of the outer-sphere complexation constant K and the rate constant for ligand interchange). (See Table 11.7, DMA, 3rd ed.)
(e) No appreciable discrimination among entering groups is observed. Unless some other group could be found which gives a very different rate, the mechanism should be regarded as I_d since the five-coordinate intermediate of a **D** mechanism should be able to discriminate among entering ligands. The observed behavior indicates that the breaking of the Co–dmso bond is rate-determining and the entering group is whatever happens to be in the second coordination sphere—the defining criteria for an I_d mechanism.

11.9 Problem: The following data refer to water exchange on $[Ln(H_2O)_8]^{3+}$.

Ln^{3+}	r (pm)	$10^{-8} k_{exch}\ (s^{-1})$	$\Delta H^{\ddagger}$ (kJ/mol)	$\Delta S^{\ddagger}$ (J/mol K)	$\Delta V^{\ddagger}$ (cm³/mol)
Tb^{3+}	109.5	5.58	12.08	−36.8	−5.7
Dy^{3+}	108.3	4.34	16.57	−23.9	−6.0
Ho^{3+}	107.2	2.14	16.36	−30.4	−6.6
Er^{3+}	106.2	1.33	18.37	−27.7	−6.9
Tm^{3+}	105.2	0.91	22.68	−16.3	−6.0
Yb^{3+}	104.2	0.47	23.39	−29.9	

Comment on the significance of these data for the exchange mechanism. (See C. Cossy, L. Helm and A. E. Merbach, *Inorg. Chem.* **1989**, *28*, 2699.)

11.9 Solution: The decreasing rates as ionic radii decrease would seem to be evidence for a **d** mechanism in which bond breaking is made more difficult by increasing charge density.

However, it is difficult to rationalize negative $\Delta V^{\ddagger}$ for a **d** reaction when the leaving group is neutral solvent. Hence, the reactions are probably **a** and involve assisted dissociation of water through H-bonding to solvent water. This provides electrostriction of solvent which should increase with the acidity of ligand water as r gets smaller.

11.10 Problem: In 1.0 M OH$^-$ solution, at least 95% of $[CoCl(NH_3)_5]^{2+}$ that undergoes base hydrolysis must be present in the pentaammine form. If 5% or more were present as the conjugate base $[CoCl(NH_2)(NH_3)_4]$, it could be detected by departure from second-order kinetics which is *not* observed to happen.

(a) For production of the conjugate base

$$[CoCl(NH_3)_5]^{2+} + OH^- \rightleftharpoons [CoCl(NH_2)(NH_3)_4]^+ + H_2O$$

$$K_h = \frac{[CoCl(NH_2)(NH_3)_4^+]}{[CoCl(NH_3)_5^{2+}][OH^-]} = \frac{K_a}{K_w}$$

Using this information and the observation described above, show that for $[CoCl(NH_3)_5]^{2+}$, $K_a \leq 5 \times 10^{-16}$.

(b) Show the form of the rate law that would result if the conjugate base were present in appreciable amount. (*Hint*: The concentration of $[CoCl(NH_3)_5]^{2+}$ used to make up the solution will be partitioned between the complex and its conjugate base.)

11.10 Solution: (a) We must have for the Cl$^-$ substitution:

$$[CoCl(NH_3)_5]^{2+} + OH^- \rightleftharpoons [CoCl(NH_2)(NH_3)_4]^+ + H_2O$$

$$K_h = \frac{[CoCl(NH_2)(NH_3)_4^+]}{[CoCl(NH_3)_5^{2+}][OH^-]} = \frac{K_a}{K_w} \leq \frac{0.05}{(0.95)(1.0)}$$

$$K_a \leq K_w \frac{0.05}{(0.95)(1.0)} = 1.0 \times 10^{-14} \times \frac{0.05}{0.95} = 5 \times 10^{-16}$$

(b) Rate = $k_{obs} [CoCl(NH_3)_5^{2+}]_{total}[OH^-]$ since the total concentration of complex is known from preparing the solution.
But $[CoCl(NH_3)_5^{2+}]_{total} = [CoCl(NH_3)_5]^{2+} + [CoCl(NH_2)(NH_3)_4]^+$
From **(a)** $[CoCl(NH_2)(NH_3)_4^+] = K_h [CoCl(NH_3)_5^{2+}][OH^-]$
If $[CoCl(NH_2)(NH_3)_4]^+$ is the reactive species.

$$\begin{aligned}
\text{rate} &= k [CoCl(NH_2)(NH_3)_4^+] \\
&= kK_h [CoCl(NH_3)_5^{2+}][OH^-] \\
&= k_{obs} [CoCl(NH_3)_5^{2+}]_{total}[OH^-] \\
&= k_{obs} \{[CoCl(NH_3)_5^{2+}] + K_h [CoCl(NH_3)_5^{2+}][OH^-]\}[OH^-]
\end{aligned}$$

$$\therefore k_{obs} = \frac{kK_h}{1 + K_h[OH^-]}$$

and deviations from second-order kinetics will appear when $K_h[OH^-] \cong 1$.

Another way of viewing this problem is as a special case of the mechanism

$$A + B \underset{k_{-1}}{\overset{k_1}{\rightleftarrows}} C + D \qquad\qquad C \overset{k_2}{\rightarrow} E + F$$

Applying an "improved steady-state approximation" to [C], we get (See D. H. McDaniel and C. R. Smoot, *J. Phys. Chem.* **1956**, *60*, 966.):

$$\frac{d[E]}{dt} = \frac{k_1 k_2 [A][B]}{k_1[B] + k_{-1}[D] + k_2}$$

In our case, A = $[CoCl(NH_3)_5]^{2+}$, B = OH^-, C = $[CoCl(NH_2)(NH_3)_4]^+$, D = H_2O, E = $[Co(NH_2)(NH_3)_4]^{2+}$ and $F^- = Cl^-$. The reaction of E with water is fast and so does not appear in the rate law. When the first step is an equilibrium established prior to the second step, k_1, k_{-1} $\gg k_2$. Since D = H_2O, $k_{-1}[D] = k'_{-1}$, and $k_1/k'_{-1} = K_h$. So,

$$\frac{d[E]}{dt} = \frac{k_1 k_2 [CoCl(NH_3)_5^{2+}][OH^-]}{k_1[OH^-] + k'_{-1}}$$

Dividing by k'_{-1}, we get

$$\frac{d[E]}{dt} = \frac{k_2 K_h [CoCl(NH_3)_5^{2+}][OH^-]}{1 + K_h[OH^-]}$$

11.11 Problem: The complexes *trans*-$[RhX(en)_2L]^+$ react with various Y^-, giving $[RhXY(en)_2]^+$. The rate law for the appearance of the product is rate = $(k_1 + k_2[Y])[Rh(en)_2LX^+]$. Do these data constitute evidence for an **a** mechanism? Why or why not? (See A. J. Poë and C. P. J. Vuik, *Inorg. Chem.* **1980**, *19*, 1771.)

11.11 Solution: Not necessarily. The form of the rate law is compatible with two parallel paths, one of which involves ion pairing. k_2 might be equal to some kK.

11.12 Problem: Anation of *trans*-$[Rh(en)_2(H_2O)_2]^{3+}$ with chloride has been studied. Two plots of k_{obs} versus concentration of species in solution are reproduced on the next page.
(a) Account for the linearity of the plots in figure *a*.
(b) What does the nonzero intercept for each of the two lines in Figure a indicate about the reaction mechanism?
(c) Account for the shape of the two curves in figure *b*.
(d) Do these curves alone allow you to distinguish what the stoichiometric mechanism is?

11.12 Solution:
(a) The reaction rate increases linearly as $1/[H^+]$ increases, i.e., as $[H^+]$ decreases. This implies

that the reactive species is a deprotonated one.

(b) Because the intercept is non-zero, a second pathway involving slower reaction of the parent (protonated) species is also operative. From this information the partial rate law is:

$$\text{rate} = \left(a + \frac{b}{[H^+]}\right)\left[Rh(en)_2(H_2O)_2^{3+}\right]$$

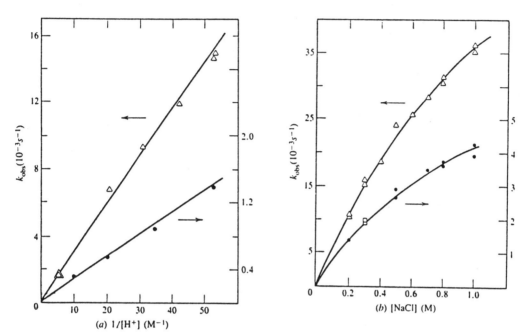

(a) Dependence of k_{obs} on reciprocal concentration of H^+ with I = 1.00 M, [complex] = 0.001 M; Δ, 65°C, [NaCl] = 0.801 M; •, 50°C, [NaCl] = 0.641 M. Arrows indicate corresponding ordinate. (b) Dependence of k_{obs} on concentration of chloride with T = 40°C, I = 1.00 M, and [complex] = 0.0190 M; Δ, $[H^+]$ = 0.0190 M; •, $[H^+]$ = 0.199 M. Arrows indicate corresponding ordinate. (Reprinted with permission from M. J. Pavelich, *Inorg. Chem.* **1975**, *14*, 982. Copyright 1975, American Chemical Society.)

(c) These curved plots indicate a transition from first-order to mixed first- and zero-order behavior in [Cl⁻]. Thus, the complete rate law has the form:

$$\text{rate} = \left(a + \frac{b}{[H^+]}\right)[Rh(en)_2(H_2O)_2]\frac{[Cl^-]}{1 + d[Cl^-]}$$

(d) The rate law is compatible with **D**, **I_d**, or **I_a** mechanisms.

11.13 Problem: (a) A linear relationship between log k's for a series of related complexes is taken to imply similarity in intimate mechanism. Plot log k for aquation vs. log k for base hydrolysis for the $[CoX(NH_3)_5]^{n+}$ complexes (given in Tables on the next page). Use linear least squares regression to fit the best straight line. How does this relationship accord with what you know about substitution on Co^III complexes?

(b) The Table below gives base hydrolysis constants for $[MClL_5]^{2+}$ complexes where M = Cr and Co.

Rate constants for acid aquation of some octahedral complexes of Co^{III} at 25°C

Complex	k (s^{-1})	Complex	k (s^{-1})
$[Co(NH_3)_5\{OP(OMe)_3\}]^{3+}$	2.5 x 10^{-4}	$[Co(NH_3)_5Cl]^{2+}$	1.8 x 10^{-6}
$[Co(NH_3)_5(NO_3)]^{2+}$	2.4 x 10^{-5}	$[Co(NH_3)_5(SO_4)]^{+}$	8.9 x 10^{-7}
$[Co(NH_3)_5I]^{2+}$	8.3 x 10^{-6}	$[Co(NH_3)_5F]^{2+}$	8.6 x 10^{-8}
$[Co(NH_3)_5(H_2O)]^{3+}$	5.8 x 10^{-6}	$[Co(NH_3)_5N_3]^{2+}$	2.1 x 10^{-9}
		$[Co(NH_3)_5(NCS)]^{2+}$	3.75 x 10^{-10}

Leaving ligand written last. Data from M. L. Tobe, *Adv. Inorg. Bioinorg. Mech.* **1984**, *2*, 1.

Rate constants for base hydrolysis of some octahedral complexes (25°C)

Complex	k ($M^{-1}s^{-1}$)	Complex	k ($M^{-1}s^{-1}$)
$[Co(NH_3)_5\{OP(OMe)_3\}]^{3+}$	79	$[Co(NH_3)_5(SO_4)]^{+}$	4.9 x 10^{-2}
$[Co(NH_3)_5(NO_3)]^{2+}$	5.5	$[Co(NH_3)_5F]^{2+}$	1.3 x 10^{-2}
$[Co(NH_3)_5I]^{2+}$	3.2	$[Co(NH_3)_5(NCS)]^{2+}$	5.0 x 10^{-4}
$[Co(NH_3)_5Cl]^{2+}$	0.23	$[Co(NH_3)_5N_3]^{2+}$	3.0 x 10^{-4}

Leaving ligand written last. From the work of Lalor, Chan, Bailar, Basolo, Pearson, Wallace, Kane-Maguire, and others.

Base hydrolysis constants for $[MClL_5]^{2+}$ at 25°C

L_5	k_{Cr} ($M^{-1}s^{-1}$)	k_{Co} ($M^{-1}s^{-1}$)
$(NH_3)_5$	1.8 x 10^{-3}	0.86
fac-(en)(dien)	1.56 x 10^{-2}	4.71
fac-(NMetn)(dien)	2.21 x 10^{-1}	7.57
$(NH_2Me)_5$	>4.3 x 10^{-1}	8 x 10^2
mer-(en)(dpt)	1.05 x 10^{-1}	2.2 x 10^3
mer-(tn)(dpt)	6.06 x 10^{-1}	2.8 x 10^3
mer-(en)(2,3-tri)	7.35 x 10^{-1}	1.24 x 10^3

NMetn = $MeNH(CH_2)_3NH_2$; dpt = $NH_2(CH_2)_3NH(CH_2)_3NH_2$;
tn = $NH_2(CH_2)_3NH_2$; 2,3-tri = $NH_2(CH_2)_2NH(CH_2)_3NH_2$

What is the relationship between the two sets of data and what does this suggest about the mechanism of base hydrolysis for Cr^{III} complexes?

(c). Acid hydrolysis rate constants for the Cr complexes in b. are given in the same order: $10^7 k_H$ = 95, 224, 37.0, 2.48, 5.34, 5.04, 2.87. Plot log k_{OH} vs. log k_H for the series of Cr complexes. What does the plot suggest about the mechanism of acid hydrolysis?

(See D. A. House, *Inorg. Chem.* **1988**, *27*, 2587.)

11.13 Solution: Since RT ln $k = -\Delta G^{\ddagger}$, a linear relationship between log k's indicates a proportionality between $\Delta G^{\ddagger}$'s and implies a similar mechanism.

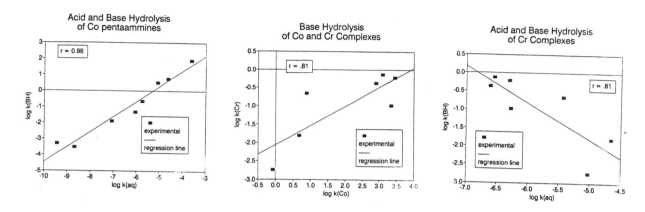

(a) A reasonable straight line is produced whose equation is: log k_{BH} = 0.9496 log k_{aq} + 5.0295 and whose correlation coefficient is 0.98. The least-squares method commonly used to fit experimental data points assumes that the abscissa points are known with no uncertainty and fits the ordinate points. This is undoubtedly not too realistic for these sets of kinetic data.

(b) A "straight" line is produced whose equation is: log k_{Cr} = 0.5284 log$_{Co}$ − 2.0637. The correlation coefficient is only 0.81. This is not convincing evidence that base hydrolysis for Cr^{III} is also **d**.

(c) The data are very scattered and the correlation coefficient is also 0.81 suggesting different mechanisms for base hydrolysis and aquation (presumably **a** for aquation).

11.14 Problem: Given below are overall volume changes $\Delta \overline{V}_{BH}$ and volumes of activation $\Delta V^{\ddagger}$ for base hydrolysis of some Co complexes:

Complex	$\Delta \overline{V}_{BH}$ (cm^3/mol)	$\Delta V^{\ddagger}$ (cm^3/mol)
[Co(NH$_3$)$_5$(Me$_2$SO)]$^{3+}$	21.2	42.0
[Co(NH$_3$)$_5$(OC(Me)NMe$_2$)]$^{3+}$	23.9	43.2
[Co(NH$_3$)$_5$Cl]$^{2+}$	9.9	33.0
[Co(NH$_3$)$_5$Br]$^{2+}$	11.1	32.5
[Co(NH$_3$)$_5$I]$^{2+}$	15.5	33.6
[Co(NH$_3$)$_5$(SO$_4$)]$^{+}$	−3.9	22.7

(See Y. Kitamura, G. A. Lawrance and R. van Eldik, *Inorg. Chem.* **1989**, *28*, 333.)

(a) Explain why the values of $\Delta V^{\ddagger}$ are qualitatively reasonable given your knowledge of the mechanism of base hydrolysis.

(b) Rationalize the trend in $\Delta V^{\ddagger}$ as a function of charge on the starting complex and on the leaving ligand.

(c) Rationalize the trend in $\Delta \overline{V}_{BH}$ as a function of charge on the starting complex and on the leaving ligand.

(d) Construct a volume profile for base hydrolysis of $[Co(NH_3)_5I]^{2+}$ analogous to the energy profile depicted in the figure.

Energy profile for
$H_2 + F \leftrightarrows H + HF$

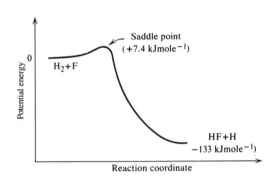

11.14 Solution: (a) All the $\Delta V^{\ddagger}$'s are positive; this is reasonable if the transition state with a very weakly bonded leaving group actually leads to a five-coordinate intermediate releasing the leaving group into solution.

(b) $\Delta V^{\ddagger}$ becomes progressively smaller as the charge on the starting complex becomes less positive reflecting greater negative charge on the leaving group. The more highly negatively charged the group which is mostly released into solution in the transition state, the more solvent electrostriction which makes a negative contribution to $\Delta V^{\ddagger}$.

(c) The metal-containing product is the same for all reactants: $[Co(OH)(NH_3)_5]^{2+}$. Thus, trends in overall volume change must reflect differences between the molar volumes of $[CoX(NH_3)_5]^{n+}$ and those of the released groups. For 3+ reactants, a 2+ complex and a neutral ligand are the products. Neither will be as tightly solvated as the reactant and H-bonded OH^- leading to a large positive $\Delta \overline{V}_{BH}$. For the 2+ reactants, the charge on the product complex is the same which leads to the expectation of minor solvation changes. However, X^- will be more highly solvated than a neutral leaving group leading to smaller values of $\Delta \overline{V}_{BH}$. The order of increasing $\Delta \overline{V}_{BH}$ for halide complexes reflects decreasing solvent electrostriction by X^- in the order $Cl^- > Br^- > I^-$. For the 1+ complex, both the product complex and SO_4^{2-} will be highly solvated and $\Delta \overline{V}_{BH}$ is smaller still (<0).

(d)

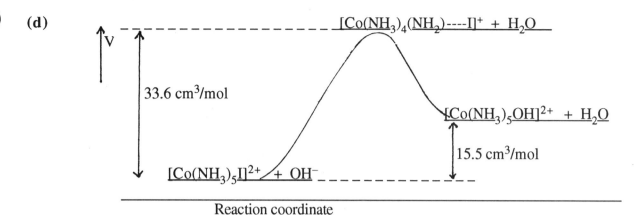

11.15 Problem: (a) By drawing the products obtained from Berry pseudorotations of the TBP isomers of (12345)M where the ligands are monodentate verify the Desargues-Levi map shown below left.
(b) Show that simplifying the diagram by making $\widehat{N_1N_2} = \widehat{N_3N_4}$ with two didentate ligands where all N's are the same gives the map shown below right. This map is equivalent to the Figure on p. 183.
(c) Modify the diagram shown here for the case of A_2B_3M TBP complexes.

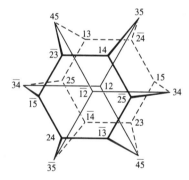

Desargues–Levi map. Vertices represent isomers indexed by their axial ligands; 13 has ligands 1 and 3 axial with 2,3,4 in clockwise order as viewed from the lower-numbered axis. $\overline{13}$ is the mirror image. Lines connecting the vertices represent pseudorotations. (Reprinted with permission from K. Mislow, *Acc. Chem. Res.* **1970**, *3*, 321. Copyright 1970, American Chemical Society.)

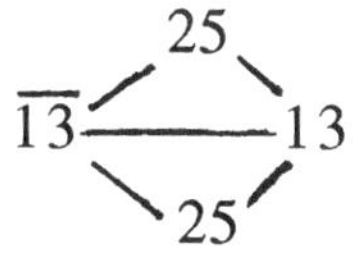

11.15 Solution (a)

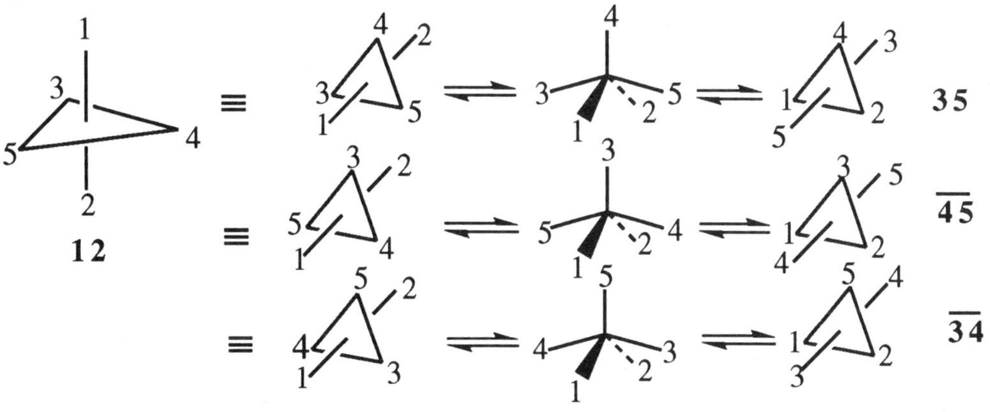

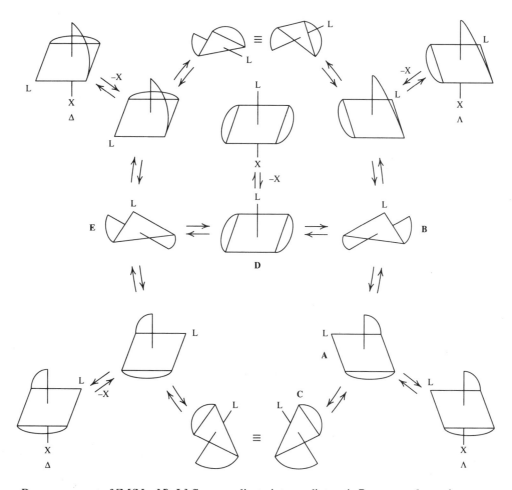

As the final example demonstrates, starting with a mirror image ($\overline{12}$ as opposed to **12**) generates mirror image isomers on pseudorotation.

Rearrangement of [M(N—N)$_2$L] five-coordinate intermediates via Berry pseudorotation.

(b) If 1 = 2 = 3 = 4, then 13 = 14 = 23 = 24 and 15 = 25 = 35 = 45, similar equivalencies holding for the mirror images. Moreover, 12, 34 and their mirror images will not exist because of the small bite of the didentate ligand. Also, if we identify 5 with L, each n5 isomer is equivalent to its mirror image n$\bar{5}$. Thus, there are only three kinds of isomers (using the lowest numbers): 13, $\overline{13}$ and 15. This introduces two C_2 axes into the Desargeues-Levi map shown above and the only non-redundant slice is

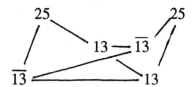

and a simpler way of representing the connectivity is

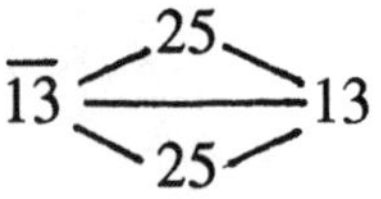

$$\overline{13} \rightleftarrows 13$$
(with 25 connections above and below)

and this is the connectivity of the figure on p. 183 which also depicts the square pyramids leading from one tbp to another.

(c) Three isomers are possible: AA, AB and BB. Because Berry pseudorotation requires that two equatorial ligands become axial, it is impossible to convert AB into AA in this manner. Thus, the map will be AA---BB---AB.

11.16 Problem: Photolysis of a square-planar complex M(ABCD) produces distortion to a tetrahedron. Relaxation back to the planar geometry leads to all possible isomers of M(ABCD). Construct a Desargues–Levi map connecting the possible isomers. (*Hint:* There are few enough isomers that their full structures can appear on the map.)

11.16 Solution:

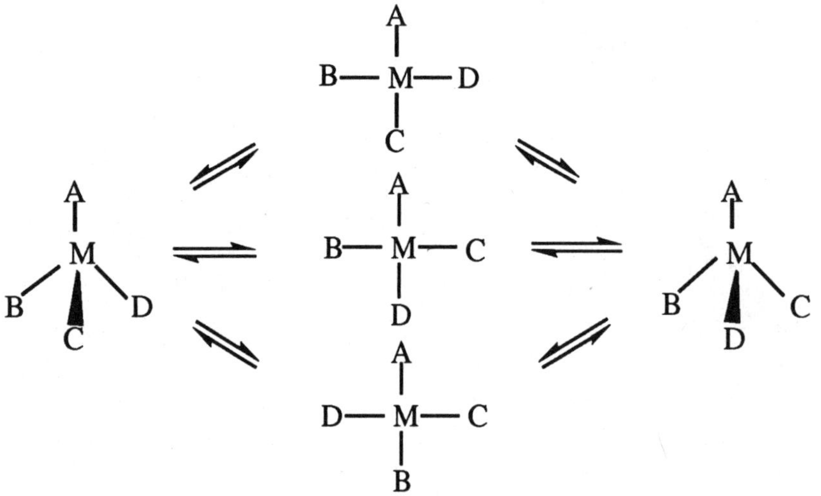

11.17 Problem: Predict the isomer distribution from associative attack of Y on the faces of *cis*-MN$_4$LX using the model shown in Section 11.3.6, DMA, 3rd ed.

11.17 Solution: Associative attack on an octahedron could be considered to form a pentagonal bipyramid. In each case, the leaving ligand X is part of the pentagonal plane and, as X departs, the pentagon relaxes to a square. The manifold of pentagonal bipyramids may be generated by edge attack.

The counterparts of attack on edges 1 and 2 are available on lower faces of the octahedron giving a statistical probability of 7:1 *cis:trans*-[L$_4$MAY]. The above argument assumes that attack on any edge which produces a pentagonal plane of ligands containing X is equally probable. Steric and electronic properties of particular ligands will modify this picture substantially.

11.18 Problem: A plot of k_{obs} versus $[X^-]$ is shown for the reactions $[PdBr(Et_4dien)]^+ + X^- \rightarrow [PdX(Et_4dien)]^+ + Br^-$. Account for the shape of the plot. In particular, what mechanism can you propose to account for the zero slope when $X^- = N_3^-$, I^-, NO_2^-, SCN^-?

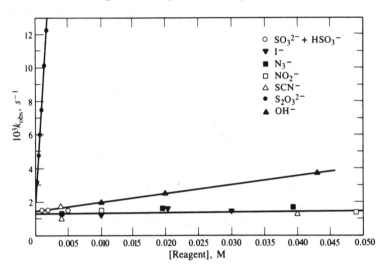

Plot of k_{obs} versus concentration of entering nucleophile for anation of $[PtBr(Et_4dien)]^+$ in water at 25°C. (Reprinted with permission from J. B. Goddard and F. Basolo, *Inorg. Chem.* **1968**, *7*, 936. Copyright 1968, American Chemical Society.)

11.18 Solution: The linear dependence on $[S_2O_3^{2-}]$ and $[OH^-]$ indicates direct attack by these species. The intercepts of all lines are the same and indicate a substitution path independent of the nature of the incoming ligand. This is presumably *not* water attack followed by water displacement by X^-. If it were, we would have to postulate that SO_3^{2-}, I^-, N_3^-, NO_2^-, and SCN^- are all much less nucleophilic toward the Pd compound than water since the rate never depends on their concentrations. This would be in marked contrast to their behavior toward Pt complexes. Probably the sterically crowded Et_4dien complex undergoes rate-determining dissociation of Br^- followed by rapid ligand attack for SO_3^{2-}, I^-, N_3^-, NO_2^-, and SCN^- which are less nucleophilic than $S_2O_3^{2-}$ and OH^-. For $S_2O_3^{2-}$ and OH^- parallel **a** and **d** paths operate.

11.19 Problem: Base hydrolysis, a well known reaction of octahedral complexes, is generally about 10^6 time faster than acid hydrolysis. Given the following facts, what can you say about the existence and importance, if any, of a special mechanism for base hydrolysis of square-planar complexes?

(a) Au^{III} complexes, which are also d^8, typically react about 10^4 times as fact as Pt^{II} complexes. The following data are representative:

$$[MCl(dien)]^{n+} + Br^- \rightarrow [MBr(dien)]^{n+} + Cl^-$$

M	n	k_1 (s^{-1})	k_2 (M^{-1}s^{-1})
Pt	1	8.0×10^{-5}	3.3×10^{-3}
Au	2	0.5	154

(b) The reactivity of $[AuCl(dien)]^{2+}$ and $[PtCl(dien)]^+$ toward anions increases in the order $OH^- \ll Br^- < SCN^- < I^-$.

(c) The replacement of chloride in $[AuCl(Et_4dien)]^{2+}$ by anions Y^- is independent of $[Y^-]$. The rate constant (25°C) increases with pH from $k = 1.9 \times 10^{-6}$ to a maximum of $1.3 \times 10^{-4} s^{-1}$.

(d) The anation rate of $[(R_5dien)Pd(H_2O)]^{2+}$ by I^- decreases from pH 5 to 9, but increases at higher pH.

11.19 Solution: (a) The data imply the usual **a** mechanism for square-planar substitution since rates increase with electrophilicity of the metal complex in the order $Pt^{II} < Au^{III}$.

(b) We can conclude that OH^- is a poor entering group toward square-planar complexes. Its mechanistic role, if any, would be confined to formation of a conjugate base. But this does not happen with the dien complexes since OH^- is the least reactive.

(c) The fact that the substitution rate of the Et_4dien complex is orders of magnitude smaller than for dien complexes and independent of $[Y^-]$ shows that rate-determining very slow dissociation of Cl^- must be occurring in this sterically crowded complex. The rate increase with pH implies that the conjugate base is more effective at dissociating Cl^- than the protonated parent complex.

(d) The decrease in rate from pH 5 to 9 indicates that the concentration of the reactive species $[(R_5dien)Pd(H_2O)]^{2+}$ is being decreased by deprotonation to $[(R_5dien)Pd(OH)]^+$. At $pH \geq 9$, either a more reactive conjugate base is being formed by deprotonation of the dien ligand or a sufficient concentration of a more reactive hydroxo complex has built up.

11.20 Problem: Rate constants for the reaction $trans\text{-}[PtCl_2(py)_2] + XC_6H_4SC_6H_4Y \rightarrow trans\text{-}[PtCl(py)_2(XC_6H_4SC_6H_4Y)]^+ + Cl^-$ were measured in methanol at 30°C. From these data calculate n_{Pt} for the following ligands.

X	Y	$k_2 \times 10^3$ $(M^{-1}s^{-1})$
NH_2	NH_2	7.25
OH	OH	2.72
CH_3	CH_3	1.84
NH_2	NO_2	0.78
NO_2	NO_2	0.096

(See J. R. Gaylor and C. V. Senoff, *Can. J. Chem.* **1971**, *49*, 2390.)

11.20 Solution:

X	Y	$n_{Pt} = \log k_2(L) - \log k_2(MeOH)$
NH_2	NH_2	$\log (7.25 \times 10^{-3}) - \log (2.7 \times 10^{-7}) = 4.43$
OH	OH	$\log (2.72 \times 10^{-3}) - \log (2.7 \times 10^{-7}) = 4.00$
CH_3	CH_3	$\log (1.84 \times 10^{-3}) - \log (2.7 \times 10^{-7}) = 3.83$
NH_2	NO_2	$\log (0.78 \times 10^{-3}) - \log (2.7 \times 10^{-7}) = 3.46$
NO_2	NO_2	$\log (0.096 \times 10^{-3}) - \log (2.7 \times 10^{-7}) = 2.55$

11.21 Problem: The following activation parameters have been measured for the reactions

$$[PtClXL_2] + py \rightarrow [PtClL_2py]^+ + X^-$$

in methanol at 30 °C.

Complex	$k_2 \times 10^3$ ($M^{-1}s^{-1}$)	$\Delta H^{\ddagger}$ (kJ/mol)	$\Delta S^{\ddagger}$ (J/molK)	$\Delta V^{\ddagger}$ (cm^3/mol)
trans-[PtCl(NO$_2$)py$_2$]	7.35	49.3	−94	−8.8
cis-[PtCl(NO$_2$)py$_2$]	0.150	55.2	−110	−19.8
trans-[PtCl$_2$(PEt$_3$)$_2$]	0.53	53.9	−100	−13.6

What mechanistic information can be extracted from these values? (See M. Kosikowski, D. A. Palmer and H. Kelm, *Inorg. Chem.* **1979**, *18*, 2555.)

11.21 Solution: A negative $\Delta S^{\ddagger}$ is consistent with an associative (**a**) mechanism since we expect degrees of motional freedom to be restricted by incipient formation of a new metal-ligand bond before the old one is broken. Likewise, the negative $\Delta V^{\ddagger}$ is compatible with the formation of a closely associated species occupying less volume than the separate individual components. The relatively small dependence of $\Delta H^{\ddagger}$ on the nature of the leaving group also points to an **a** mechanism.

11.22 Problem: Activation volumes for acid hydrolysis of cis-[PtCl$_2$(NH$_3$)$_2$] and [PtCl$_2$en] are −9.5 (at 45°C) and −9.2 (at 41°C) cm^3/mol, respectively. How are these values consistent with what you know about the mechanism for square-planar substitution?

11.22 Solution: Since the mechanism of square-planar substitution is generally considered to be **a**, important bond-making in the transition state should coordinate the entering group to the reacting complex and thereby reduce the total volume as compared to the starting materials where both complex and entering group are not associated and both possess an outer sphere of coordinated solvent molecules.

11.23 Problem: Explain how the following data for the reaction

$$cis\text{-}[PtCl_2(N\text{-}N)] + Me_2SO \rightarrow cis\text{-}[PtCl(N\text{-}N)(Me_2SO)]^+ + Cl^-$$

are in accord with an **a** mechanism for substitution in square-planar Pt complexes.

N—N Ligand	$10^4 k_{meso}$	$10^5 k_{d,l}$
NH$_2$CHMeCHMeNH$_2$	1.39	9.3
C$_6$H$_{10}$(CH$_2$CH$_2$NH$_2$)$_2$	1.06	9.1
MeNHCH$_2$CH$_2$NHMe	0.92	7.7
i-PrNHCH$_2$CH$_2$NHi-Pr	0.47	3.0

See F. P. Fanizzi, F. P. Intini, L. Mareaca, G. Natile and G. Ucello-Barretta, *Inorg. Chem.* **1990**, *29*, 33.

11.23 Solution: In *meso* ligands the alkyl substituents are all on the same side of the coordination plane, whereas they are staggered above and below the plane in the *d,l*. Hence, one side of the plane is completely free for attack in the *meso* isomers and they react faster than the *d,l* isomers. Increasing the steric bulk of alkyl substituents slows the rate by retarding attack and the effect is larger when the alkyl groups are on N bonded to Pt than when they are further removed.

11.24 Problem: Show that for a redox reaction of the type $A^+ + B \overset{K}{\rightleftarrows}$ Intermediate $\overset{k}{\rightarrow} A + B^+$, where B is present in excess, the rate law will be of the form

$$\text{rate} = \frac{a[A^+]_0[B]}{1 + b[B]}$$

whether the intermediate is a precursor or successor complex. (*Hint:* $[A]_0$ will be partitioned between free A and the intermediate.)

11.24 Solution: $[A^+]_0 = [A^+] + [\text{Intermediate}]$
If B is in excess, its concentration will not be appreciably diminished by being tied up in the intermediate.

$$[\text{Intermediate}] = K[A^+][B]$$
$$\text{rate} = k_{obs}[A^+]_0[B] = k[\text{Intermediate}]$$
$$\text{rate} = k_{obs}([A^+] + K[A^+][B])[B] = kK[A^+][B]$$
$$k_{obs} = \frac{kK}{1 + K[B]} \quad \text{and} \quad \text{rate} = \frac{kK}{1 + K[B]}[A^+]_0[B]$$

where $kK = a$ and $K = b$.
Again, this is a special case of the mechanism discussed in Problem 11.10.

11.25 Problem: Calculate predicted rate constants for the following outer-sphere redox reactions from the information provided. Measured values of k_{12} are given for comparison

Reaction	k_{11} ($M^{-1}s^{-1}$)	k_{22} ($M^{-1}s^{-1}$)	E^0 (volts)	k_{12}^{obs} ($M^{-1}s^{-1}$)
$Cr^{2+} + Fe^{3+}$	$\leq 2 \times 10^5$	4.0	1.18	2.3×10^3
$[W(CN)_8]^{4-} + Ce^{IV}$	$> 4 \times 10^4$	4.4	0.90	$> 10^8$
$[Fe(CN)_6]^{4-} + MnO_4^-$	7.4×10^2	3×10^3	0.20	1.7×10^5
$[Fe(phen)_3]^{2+} + Ce^{IV}$	$> 3 \times 10^7$	4.4	0.36	1.4×10^5

11.25 Solution: K_{eq} is obtained from E^0 through the Nernst equation $\Delta G^0 = -nFE^0 = -RT \ln K_{12}$ or $\log K_{12} = \frac{E^0}{0.059}$ at 25°. We calculate f_{12} from the following formula

$$f_{12} = \frac{(\log K_{12})^2}{4 \log \frac{k_{11} k_{22}}{z^2}}$$ where z is the collision frequency at 25°C ($\approx 10^{11} M^{-1} s^{-1}$). Substitution in

the simplified Marcus Equation ($k_{12} = \sqrt{k_{11} k_{22} K_{12} f_{12}}$) gives the following results:

	f	k_{12} (calc)
$Cr^{2+} + Fe^{3+}$	1.5×10^{-4}	$\leq 1 \times 10^6$
$[W(CN)_8]^{4-} + Ce^{IV}$	3.4×10^{-4}	$> 3 \times 10^8$
$[Fe(CN)_6]^{4-} + MnO_4^-$	6.6×10^{-1}	6×10^4
$[Fe(phen)_3]^{2+} + Ce^{IV}$	2.1×10^{-1}	$>6 \times 10^6$

11.26 Problem: Use the value of $1.0 \times 10^{-3} M^{-1} s^{-1}$ for $Fe^{3+/2+}$ self-exchange to estimate k_{12} for the Fe^{2+}–$[Fe(phen)_3]^{3+}$ cross reaction in the table below.

Comparison between calculated and observed rate constants for outer-sphere reactions

Reaction	k_{12} obsd. ($M^{-1}s^{-1}$)	k_{12} calcd. ($M^{-1}s^{-1}$)	k_{11} ($M^{-1}s^{-1}$)	k_{22} ($M^{-1}s^{-1}$)	$E_1 V(K_{12})$	f
Fe^{2+}–$[Fe(phen)_3]^{3+}$	3.7×10^4	$>5 \times 10^6$	4.0	$>3 \times 10^7$	$0.35(7.6 \times 10^5)$	2.4×10^{-1}
Fe^{2+}–Ce^{IV}	3.7×10^4	5×10^5	4.0	4.4	$0.71(8.3 \times 10^{11})$	2.0×10^{-2}
Fe^{2+}–$[IrCl_6]^{2-}$	3.7×10^4	2×10^4	4.0	2.3×10^5	$0.16(5.0 \times 10^2)$	8.9×10^{-1}

k_{11} and k_{22} are rate constants for self-exchange for reductant and oxidant, respectively, for the redox pair. Data from D. A. Pennington, in *Coordination Chemistry*, Vol. 2, A. E. Martell, Ed., American Chemical Society, Washington, D. C., 1978.

11.26 Solution:
$\log f_{12} = (\log 7.6 \times 10^5)^2 / 2 \log \{(1 \times 10^{-3})(3 \times 10^7)/1 \times 10^{22}\}$
$= 34.58/4(-17.5) = -0.494$
$f_{12} = 0.321$

$k_{12} = \sqrt{k_{11} k_{22} K_{12} f_{12}}$
$= \sqrt{(1 \times 10^{-3})(3 \times 10^7)(7.6 \times 10^5)(0.321)}$
$= 9 \times 10^4 M^{-1} s^{-1}$

Agreement with experiment is improved, but this will not be so for the other reactions in the Table.

11.27 Problem: Assign an outer- or inner-sphere mechanism for each of the following:
(a) The main product of the reaction between $[Cr(NCS)F(H_2O)_4]^+$ and Cr^{2+} is $[CrF(H_2O)_5]^{2+}$.

(See F. N. Welch and D. E. Pennington, *Inorg. Chem.* **1976**, *15*, 1515.)

(b) When [(VO)(edta)]$^{2-}$ reacts with [V(edta)]$^{2-}$, a transient red color is observed. (See F. J. Kristine, D. R. Gard and R. E. Shepherd, *J. Chem. Soc., Chem. Commun.* **1976**, 944.)

(c) The rates of reduction of [Co(NH$_3$)$_5$(py)]$^{3+}$ by [Fe(CN)$_6$]$^{4-}$ are insensitive to substitution on py. (See A. J. Miralles, R. E. Armstrong and A. Haim, *J. Am. Chem. Soc.* **1977**, *99*, 1416.)

(d) The rate of reduction of [Co(NCS)(NH$_3$)$_5$]$^{2+}$ by Ti^{3+} is 36,000 times smaller than the rate of [Co(N$_3$)(NH$_3$)$_5$]$^{2+}$ reduction. (See J. P. Birk, *Inorg. Chem.* **1975**, *14*, 1724.)

(e) Activation parameters for some reductions by V^{2+} are:

Complex	$\Delta H^{\ddagger}$ (kJ/mol)	$\Delta S^{\ddagger}$ (J/mol K)
[CoF(NH$_3$)$_5$]$^{2+}$	46.4	−77.4
[CoCl(NH$_3$)$_5$]$^{2+}$	31.4	−120
[CoBr(NH$_3$)$_5$]$^{2+}$	30.1	−115
[CoI(NH$_3$)$_5$]$^{2+}$	30.5	−103
[Co(N$_3$)(NH$_3$)$_5$]$^{2+}$	48.9	−58.5
[Co(SO$_4$)(NH$_3$)$_5$]$^{+}$	48.5	−54.8

(See M. R. Hyde, R. S. Taylor and A. G. Sykes, *J. Chem. Soc., Dalton Trans.* **1973**, 2730.)

(f) A series of Co^{III} carboxylate complexes is known to be reduced by Cr^{2+} in an inner-sphere mechanism. The log-log plot comparing specific rates of reductions of carboxylatopentaammine Co^{III} complexes [R(NH$_3$)$_5$Co]$^{2+}$ by Eu^{2+} and Cr^{2+} at 25°C is shown below. (Reprinted with permission from F. R. Fan and E. S. Gould, *Inorg. Chem.* **1974**, *13*, 2639. Copyright 1974, American Chemical Society.)

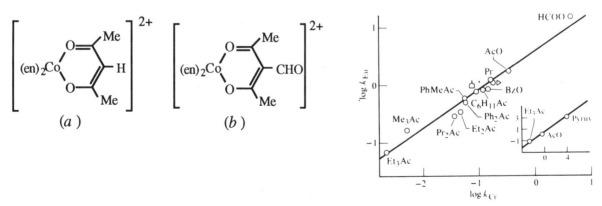

(g) Reduction of (a) by Cr^{2+} is much slower than reduction of (b) (formulas above). (See R. J. Balahura and N. A. Lewis, *Can. J. Chem.* **1975**, *53*, 1154.)

(h) The reduction rates of [(NH$_3$)$_5$CoOC(O)R]$^{2+}$ (R = Me, Et) by Eu^{2+}, V^{2+} and Cr^{2+} decrease as the pH decreases. (See J. C. Thomas, J. W. Reed and E. S. Gould, *Inorg. Chem.* **1975**, *14*, 1696.)

(i) On mixing $\left[(NH_3)_5CoOC(O)C_5H_4N\right]^{2+}$ and Cr^{2+}, a transient ESR signal could be observed. The g-value indicated that the odd electron resided mainly in the aromatic ring. (See H. Spiecker and K. Wieghardt, *Inorg. Chem.* **1977**, *16*, 1290.)

11.27 Solution: See DMA, 3rd ed., Table 11.29.
(a) Inner sphere. As expected for an inner-sphere mechanism, the bridging F ligand is transferred to the product, $[CrF(H_2O)_5]^{2+}$.
(b) Inner sphere. The transient red color could be that of the precursor or successor complex which disappears on account of electron transfer or breakup of the complex to products.
(c) Outer sphere. Py has no lone pair available for bridge formation. Moreover, both complexes are substitution-inert. If the lone pair on CN^- were to bridge, we might expect the stability of the bridged intermediate to be affected by the electron density on Co which is influenced by the substituent on py. Since all py complexes react at a similar rate, it is likely that none forms the bridged complex required for an inner-sphere mechanism.
(d) Inner sphere. The difference in the ability of S and N to form a bridge with Ti^{3+} could account for the observed difference in rates. This difference is in the direction expected since Ti^{3+} is a hard acid and N a hard base. (See DMA, 3rd. ed., Chapter 7.)
(e) The data divide into two groups—one with $\Delta S^{\ddagger} \simeq -55$ J/molK and one with more negative $\Delta S^{\ddagger}$. Presuming that these differences indicate a difference in mechanism, we would ascribe an inner-sphere path to all the complexes except the sulfato and azido on account of more negative $\Delta S^{\ddagger}$. See Problem 11.21.
(f) The linear plot indicates that $\log k_{Eu} = 0.67 \log k_{Cr} + 0.54$ or in other words

$$\log\left(\frac{3k_{Eu}}{2k_{Cr}}\right) = 0.54 \quad \text{and} \quad k_{Eu} = 2.31 k_{Cr}$$

A constant ratio exists between the constants for reduction by both Eu^{2+} and Cr^{2+}. This indicates a proportionality between activation parameters and thus suggests similarity of mechanism. Since the Cr^{2+} reductions are known to be inner-sphere, the Eu^{2+} ones are also inner-sphere. (See DMA, Section 11.8.4.)
(g) The C=O: group in **b** should make the acac ligand a better bridging group because of its available lone pairs. The fact that the rate goes up as the bridging ability of the ligand should increase is evidence that the reduction of **b** is inner-sphere.
(h) As pH decreases, $[H^+]$ increases and protons can compete with Eu^{2+}, V^{2+}, or Cr^{2+} for the lone pairs on the carbonyl O. Hence, an inner-sphere mechanism is indicated.
(i) The transfer of an electron to the ring indicates an inner-sphere radical ion mechnism. The ESR signal decays as the electron is transferred to Co^{3+}.

11.28 Problem: The following rate constants (in $M^{-1}s^{-1}$) were measured for reduction of Co complexes:

Complex	$[Ru(NH_3)_6]^{2+}$	$[Ru(NH_3)_5(H_2O)]^{2+}$
$[Co(N_3)(CN)_5]^{3-}$	1.71	2.06
$[Co(NCS)(CN)_5]^{3-}$	0.81	0.99

Comment on the inner- or outer-sphere mechanism of these four reactions. (See A. B. Ageboro, J. F. Ojo, O. Olubuyide and O. T. Sheyin, *Inorg. Chim. Acta* **1987**, *131*, 247.

11.28 Solution: Reductions with $[Ru(NH_3)_6]^{2+}$ are outer-sphere since no electrons are available for bridging on ammonia. Water does have available lone pairs, so $[Ru(NH_3)_5(H_2O)]^{2+}$ is a possible inner-sphere reductant. However, the ratio $k_{azide}/k_{thiocyanate} \simeq 2.1$ for both indicating that all reductions are outer-sphere. (See also Problem 11.27f.)

11.29 Problem: Using the data below, calculate equilibrium constants for the reaction

$[Ru^*(bipy)_3]^{2+} + [Ru(NH_3)_6]^{3+} \rightarrow [Ru(bipy)_3]^{3+} + [Ru(NH_3)_6]^{2+}$

	E^0
$[Ru(bipy)_3]^{3+} + e \rightarrow [Ru(bipy)_3]^{2+}$	–0.84
$[Ru(NH_3)_6]^{3+} + e \rightarrow [Ru(NH_3)_6]^{2+}$	0.051

11.29 Solution:

$E^0 = E^0 \{[^*Ru(bipy)_3]^{2+} \rightarrow [Ru(bipy)_3]^{3+}\} + E^0 \{[Ru(NH_3)_6]^{3+/2+}\}$
$= +0.84 + (0.051)$ V $= 0.891$ V.
$\Delta G^0 = -16.9 (0.891)$ kJ/mol $= -15.1$ kJ/mol

11.30 Problem: The complex $[(bipy)_2ClRu-N\bigcirc N-Ru(bipy)_2Cl]^{5+}$ displays an intervalence transfer absorption band at 7.79 x 10^3 cm^{-1}. Estimate the energy barrier to thermal electron transfer. If E_{th} can be very approximately equated to $\Delta G^{\ddagger}$ for thermal electron transfer, estimate the rate constant. The thermal rate constant has been reported to be ≤ 3 x $10^{10} s^{-1}$.

11.30 Solution: $E_{op} = \frac{1}{4}E_{th} = \frac{1}{4}(7.69 \times 10^3$ cm$^{-1}) = 1.92 \times 10^3$ cm^{-1}

For a mole of complex ions:

$\Delta G^{\ddagger} \simeq NE_{op}$
$= 6.022 \times 10^{23}$ mol^{-1} (1.92 x 10^3 cm^{-1})(6.63 x 10^{-34} Jsec) x (3.00 x 10^{10} cm sec^{-1}) = 2.30 x 10^4 J/mol

From activated complex theory:

$$k = \frac{\kappa T e^{-\Delta G^{\ddagger}/RT}}{h}$$

$$= \frac{(1.38 \times 10^{-23} \text{ JK}^{-1})(300K)}{6.63 \times 10^{-34} \text{J s}} \exp \frac{(2.30 \times 10^4 \text{J/mol})}{8.33 \text{J/mol K} \times 300K}$$

$6.24 \times 10^{12} s^{-1} \exp(-9.20) = 6.24 \times 10^{12} s^{-1} (1.01 \times 10^{-4})$

$= 6.3 \times 10^8 s^{-1}$, not in very agreement with the quoted estimate.

11.31 Problem: How might you account for the observation that the rate of hydrolysis of [Co(en)$_2$F$_2$]NO$_3$ is greatly enhanced by the either acid or base—it increases linearly with [H$^+$] below pH = 2 or with [OH$^-$] above pH = 6?

11.31 Solution: The linear increase with [OH$^-$] above pH is base hydrolysis. Below pH 2, [H$^+$] is sufficient to protonate an F ligand assisting its departure as HF. The reaction has a three-term rate law in which different terms become predominant in different pH regions.

$$\text{rate} = k_H\left[\text{Co(en)}_2\text{F}_2^+\right][\text{H}^+] + k\left[\text{Co(en)}_2\text{F}_2^+\right] + k_{OH}\frac{\left[\text{Co(en)}_2\text{F}_2^+\right]}{[\text{H}^+]}$$

11.32 Problem: What mechanistic implications may be drawn from the observation that base hydrolysis of Λ-*cis*-[Co(en)$_2$(NH$_3$)Cl]$^{2+}$ yields 84% *cis*- and 16% *trans*-[Co(en)$_2$(NH$_3$)(OH)]$^{2+}$ whereas acid hydrolysis in the presence of AgClO$_4$ yields 75% *cis*- and 25% *trans*-isomer?

11.32 Solution: The 84%:16% *cis:trans* ratio is approximately the product distribution which would be expected if the square pyramidal species can undergo substitution before rearrangement and also rearrange to all possible trigonal bipyramids before substitution with equal probablilty. See the figure in Problem 11.15. The 75%:25% *cis:trans* ratio would result if **A** in the figure either rearranged only to **B, C,** and **D** or reverted to starting complex with equal probability.

11.33 Problem: A large number of the examples of octahedral ligand substitution reactions involve ammine or amine complexes such as [Co(NH$_3$)$_5$X]$^{n+}$ and [Co(en)$_2$LCl]$^{2+}$. The ammine ligands are almost never the ones replaced. For example, k_1 for aquation of [Co(NH$_3$)$_6$]$^{3+}$ is ~10^{-12} s^{-1}. Suggest a reason why this may be so.

11.33 Solution: The ammine or amine ligands have only one lone pair which is used in coordination to the metal cation. More labile leaving ligands such as water or Cl$^-$ have additonal lone pairs. Their departure may be assisted by interaction with electrophilic species present in solution such as H$^+$ or the positive end of the water dipole.

11.34 Problem: The presence of excess L = PMe$_2$Ph catalyzes the isomerization of *cis*-[Pt(Cl)$_2$L$_2$] to *trans*-[Pt(Cl)$_2$L$_2$]. Suggest a mechanism for this reaction. (See D. G. Cooper and J. Powell, *J. Am. Chem. Soc.* **1973**, *95*, 1102.)

11.34 Solution: Since attack by L could form an 18*e* five-coordinate complex, such a complex is likely involved in the isomerization. One possible mechanism would involve sequential displacement of Cl$^-$ by L and L by Cl$^-$. (L displacement of L occurs with retention and so cannot lead to isomerization.)

Alternatively, Berry pseudorotation of the five-coordinate species (if it survives long enough) can lead to isomerization.

V. ORGANOMETALLIC CHEMISTRY
12
General Principles of Organometallic Compounds

12.1 Problem: Give the valence electron counts for the following species. Which ones obey the 18-electron rule?

(a) $W(CO)_6$
(b) $Cr(CNMe)_6$
(c) $Co_2(\mu\text{-}CO)_2(CO)_6$
(d) $[Fe(CN)_6]^{4-}$
(e) $Mn(CO)_5CH_2C_6H_5$
(f) $Fe(CO)_4Br_2$
(g) $[Mn(CO)_5]^-$
(h) $HRh(CO)_4$
(i) $Ru_3(CO)_{12}$
(j) $[Co(CN)_5]^{3-}$

12.1 Solution:

(a) $W(CO)_6$ $W^0 + 6\,CO$
     $6e + 6 \times 2e = 18e$

(b) $Cr(CNMe)_6$ $Cr^0 + 6\,CNMe$
     $6e + 6 \times 2e = 18e$

(c) For each Co in $Co_2(\mu\text{-}CO)_2(CO)_6$
$Co^0 + 3 \times CO + 2\,\mu\text{-}CO + Co\text{–}Co$
$9e + 3 \times 2e + 2 \times 1e + 1e = 18e$

(d) $[Fe(CN)_6]^{4-}$
$Fe^{II} + 6\,CN^-$
$6e + 6 \times 2e = 18e$

(e) $Mn(CO)_5CH_2C_6H_5$
$Mn^I + 5\,CO + \text{benzyl}^-$
$6e + 5 \times 2e + 2e = 18e$

(f) $Fe(CO)_4Br_2$ $Fe^{II} + 4\,CO + 2\,Br^-$
     $6e + 4 \times 2e + 2 \times 2e = 18e$

(g) $[Mn(CO)_5]^-$ $Mn^{-I} + 5\,CO$
     $8e + 5 \times 2e = 18e$

(h) $HRh(CO)_4$
$Rh^I + 4\,CO + H^-$
$8e + 4 \times 2e + 2e = 18e$

(i) For each Ru in $Ru_3(CO)_{12}$
$Ru^0 + 4\,CO + 2\,Ru\text{-}Ru$
$8e + 4 \times 2e + 2 \times 1e = 18e$

(j) $[Co(CN)_5]^{3-}$
$Co^{II} + 5\,CN^-$
$7e + 5 \times 2e = 17e$

12.2 Problem: Name each of the species in Problem 12.1.

12.2 Solution:
(a) hexacarbonyltungsten
(b) hexakis(methyl isocyanide) chromium
(c) di-μ-carbonylhexacarbonyldicobalt
(d) hexacyanoferrate(4–) or ferrocyanide
(e) benzylpentacarbonylmanganese
(f) dibromotetracarbonyliron
(g) pentacarbonylmanganate(1–)
(h) tetracarbonylhydridorhodium
(i) dodecacarbonyltriruthenium
(j) pentacyanocobaltate(3–)

12.3 Problem: The shapes of the binary carbonyls can be rationalized using a VB model in which each CO donates an electron pair to the metal in a dative σ bond and the 18-e rule applies. Metal lone pairs are stereochemically inactive and the geometry is fixed by the hybridization of sigma bonds alone. Use this model to predict the structure of (a) $Ni(CO)_4$, (b) $Fe(CO)_5$, (c) the simplest Co carbonyl.

12.3 Solution:
(a) Ni requires four σ bonds; these will use sp^3 hybrids and lead to a tetrahedral geometry.

(b) Five σ bonds are required for Fe; these can be supplied by dsp^3 hybrids leading to TBP geometry.

(c) Co (d^9) requires nine more electrons to fill its d, s and p orbitals and cannot achieve this by adding pairs of electrons, and so forms dimeric molecules with a Co-Co bond, with $Co_2(CO)_8$ the simplest. This leads to an expected hybridization of dsp^3. For a dimeric species with equivalent Co atoms, metal-metal bonds join either axial or equatorial positions with the expected TBP Co coordination. An alternative would be d^2sp^3 hybridization with both a metal-metal bond and two bridging carbonyl groups.

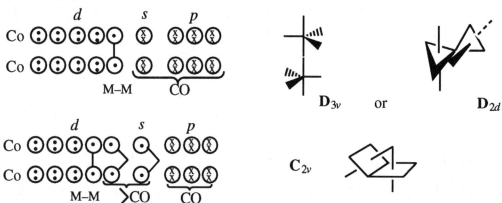

12.4 Problem:
(a) Explain why the 18-e rule applies to transition metal complexes and particularly to those in the middle of the series.
(b) Why does the 16-e configuration become more stable at the end of the transition series?

12.4 Solution:
(a) At the middle of the transition series the $(n-1)d$, ns and np orbitals are all of similar energy and so are all valence orbitals. Since there are nine valence orbitals, eighteen electrons can be accommodated as bonding or lone pairs.
(b) Since p orbitals drop less rapidly in energy with increasing atomic number than s or d orbitals, an energy trade-off develops between employing high-energy orbitals in bonding and forming additional chemical bonds. At the end of the series, this can result in one (or sometimes even two) high-energy orbitals remaining unused in bonding leading to a sixteen- (or sometimes fourteen-) electron configuration.

12.5 Problem: Give the valence electron count for the following species. Which ones obey the 18-e rule?

(a) Ni(PPh$_3$)$_4$
(b) H$_2$Fe(CO)$_4$
(c) [Cr(CO)$_5$]$^{2-}$
(d) [Co(CNMe)$_5$]$^{2+}$
(e) Mo(CO)$_5$(Pn-Bu$_3$)
(f) [Ta(CO)$_6$]$^-$
(g) [FeRu(CO)$_8$]$^{2-}$
(h) [Fe(CN)$_5$NO]$^{2-}$
(i) [W(CO)$_4$]$^{4-}$
(j) V(CO)$_5$NO
(k) [Ir(PPh$_3$)$_2$(CO)Cl(NO)]$^+$
(l) Hf(CO)$_2$(Me$_2$PCH$_2$CH$_2$PMe$_2$)$_2$I$_2$

12.5 Solution:

(a) Ni(PPh$_3$)$_4$
Ni^0 + 4 PPh$_3$
10e + 4 x 2e = 18e

(b) H$_2$Fe(CO)$_4$
Fe^{II} + 2 H$^-$ + 4 CO
6e + 2 x 2e + 4 x 2e = 18e

(c) [Cr(CO)$_5$]$^{2-}$
Cr^{-II} + 5 CO
8e + 5 x 2e = 18e

(d) [Co(CNMe)$_5$]$^{2+}$
Co^{II} + 5 CNMe
7e + 5 x 2e = 17e

(e) Mo(CO)$_5$(Pn-Bu$_3$)
Mo^0 + 5 CO + (Pn-Bu$_3$)
6e + 5 x 2e + 2e = 18e

(g) For Fe and Ru in [FeRu(CO)$_8$]$^{2-}$
M^{-I} + 4 CO + M–M
9e + 4 x 2e + 1e = 18e

(h) [Fe(CN)$_5$NO]$^{2-}$
Fe^{II} + 5 CN$^-$ + NO$^+$
6e + 5 x 2e + 2e = 18e

(i) [W(CO)$_4$]$^{4-}$
W^{-IV} + 4 CO
10e + 4 x 2e = 18e

(j) V(CO)$_5$NO
V^{-I} + NO$^+$ + 5 CO
6e + 2e + 5 x 2e = 18e

(k) [Ir(PPh$_3$)$_2$(CO)Cl(NO)]$^+$
Ir^I + 2 PPh$_3$ + CO + Cl$^-$ + NO$^+$
8e + 2 x 2e + 2e + 2e + 2e = 18e

(f) $[Ta(CO)_6]^-$
Ta^{-I} + 6 CO
6e + 6 × 2e = 18e

(l) $Hf(CO)_2(Me_2PCH_2CH_2PMe_2)_2I_2$
Hf^{II} + 2 CO + 2 dppe + 2I$^-$
2e + 2 × 2e + 2 × 4e + 2 × 2e = 18e

12.6 Problem: Name each species in Problem 12.5.

12.6 Solution:
(a) tetrakis(triphenylphosphine)nickel
(b) tetracarbonyldihydridoiron
(c) pentacarbonylchromate(2–)
(d) pentakis(methyl isocyanide)cobalt(2+)
(e) pentacarbonyl(tri-n-butylphosphine)molybdenum
(f) hexacarbonyltantalate(1–)
(g) octacarbonylferrateruthenate(Fe–Ru)(2–)
(h) pentacyanonitrosylferrate(2–) or nitroprusside
(i) tetracarbonyltungstate(4–)
(j) pentacarbonylnitrosylvanadium
(k) carbonylchloronitrosylbis(triphenylphosphine)iridium(1+)
(l) dicarbonylbis(1,2-dimethylphosphinoethane)diiodohafnium

12.7 Problem: Transition metal compounds differ from those of the s and p block elements in that nonbonding d electrons do not participate in the σ-bond hybridization in the VB model. Compare changes in the σ-bond hybridization and changes in geometry in the following sets of reactions:

(a) $H^+ + PH_3 \rightarrow PH_4^+$

$H^+ + [Co(CO)_4]^- \rightarrow HCo(CO)_4$

(b) $SO_3 + 2e \rightarrow SO_3^{2-}$

$[Ni(CN)_4]^{2-} + 2e \rightarrow [Ni(CN)_4]^{4-}$

12.7 Solution:
(a) In phosphine, the bonding and lone pair electrons are in sp^3 orbitals. PH_3 is a trigonal pyramid and PH_4^+ is tetrahedral. The four Co–C bonds employ sp^3 hybrids so that $[Co(CO)_4]^-$ is tetrahedral. On protonation, a formerly stereochemically inactive lone pair on Co is now hybridized to give dsp^3 trigonal bipyramidal $HCo(CO)_4$.
(b) SO_3 is trigonal planar with sp^2 hybridization. Addition of an electron pair gives sp^3 hybridization and a pyramidal geometry for SO_3^{2-}. d^8-$[Ni(CN)_4]^{2-}$ is square planar with dsp^2 hybridization. Adding an additional electron pair fills the empty Ni d orbital, removing it from hybridization. We thus have d^{10}, tetrahedral (sp^3) $[Ni(CN)_4]^{4-}$.

12.8 Problem: The Figure on the next page shows a plot of v_{CO} vs. E^0 for CpFeL(CO)Me where L is a phosphine. E^0 measures the ease of the reaction CpFeL(CO)Me + $e \rightarrow$ [CpFeL(CO)Me]$^-$; the more positive E^0, the more favorable the reduction.
(a) What is the significance of the linear relationship displayed for the ligands represented by filled circles?

(b) What is the significance of the position of the open-circle ligands above the line?

12.8 Solution:
(a) ν_{CO} values reflect the net donation of electrons to Fe by the phosphine ligands. Thus, as phosphines become better donors of electrons through the σ-system the electrons are delocalized by back donation into π^* of CO and ν_{CO} moves to lower energy. This same trend is reflected in a decreased tendency for the metal to accept an additional electron from an external source and be reduced. The linear relationship indicates that ease of reduction and σ-donor ability are correlated.

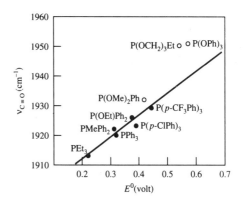

(b) The open circles correspond to phosphite ligands which are not only σ-donors, but also better π-acceptors than phosphines. Thus, for a given value of E^0 (which reflects the σ-donor ability of the ligand), ν_{CO} is "too high" reflecting the fact that electron density can be delocalized by donation into phosphite π^* orbitals. The effect increases as the number of P–O bonds increases. Compare $P(OMe)_2Ph$ with $P(OPh)_3$.

12.9 Problem: Explain in your own words how the MO and VB descriptions of $C_3H_5^-$ are equivalent.

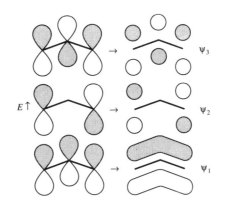

MO's for $C_3H_5^-$

12.9 Solution: The two VB resonance structures indicate the equivalence of the two end C's and of both C–C bonds. In the MO description, occupation of ψ_1 and ψ_2 leads to equivalence of C–C bonds. Implicit in the construction of the MOs is the equivalence of end C's which is consequence of the C_{2v} local symmetry.

12.10 Problem: Give the valence electron count for the following species. Which ones obey the 18-*e* rule?

(a) $Cp_2Ru_2(CO)_4$
(b) $(\eta^3\text{-}CH_2CMeCH_2)_2Ni$
(c) $(\eta^4\text{-cot})Fe(CO)_3$
(d) $CpCr(NO)_2Me$
(e) $[(\eta^4\text{-cod})Ru(\mu\text{-Cl})]_2$
(f) $Ru(Me_2PCH_2CH_2PMe_2)_2$
(g) $Cp^*_2ZrCl_2$
(h) $Rh(C_2H_4)(PPh_3)_2Cl$
(i) $(\eta^6\text{-}C_6H_6)_2Mo$
(j) $MeC(O)Re(CO)_4(CNPh)$

12.10 Solution:

(a) For each Ru in $Cp_2Ru_2(CO)_4$
$Ru^I + Cp^- + Ru\text{-}Ru + 2\,CO$
$7e + 6e + 1e + 2 \times 2e = 18e$

(b) $(\eta^3\text{-}CH_2CMeCH_2)_2Ni$
$Ni^{II} + 2\,\text{allyl}^-$
$8e + 2 \times 4e = 16e$

(c) $(\eta^4\text{-cot})Fe(CO)_3$
$Fe^0 + 3\,CO + \text{cot}$
$8e + 3 \times 2e + 4e = 18e$

(d) $CpCr(NO)_2Me$
$Cr^0 + Cp^- + 2\,NO^+ + Me^-$
$6e + 6e + 2 \times 2e + 2e = 18e$

(e) For each Ru in $[(\eta^4\text{-cod})Ru(\mu\text{-Cl})]_2$
$Ru^I + Ru\text{-}Ru + \text{cod} + \mu\text{-Cl} + Cl^-$
$7e + 1e + 4e + 2e + 2e = 16e$

(f) $Ru(Me_2PCH_2CH_2PMe_2)_2$
$Ru^0 + 2\,\text{dmpe}$
$8e + 2 \times 4e = 16e$

(g) $Cp^*_2ZrCl_2$
$Zr^{IV} + 2\,Cp^{*-} + 2\,Cl^-$
$0e + 2 \times 6e + 2 \times 2e = 16e$

(h) $Rh(C_2H_4)(PPh_3)_2Cl$
$Rh^I + C_2H_4 + 2\,PPh_3 + Cl^-$
$8e + 2e + 2 \times 2e + 2e = 16e$

(i) $(\eta^6\text{-}C_6H_6)_2Mo$
$Mo^0 + 2\,C_6H_6$
$6e + 2 \times 6e = 18e$

(j) $MeC(O)Re(CO)_4(CNPh)$
$Re^I + MeC(O)^- + 4\,CO + CNPh$
$6e + 2e + 4 \times 2e + 2e = 18e$

12.11 Problem: Name each species in Problem 12.10.

12.11 Solution:
(a) tetracarbonylbis(cyclopentadienyl)diruthenium
(b) bis(2-methylallyl)nickel
(c) tricarbonyl(η^4-cyclooctatetraene)iron
(d) cyclopentadienyl(methyl)dinitrosylchromium
(e) di-μ-chlorobis(η^4-cyclooctadiene)diruthenium
(f) bis(1,2-dimethylphosphinoethane)ruthenium
(g) dichlorobis(pentamethylcyclopentadienyl)zirconium
(h) chloro(ethylene)bis(triphenylphosphine)rhodium
(i) bis(benzene)molybdenum

(j) acetyltetracarbonyl(phenyl isocyanide)rhenium

12.12 Problem: Give the formula of the most stable compound of the type $M(olefin)(CO)_x$ expected for each of the following metals with each olefin listed.

Metals	Olefins
Cr, Mn, Fe	$C_3H_5^-$, Cp^-, C_6H_6, $C_7H_7^+$

12.12 Solution: The most stable compounds are expected to be those which conform to the EAN rule. Formulas which satsify this criterion are:

$[(C_3H_5)Cr(CO)_4]_2$ $[(C_6H_6)Mn(CO)_3]^+$
$[CpCr(CO)_3]_2$ $[(C_7H_7)Mn(CO)_3]^{2+}$ (unlikely because of high + charge)
$(C_6H_6)Cr(CO)_3$ $[(C_3H_5)Fe(CO)_3]_2$
$[(C_7H_7)Cr(CO)_3]^+$ $[CpFe(CO)_2]_2$
$(C_3H_5)Mn(CO)_4$ $(C_6H_6)Fe(CO)_2$
$CpMn(CO)_3$ $[(C_7H_7)Fe(CO)_2]^+$

12.13 Problem: Given the following pairs of π-donor rings, choose the proper d^6 ion from the following list to form a neutral mixed sandwich compound: Cr^0, Mn^I, Fe^{II}, Co^{III}.

12.13 Solution: Each pair of rings must have a total of twelve π electrons. We must also choose the set with correct total negative charge to give a neutral complex: $(\eta^5\text{-}C_5H_5^-)(\eta^7\text{-}C_7H_7^+)Cr^0$, $(\eta^5\text{-}C_5H_5^-)(\eta^6\text{-}C_6H_6)Mn^I$, $(\eta^5\text{-}C_5H_5^-)_2Fe^{II}$, and $(\eta^4\text{-}C_4H_4^{2-})(\eta^5\text{-}C_5H_5^-)Co^{III}$.

12.14 Problem: The complex ion $[(\eta^3\text{-}C_3H_5)Ir(PPh_3)_2(NO)]^+$ can be crystallized with either linear or bent NO depending on the counterion. Give the electron count for each isomer.

12.14 Solution:
Linear isomer:
$\quad Ir^I \;+\; NO^+ \;+\; \eta^3\text{-allyl}^- \;+\; 2\,PPh_3$
$\quad 8e \;+\; 2e \;\;+\; 4e \;\;\;\;\;\;+\; 2 \times 2e \;\;= 18e$
Bent isomer:
$\quad Ir^{III} \;+\; NO^- \;+\; \eta^3\text{-allyl}^- \;+\; 2\,PPh_3$
$\quad 6e \;+\; 2e \;\;+\; 4e \;\;\;\;\;\;+\; 2 \times 2e \;\;= 16e$
In bent NO^- an electron pair formally extracted from Ir^I is localized on N.

12.15 Problem: Using the MO energy-level diagram for metallocenes below, write electron configurations that account for the magnetic properties of Cp_2V (3 unpaired e), Cp_2V^+ (2 unpaired

e), Cp$_2$Cr (2 unpaired e), Cp$_2$Mn (5 unpaired e), Cp$_2$Fe (0 unpaired e), Cp$_2$Fe$^+$ (1 unpaired e), Cp$_2$Co (1 unpaired e), Cp$_2$Co$^+$ (0 unpaired e), Cp$_2$Ni (2 unpaired e), and Cp$_2$Ni$^+$ (1 unpaired e).

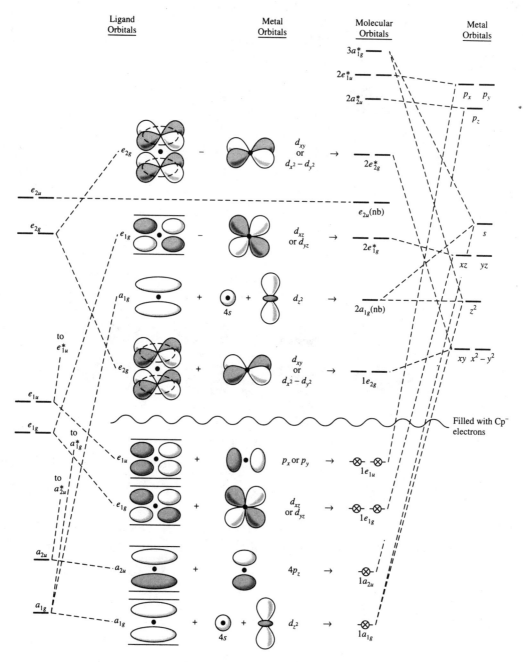

MO energy-level diagram for metallocenes. (From B. E. Douglas and C. A. Hollingsworth, *Symmetry in Bonding and Spectra*, Academic Press, New York, 1985.)

12.15 Solution: All compounds have $(1a_{1g})^2(1a_{2u})^2(1e_{1g})^4(1e_{1u})^4\cdots$

Cp$_2$V $\cdots(1e_{2g})^2(2a_{1g})^1$ Cp$_2$Fe$^+\cdots(1e_{2g})^4(2a_{1g})^1$

Cp$_2$V$^+\cdots(1e_{2g})^2$ Cp$_2$Co $\cdots(1e_{2g})^4(2a_{1g})^2(2e_{1g})^1$

$$Cp_2Cr \cdots (2a_{1g})^2 (1e_{2g})^2 \qquad Cp_2Co^+ \cdots (1e_{2g})^4 (2a_{1g})^2$$
$$Cp_2Mn \cdots (1e_{2g})^2 (2a_{1g})^1 (2e_{1g})^2 \qquad Cp_2Ni \cdots (1e_{2g})^4 (2a_{1g})^2 (2e_{1g})^2$$
$$Cp_2Fe \cdots (1e_{2g})^4 (2a_{1g})^2 \qquad Cp_2Ni^+ \cdots (1e_{2g})^4 (2a_{1g})^2 (2e_{1g})^1$$

In order to account for the observed magnetic properties, we must assume that electron pairing energies are comparable to orbital separations so that orbitals are sometimes filled "out of order".

12.16 Problem: Give the valence electron count for the following species. Which ones obey the 18-e rule?

(a) $[CpCo(NO)]_2$
(b) $Cp^*_2Re(\eta^2\text{-EtC}\equiv\text{CEt})Cl_2$
(c) $Cp_2Co_2(\mu\text{-NO})_2$
(d) $[CpTi(CO)_4]^-$
(e) $Cr(CH_2CMe_3)_2[i\text{-Pr}_2PCH_2CH_2Pi\text{-Pr}_2]_2$
(f) $Cp^*(CO)Ir(\mu\text{-CO})_2Re(CO)Cp$

12.16 Solution:

(a) For each Co in $[CpCo(NO)]_2$
$Co^0 + Cp^- + NO^+ + Co\text{-}Co$
$9e + 6e + 2e + 1e = 18e$

(d) $[CpTi(CO)_4]^-$
$Ti^0 + Cp^- + 4\,CO$
$4e + 6e + 4 \times 2e = 18e$

(b) $Cp^*Re(\eta^2\text{-EtC}\equiv\text{CEt})Cl_2$
$Re^{III} + Cp^{*-} + C\equiv C + 2\,Cl^-$
$4e + 6e + 2e + 2 \times 2e = 16e$

(e) $Cr(CH_2CMe_3)_2[i\text{-Pr}_2PCH_2CH_2Pi\text{-Pr}_2]_2$
$Cr^{II} + 2\,\text{alkyl}^- + 2\,\text{diphosphine}$
$4e + 2 \times 2e + 2 \times 4e = 16e$

(c) For each Co in $Cp_2Co_2(\mu\text{-NO})_2$
$Co^I + Cp^- + \mu\text{-NO} + \mu\text{-NO} + Co\text{-}Co$
$8e + 6e + 1e + 2e + 1e = 18e$

(f) $Cp^*(CO)Ir(\mu\text{-CO})_2Re(CO)Cp$
$Ir^I + Cp^{*-} + 2\,\mu\text{-CO} + CO$
$8e + 6e + 2 \times 1e + 2e = 18e$
$Re^I + Cp^- + 2\,\mu\text{-CO} + CO$
$6e + 6e + 2 \times 1e + 2e = 16e$ (An Ir → Re dative bond would increase the count to 18; but its presence needs to be established by structural data.)

12.17 Problem: Name each of the species in Problem 12.16.

12.17 Solution:
(a) dicyclopentadienyldinitrosyldicobalt
(b) dichloro(η^2-3-hexyne)pentamethylcyclopentadienylrhenium
(c) dicyclopentadienyldi-μ-nitrosyldicobalt
(d) tetracarbonylcyclopentadienyltitanate(1–)
(e) di(*neo*-pentyl)bis(1,2-di-*iso*-propylphosphinoethane)chromium
(f) di-μ-carbonyldicarbonylcyclopentadienylpentamethylcyclopentadienyliridiumrhenium

12.18 Problem: Give the valence electron count for the following species. Which ones obey the 18-e rule?

(a) $(CO)_5CrC(OEt)(NMe_2)$
(b) $CpMn(CO)_2CPh_2$
(c) *cis*-$Cl_2Pt[C(Oi\text{-Pr})(Me)]_2$
(d) $[CpFe(CO)(PPh_3)(CF_2)][BF_4]$
(f) $(Me_3CO)_3W(CCMe_3)$
(g) $CpMo[P(OMe)_3]_2(CCH_2CMe_3)$
(h) $Br(CO)_4Cr(CPh)$
(i) $CpIr(PMe_3)(CH_2)$

(e) TaCl$_3$(PEt$_3$)$_2$(CHCMe$_3$) (j) Cp*Ta(PMe$_3$)$_2$ClCPh

12.18 Solution:
(a) (CO)$_5$CrC(OEt)(NMe$_2$)
Cr^0 + 5 CO + =C(OEt)(NMe$_2$)
6e + 5 x 2e + 2e   = 18e

(f) (Me$_3$CO)$_3$W(CCMe$_3$)
W^{VI} + 3 Me$_3$CO$^-$ + ≡CCMe$_3^{3-}$
0e + 3 x 2e + 6e = 12e

(b) CpMn(CO)$_2$CPh$_2$
Mn^I + Cp$^-$ + 2 CO + =CPh$_2$
6e + 6e + 2 x 2e + 2e = 18

(g) CpMo[P(OMe)$_3$]$_2$(CCH$_2$CMe$_3$)
Mo^0 + Cp$^-$ + 2 PR$_3$ + ≡CCH$_2$CH$_3^+$
6e + 6e + 2 x 2e + 2e = 18e

(c) cis-Cl$_2$Pt[C(OiPr)(Me)]$_2$
Pt^{II} + 2 Cl$^-$ + 2 =C(Oi-Pr)(Me)
8e + 2 x 2e + 2 x 2e = 16e

(h) Br(CO)$_4$Cr(CPh)
Cr^0 + Br$^-$ + ≡CPh$^+$ + 4 CO
6e + 2e + 2e + 4 x 2e = 18e

(d) [CpFe(CO)(PPh$_3$)(CF$_2$)][BF$_4$]
Fe^{II} + Cp$^-$ + CO + PPh$_3$ + =CF$_2$
6e + 6e + 2e + 2e + 2e = 18e

(i) CpIr(PMe$_3$)(CH$_2$)
Ir^I + Cp$^-$ + PMe$_3$ + =CH$_2$
8e + 6e + 2e + 2e = 18e

(e) TaCl$_3$(PEt$_3$)$_2$(CHCMe$_3$)
Ta^V + 3 Cl$^-$ + 2 PEt$_3$ + =CHCMe$_3^{2-}$
0e + 3 x 2e + 2 x 2e + 4e   = 14e

(j) Cp*Ta(PMe$_3$)$_2$ClCPh
Ta^V + Cp^{*-} + Cl$^-$ + ≡CPh^{3-} + 2PMe$_3$
0e + 6e + 2e + 6e + 2 x 2e = 18e

The total electron count is not changed if carbenes are considered as alkylidenes or if carbynes are considered as alkylidynes. The difference in ligand electron count is compensated for by a different oxidation state computed for the metal. For example,

Cp*Ta(PMe$_3$)$_2$ClCPh
Ta^I + Cp^{*-} + Cl$^-$ + ≡CPh$^+$ + 2 PMe$_3$
4e + 6e + 2e + 2e + 2 x 2e = 18e

12.19 Name each species in problem 12.18.

12.19 Solution: Carbenes and carbynes are named as alkylidenes and alkylidynes, respectively.
(a) pentacarbonyl[(dimethylamino)(ethoxy)alkylidene]chromium
(b) dicarbonylcyclopentadienyl(diphenylalkylidene)manganese
(c) cis-dichlorobis[(methyl)(iso-propoxy)alkylidene]platinum
(d) [carbonyl(η^5-cyclopentadienyl)(difluoroalkylidene)(triphenylphosphine)iron] tetrafluoroborate
(e) trichloro(neo-pentylidene)bis(triethylphosphine)tantalum
(f) tris(tert-butoxy)(neo-pentylidyne)tungsten
(g) (η^5-cyclopentadienyl)(hexylidyne)bis(trimethylphosphite)molybdenum
(h) bromotetracarbonyl(phenylalkylidyne)chromium
(i) (η^5-cyclopentadienyl)(methylene)(trimethylphosphine)iridium
(j) chloro(η^5-cyclopentadienyl)(phenylalkylidyne)bis(trimethylphosphine)tantalu

12.20 Problem: The first carbene complex was prepared in 1915 by Tschugaev, although it was not recognized as such until 1970. The reaction involved $[(MeNC)_4Pt]^{2+}$ and hydrazine, N_2H_4, followed by deprotonation of the product by excess hydrazine. Draw the (several) canonical forms for the product. *Hint:* Consider the equation:

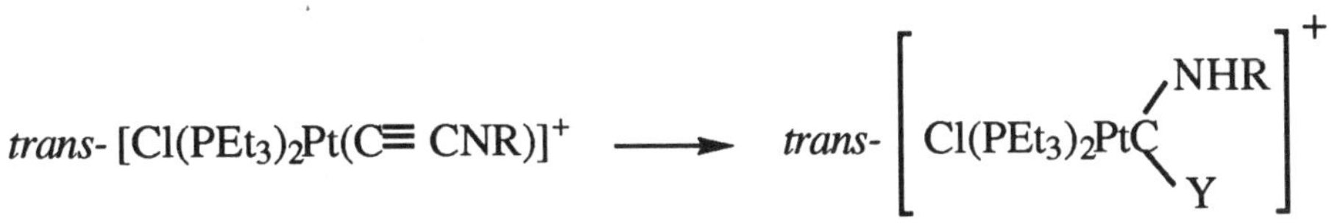

$$trans\text{-}[Cl(PEt_3)_2Pt(C\equiv CNR)]^+ \longrightarrow trans\text{-}\left[Cl(PEt_3)_2PtC\begin{array}{c}NHR\\Y\end{array}\right]^+$$

$$HY = ROH, NH_2R, NHR_2$$

12.20 Solution: The formation of carbenes by attack of HY on coordinated isocyanides is shown above. H_2NNH_2 has acidic H on both N's, so it can add across two coordinated isocyanides. Several canonical structures can be drawn for the final product.

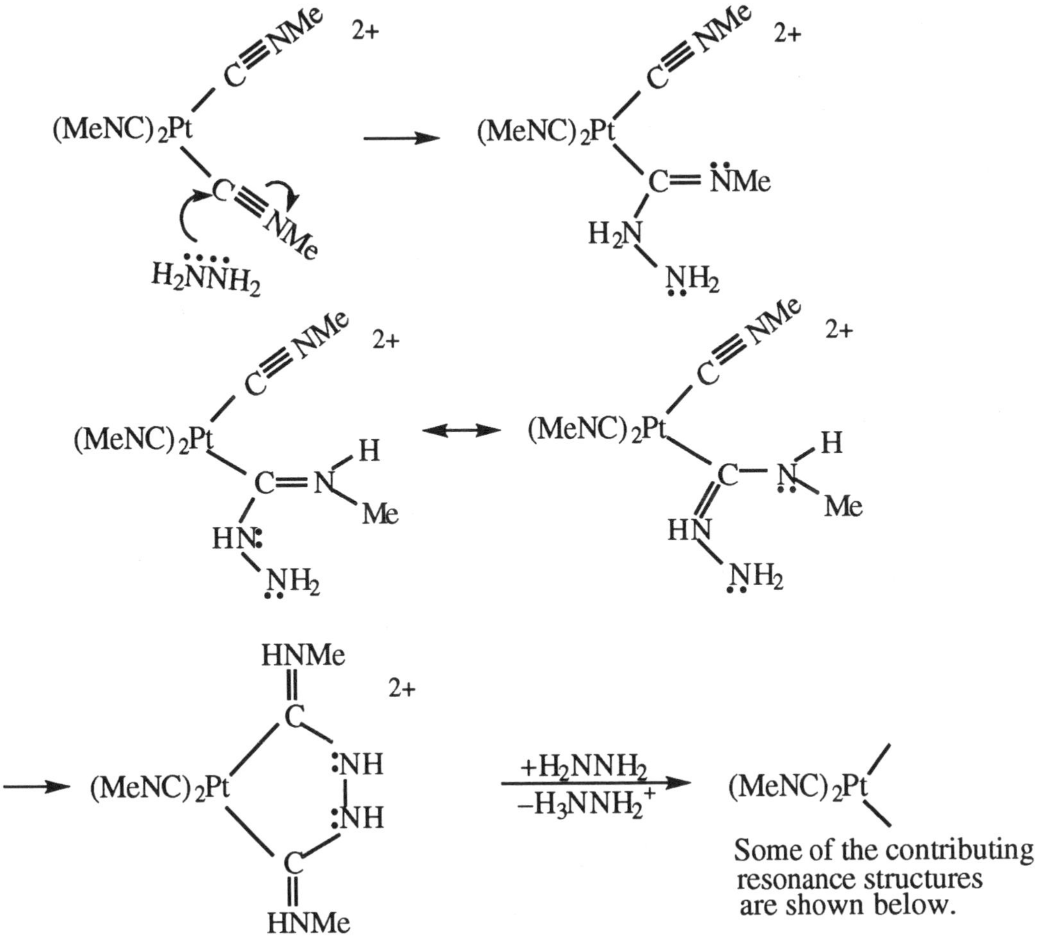

Some of the contributing resonance structures are shown below.

12.21 Problem: Rationalize the trends in the following sets of IR-active CO stretching frequencies (in cm^{-1}):

(a) (η^6-C$_6$H$_6$)Cr(CO)$_3$	1980, 1908		(d) Ni(CO)$_4$	2040
CpMn(CO)$_3$	2027, 1942		[Co(CO)$_4$]$^-$	1890
(b) CpV(CO)$_4$	2030, 1930		[Fe(CO)$_4$]$^{2-}$	1730
CpMn(CO)$_3$	2027, 1942		[Mn(CO)$_4$]$^{3-}$	1670
[CpFe(CO)$_3$]$^+$	2120, 2070			
(c) W(CO)$_5$(Pn-Bu$_3$)	2068, 1936, 1943			
W(CO)$_5$(PPh$_3$)	2075, 1944, 1944			
W(CO)$_5$[P(OPh)$_3$]	2079, 1947, 1957			

12.21 Solution: (a) Mn^I has a higher effective nuclear charge than Cr^0. Hence, it attracts electrons more strongly and less electron density can be back-donated into CO π^* orbitals. Hence, the C—O bond in the Mn compound is weakened to a lesser extent and ν_{C-O} is higher.

(b) Same argument as in (a). The effective nuclear charge of Fe^{II} is greater than that in V^I and Mn^I.

(c) The Lewis basicity of the phosphine ligands lies in the order P(OPh)$_3$ < PPh$_3$ < P(n-Bu)$_3$. The electron density supplied to W increases in this order. The metal has increasing electron density to back-donate to CO. Thus, ν_{C-O} decreases in the observed order.

(d) The isoelectronic complexes have increasing negative charges to delocalize in the order Ni(CO)$_4$ < [Co(CO)$_4$]$^-$ < [Fe(CO)$_4$]$^{2-}$ < [Mn(CO)$_4$]$^{3-}$. The method of delocalization is back donation into π^* CO which causes a shift of ν_{CO} to progressively lower energies in the order given above.

12.22 Problem:
(a) Draw at least two possible structures of Os$_3$(CO)$_9$(PPh$_3$)$_3$.

(b) The IR spectrum of this compound in CH$_2$Cl$_2$ has CO stretches at 1962 and 1917 cm^{-1}. How does this knowledge help to narrow the possible structures?

12.22 Solution:
(a) Some possibilities we can write are:

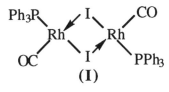

(b) Structure (**II**) is not possible because no stretching frequencies are seen in the μ-CO region. Structure (**I**) would display a very low frequency terminal CO stretch due to the unique CO. Structure (**III**) is consistent with the two frequencies observed.

12.23 Problem: The IR spectrum of $Rh_2I_2(CO)_2(PPh_3)_2$ has CO stretches at 2061 and 2005 cm^{-1}. Suggest a structure consistent with this.

12.23 Solution:

| Symmetric and antisymmetric CO stretches could account for the two observed bands; but the symmetric stretch would have low intensity. | (**II**) is also a possibility from the IR data. The two stretches could be assigned as symmetric and antisymmetric. However, it is less likely to exist than (**I**) because of less favorable entropy. | (**III**) is also a possibility, but unlikely for same reasons as (**II**). |

Structures with one or two μ-CO are ruled out because all stretches observed are in the terminal CO stretching region.

12.24 Problem: The IR spectrum of $(CO)_5Cr=C(OMe)Ph$ shows a band belonging to the CO stretch for the CO *trans* to the carbene at 1953 cm^{-1}. For comparison, the totally symmetric A_{1g} mode in $Cr(CO)_6$ is seen in the Raman at 2108 cm^{-1}. Compare the π-acid strengths of CO and the carbene ligand.

12.24 Solution: When only other CO ligands compete for electron density in $Cr(CO)_6$ ν_{CO} occurs as higher energy indicating a smaller degree of back-donation into carbonyl π* orbitals. The shift to lower energy on substituting a carbene for a carbonyl indicates that the carbene is a less effective π-acid than CO.

12.25 Problem: Treating $W(CO)_6$ with $LiC_6H_4$4-CMe_3 gives an anion of composition $[(CO)_6WC_{13}H_{13}]^-$ which can be crystallized as the NMe_4^+ salt. What is the structure of anion from its spectroscopic data? (See K. A. Belsky, M. F. Asaro, S. Y. Chan, and A. Mayr, *Organometallics*, **1992**, *11*, 1926.)

IR: 2042 cm^{-1} (weak), 1901 cm^{-1} (strong), ~1600 cm^{-1}
^{1}H NMR δ (relative intensity): 7.54 (2H), 7.35 (2H), 1.31 (9H)
^{13}C NMR δ: 279.3, 209.0, 204.7, 155.5, 152.2, 126.4, 125.2, 31.7

12.25 Solution: See Tables 12.8 and 12.9 in DMA, 3rd ed. for spectroscopic data. The IR data show the presence of C=O. The presence of two terminal CO stretches is consistent with a square-pyramidal arrangement of carbonyls in LW(CO)$_5^-$. The ^{13}C NMR spectrum has signals at δ 209.0 and 204.7, corresponding to two different types of terminal CO's. The δ 279.3 signal is in the region expected for C=O while those at 155.5, 152.2, 126.4, and 125.2 represent four different types of aromatic C's. Methyl C's resonate at 31.7. Since no H is attached to the *t*-butyl tertiary C, it has no ^{13}C NMR signal. The ^{1}H NMR is consistent with 4 aromatic protons (δ 7.54 and 7.35) and 9 *t*-butyl protons (1.31). A structure consistent with these results is

12.26 Problem: In the ^{13}C NMR spectrum of CH$_3^*$CMn(CO)$_5$, the CO's *cis* to Me absorb at 213.8 ppm and the *trans* CO absorbs at 211.3 ppm. When this labeled sample was heated, CH$_3$Mn(CO)$_5$ was produced. The ^{13}C NMR spectrum of the product showed dramatic signal enhancement at only the 213.8-ppm position. What conclusions can you draw about the mechanism of the CO loss? (See T. C. Flood, J. E. Jensen, and J. A. Statler, *J. Am. Chem. Soc.* **1981**, *103*, 4410.)

12.26 Solution: The CO lost is not the labeled (^{13}C) one originally in the acetyl group. The acetyl *C=O is converted into *C≡O in the product and the methyl group now occupies a position *cis* to the original acetyl position.

12.27 Problem: With what organic compound is each of the following isolobal?

(a)
```
   CO      CO
    |       |
  CpRh────RhCp
     \    /
      \  /
      CH₂
```

(b) [CpFe(CO)₂]₂

(c)
```
  (OC)₃Co────────Co(CO)₃
        \    CH  /
         \  |  /
          \ | /
          Co(CO)₃
```

(d)
```
  H₂C────CH₂
   |      |
 (OC)₄Os──Os(CO)₄
```

(e)
```
        Me₂
         As
        /  \
  (OC)₃Fe──Co(CO)₃
```

12.27 Solution:
(a) CpRh(CO) ↔ CH₂; hence the compound is isolobal with cyclopropane (CH₂)₃.
(b) CpFe(CO)₂ ↔ CH₃; hence the dimer is isolobal with ethane.
(c) Co(CO)₃ ↔ CH; hence the compound is isolobal with tetrahedrane (CH)₄.
(d) Os(CO)₄ ↔ CH₂; hence the compound is isolobal with cyclobutane.
(e) AsMe₂ is isolobal with CH_2^- while Fe(CO)₄ is isolobal with CH₂ and Co(CO)₃ with CH or CH_2^+. Hence, the compound is isolobal with (CH₂)₃, cyclopropane.

12.28 Problem: Dihydrogen complexes of Mo and W can be prepared via the following route: M(CO)₃(PCy₃)₂ + H₂ → M(CO)₃(PCy₃)₂(η²-H₂) where M = Mo, W and Cy = cyclohexyl. Why is the above route better for preparing dihydrogen complexes than, for example, thermal CO replacement as in W(CO)₆ + H₂ → (CO)₅W(η²-H₂) + CO?

12.28 Solution: H₂ is very easily displaced by other ligands in solution such as CO. Hence, very high pressures of H₂ would have to be employed for thermal displacement. Also, the agostic interaction of the C-H bond with the metal is undoubtedly much weaker than the CO bond, so the reaction will be thermodynamically more favorable starting with the agostic complex.

12.29 Problem: [Rh(PP₃)(H₂)]⁺ where PP₃ = (Ph₂PCH₂CH₂)₃P reacts with D₂ giving [(PP₃)Rh(D₂)]⁺ but no [(PP₃)Rh(HD)]⁺. On the other hand, W(CO)₃(Pi-Pr₃)₂(H₂) + D₂ → W(CO)₃(Pi-Pr₃)₂(HD) even in the solid state. What significance do these results have for the formulation of the Rh and W complexes as dihydrides or dihydrogen species? (See G. J. Kubas, C. J. Unkefar, B. I Swanson and E. Fukushima, *J. Am. Chem. Soc.* **1986**, *108*, 7000; C. Bianchini, C. Mealli, M. Peruzzini and F. Zanobini, *J. Am. Chem. Soc.* **1987**, *109*, 5548.)

12.29 Solution: If either complex exists as a dihydride, this involves formal oxidative addition of dihydrogen. Coordination of D₂ to this dihydride complex could afford a pathway for H-D scrambling by oxidative addition of D₂ and reductive elimination of HD. That no scrambling occurs with the Rh complex indicates that the H's remain bonded together as H₂. Likewise, the scrambling that takes place with the W compound is evidence that an equilibrium with the dihydride form may be occurring.

12.30 Problem: The x-ray structure of W(CO)₃(PCy₃)₂ shows an agostic interaction between W and cyclohexyl C–H. Give electron counts for W(CO)₃(PCy₃)₂ both with and without the agostic interaction.

12.30 Solution:

Without agostic interaction:
$W^0 + 2\ PCy_3 + 3\ CO$
$6e\ + 2 \times 2e\ + 3 \times 2e\ = 16e$

With agostic interaction:
$W^0 + 2\ PCy_3 + C\text{-}H + 3\ CO$
$6e\ + 2 \times 2e\ + 2e\ + 3 \times 2e\ = 18e$

12.31 Problem: Predict the products of the following substitution reactions:
(a) $W(CO)_6 + PPh_3 \rightarrow$
(b) $Fe(CO)_5 + CF_2{=}CF_2 \xrightarrow{h\nu}$
(c) $CpFe(CO)_2Cl + PPh_3 \rightarrow$
(d) $Cr(CO)_6 + NO \rightarrow$
(e) $Mn(CO)_5Br + CNt\text{-}Bu \rightarrow$

12.31 Solution:
(a) $W(CO)_{6-n}(PPh_3)_n + CO$; n depends on stoichiometric mixture, but will be ≤ 3.
(b) $(\eta^2\text{-}CF_2{=}CF_2)Fe(CO)_4 + CO$
(c) $CpFe(CO)(PPh_3)Cl + CO$ or $[CpFe(CO)_2(PPh_3)]Cl$
(d) $Cr(NO)_4 + 6\ CO$
(e) $Mn(CO)_{5-n}(CNt\text{-}Bu)_n Br$; n depends on stoichiometric mixture, but will probably be ≤ 3.

12.32 Problem: Predict the products of the following oxidations:
(a) $Re_2(CO)_{10} + Br_2 \rightarrow$
(b) $[CpFe(CO)_2]_2 + I_2 \rightarrow$
(c) $Pt(PPh_3)_4 + Cl_2 \rightarrow$

12.32 Solution:
(a) $Re(CO)_5Br$
(b) $CpFe(CO)_2I$
(c) $Pt(Cl)_2(PPh_3)_4$

12.33 Problem: Predict the products of the following reductions:
(a) $[CpFe(CO)_2]_2 + Na/Hg \rightarrow$
(b) $Re_2(CO)_{10} + Li[HBEt_3] \rightarrow$
(c) $[CpMo(CO)_3]_2 + KH \rightarrow$

12.33 Solution:
(a) $[CpFe(CO)_2]^- + Na^+$
(b) $[Re(CO)_5]^- + Li^+ + \frac{1}{2}H_2 + BEt_3$
(c) $[CpMo(CO)_3]^- + K^+ + \frac{1}{2}H_2$

12.34 Problem: Identify the following reactions by type and predict the products, e. g. $Mo(CO)_6 + PPh_3$ is CO displacement by a nucleophile (PPh_3) giving $Mo(CO)_5(PPh_3) + CO$.

(a) $Re_2(CO)_{10} + Na/Hg$
(b) $W(CO)_6 + (n\text{-}Bu_4N)I$
(c) $CpCo(CO)_2 + PPh_3$
(d) $Cp_2Co_2(NO)_2 + Na/Hg$
(e) $Fe(CO)_5 + EtCO_2CH=CHCO_2Et \xrightarrow{h\nu}$
(f) $Mo(CO)_3(PCy_3)_2(\eta^2\text{-}H_2) + CO$
(g) $[CpMo(CO)_3]^- + C_3H_5Br \rightarrow 1 \xrightarrow{h\nu} 2$
(h) $[Re(CO)_5]^- + MeI$

12.34 Solution:

(a) Reduction; $Na^+[Re(CO)_5]^-$

(b) Substitution; $(n\text{-}Bu_4N)^+[W(CO)_5I]^- + CO$

(c) CO substitution; $CpCo(PPh_3)(CO) + CO$

(d) Reduction; $Na^+[CpCo(NO)]^-$

(e) Substitution; $(\eta^2\text{-}EtCO_2C=CCO_2Et)Fe(CO)_4 + CO$

(f) H_2 substitution; $Mo(CO)_4(PCy_3)_2 + H_2$

(g) Halide displacement, CO displacement; $CpMo(CO)_3CH_2CH=CH_2 + Br^- \rightarrow CpMo(CO)_2(\eta^3\text{-}CH_2CHCH_2) + CO$

(h) Halide displacement; $CH_3Re(CO)_5 + I^-$
The products are all reasonable in that they obey the 16-, 18-e rule.

12.35 Problem: $[(\eta^7\text{-}C_7H_7)Fe(CO)_3]^-$ was found to react with $CpPd\;[\eta^4\text{-}C_8H_{12}]^+$ to give $(C_7H_7)[Fe(CO)_3][PdCp]$ in which both the Fe and Pd-containing groups are coordinated to the seven-membered ring. Formulate reasonable structures for the product. (See M. Airoldi, G. Daganello, G. Gennaro, M. Moret, and A. Sironi, *Organometallics* **1993**, *12*, 3964.)

12.35 Solution: Reasonable structures should lead to an 18-*e* configuration for Fe and either a 16- or 18-*e* configuration for Pt. We first consider how to formulate the reactants. A neutral $Fe(CO)_3$ fragment has $8 + 3 \times 2 = 14\;e$ and requires four more *e* to obey the 18-*e* rule. Hence, we must have $C_7H_7^-$ acting as a 4-*e* donor. Two possibile structures are given below:

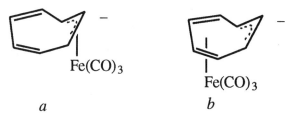

$\qquad a \qquad\qquad\qquad b$

In the Pd cation we have d^8 Pd^{II} and Cp^-, a 6-*e* donor; thus $CpPd^+$ has 14 *e*. $\eta^4\text{-}C_8H_{12}$ (cod) is a neutral 4-*e* donor giving an 18-*e* count around Pd.

Thus, two possible product structures where the ring acts as a 4-e donor to both Fe^0 and Pd^{II} are:

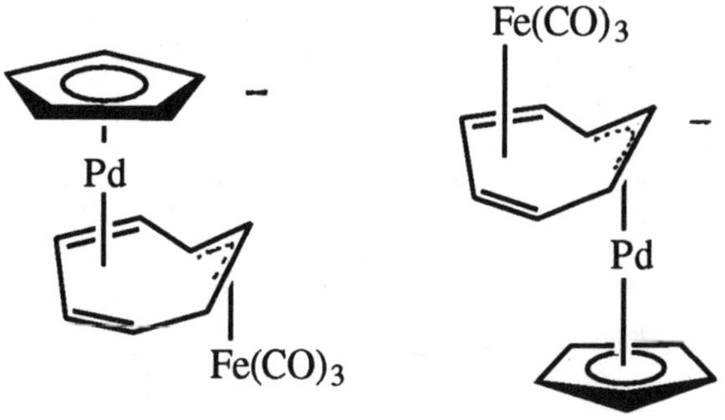

c   *d*

Alternatively, electron transfer could occur from Fe^0 to Pd^{II} giving Fe^I and Pd^I which would require an Fe—Pd bond to achieve 18-e configuration.

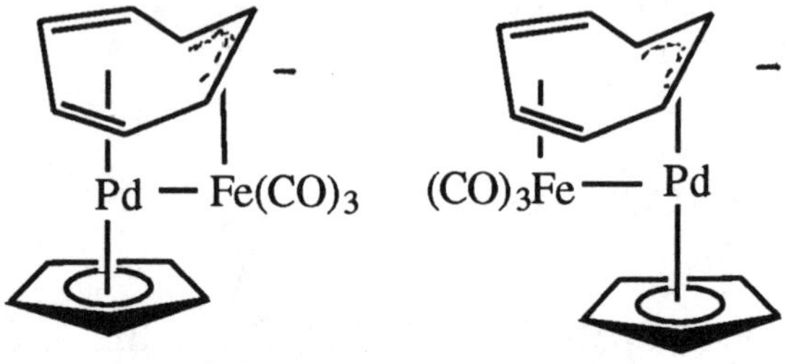

e   *f*

Further possibilities, in which Pd has a 16-e configuration are:

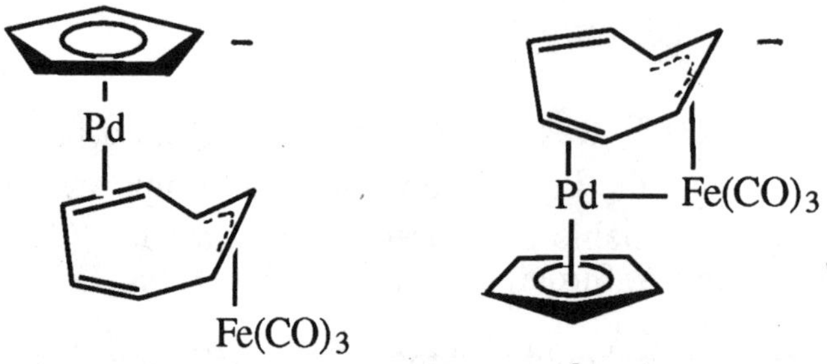

g   *h*

A decision among the possibilites *c – h* is only possible on the basis of spectroscopic on x-ray structural data. Studies indicate that *h* is the actual structure.

12.36 Problem: The following reaction was recently found to occur. The structural formulas given here show atom connectivity.

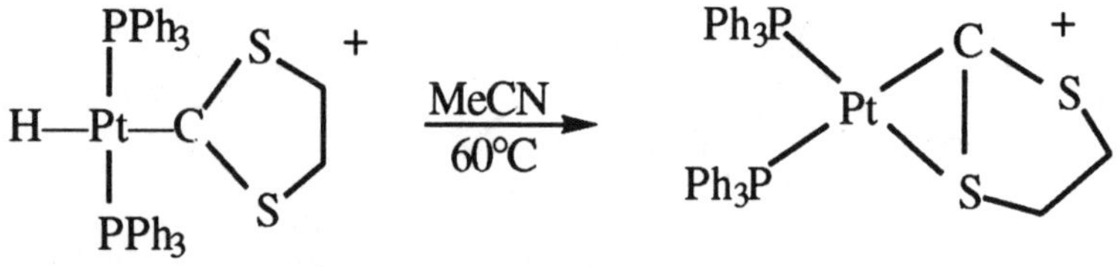

Give the electron counts for both species. (See R. A. Michelin, R. Bertani, M. Mozzon, G.

Bombieri, F. Benetello, M. Guedes da Silva, and A. J. L. Pombeiro, *Organometallics* **1993**, *12*, 2372.)

12.36 Solution:
Reactant

$$Pt^{II} + 2\,PPh_3 + H^- + =C\begin{matrix}S\\ \\S\end{matrix}\rangle \quad \text{(carbene)}$$

$$8e\ +\ 2\times 2\,e\ +2e\ +2e\ =\ 16e$$

In the product, the C bonded to Pt is saturated and is now regarded as the C of a substituted alkyl ligand. S is bonded through donation of a lone pair; lone pairs are not always shown explicity in formulas. Thus the electron count is

$$Pt^{II} + 2\,PPh_3 + \text{alkyl}^- + :S$$
$$8e\ +\ 2\times 2e\ +\ 2e\ +\ 2e\ =\ 16e$$

In a formal sense, the negative charge on the alkyl-ligand is provided by migration of H^- from Pt^{II} to a neutral carbene ligand.

12.37 Problem: $[CpTi(CO)_4]^-$ has CO stretches in the IR at 1921 and 1777 cm^{-1}. When this anion is allowed to react with Ph_3SnCl, the product $CpTi(CO)_4SnPh_3$ with CO stretches at 2020 and 1960 cm^{-1}. Account for the change in ν_{CO} (See J. E. Ellis, S. R. Frerichs, and B. K. Stein, *Organometallics* **1993**, *12*, 1048.)

12.37 Solution: The product results from nucleophilic displacement of Cl^- from Ph_3SnCl by the carbonylate anion. The Ph_3Sn^+ group is electrophilic. Hence it reduces the electron density on Ti. This reduces the ability of Ti to back donate electrons into the π^* CO orbital, thereby strengthening the CO bond and increasing its stretching frequency.

V. ORGANOMETALLIC CHEMISTRY
13
Survey of Organometallic Compounds

13.1 Problem: Give the valence electron count for the following species. Which ones obey the 18-e rule?

(a) $CpMo(CO)_3(\eta^1-C_3H_5)$
(b) $Ir(PPh_3)_2(CO)Cl(\eta^2-C_3H_6)$
(c) $Cp_2Ti(CO)(\eta^2-C_2Ph_2)$
(d) $[(\eta^4-C_4Ph_4)PdCl_2]_2$
(e) $[(\eta^5-C_7H_9)Ru(\eta^6-C_7H_8)]^+$
(f) $CpMn(\eta^6-C_6H_6)$
(g) $Pb[CH(SiMe_3)_2]_2$
(h) $[Me_2Al(\mu-NMe_2)]_2$
(i) Cp^*_2Sn
(j) $[Li(thf)_4]_2[Mn_2(\mu-Ph)_2Ph_4]$
(k) $Li[CuMe_2]$
(l) $[(\eta^3-C_3H_5)_2(\mu-Cl)Rh]_2$

13.1 Solution:

(a) $CpMo(CO)_3(\eta^1-C_3H_5)$
$Mo^{II} + Cp^- + 3\,CO + \eta^1-C_3H_5^-$
$4e\ \ +6e\ +3 \times 2e + 2e\ \ = 18e$

(b) $Ir(PPh_3)_2(CO)Cl(\eta^2-C_3H_6)$
$Ir^I\ +2\,PPh_3 + CO + Cl^- + C_3H_6$
$8e\ +2 \times 2e\ +2e\ +2e\ +2e\ =18e$

(c) $Cp_2Ti(CO)(\eta^2-C_2Ph_2)$
$Ti^{II} + CO + 2\,Cp^- + PhC\equiv CPh$
$2e\ +2e\ +2 \times 6e + 2e\ =18e$

(g) $Pb[CH(SiMe_3)_2]_2$
$Pb^{II} + 2\,CH(SiMe_3)_2^-$
$2e\ +2 \times 2e\ \ \ \ =6e$

(h) $[Me_2Al(\mu-NMe_2)]_2$
$Al^{III} + 2\,Me^- + 2\,\mu-NMe_2$
$0e\ +2 \times 2e + 2 \times 3e\ =10e$

(i) Cp^*_2Sn
$Sn^{II} + 2\,Cp^{*-}$
$2e\ +2 \times 6e\ =14e$

(d) $[(\eta^4\text{-}C_4Ph_4)PdCl_2]_2$
For each Pd
$Pd^{IV} + C_4Ph_4^{2-} + 2\,Cl^- + \mu\text{-}Cl$
$6e \quad + 6e \quad\quad + 2 \times 2e + 2e = 18e$

(e) $[(\eta^5\text{-}C_7H_9)Ru(\eta^6\text{-}C_7H_8)]^+$
$Ru^{II} + \eta^5\text{-}C_7H_9^- + \eta^6\text{-}C_7H_8$
$6e + 6e \quad\quad + 3 \times 2e = 18e$

(f) $CpMn(\eta^6\text{-}C_6H_6)$
$Mn^I + Cp^- + \eta^6\text{-}C_6H_6$
$6e + 6e + 6e = 18e$

(j) $[Li(thf)_4]_2[Mn_2(\mu\text{-}Ph)_2Ph_4]$
For each Mn in $[Mn_2(\mu\text{-}Ph)_2Ph_4]^{2-}$
$Mn^{II} + 2\,\mu\text{-}Ph^- + 2\,Ph^- + Mn\text{–}Mn$
$5e \quad + 2 \times 1e + 2 \times 2e + 1e = 12e$

(k) $Li[CuMe_2]$
$Cu^I + 2\,Me^-$
$10e + 2 \times 2e = 14e$

(l) $[(\eta^3\text{-}C_3H_5)_2(\mu\text{-}Cl)_2Rh]_2$
For each Rh
$Rh^{III} + 2\,\eta^3\text{-}C_3H_5 + Cl^- + \mu\text{-}Cl$
$6e \quad + 2 \times 4e \quad\quad + 2e + 2e = 18e$

13.2 Problem: Name each species in Problem 13.1.

13.2 Solution:
(a) allyltricarbonylcyclopentadienylmolybdenum
(b) carbonylchloro(propene)bis(triphenylphosphine)iridium
(c) carbonylbis(η^5-cyclopentadienyl)(diphenylacetylene)titanium
(d) di-μ-chlorodichlorobis(tetraphenylcyclobutadiene)dipalladium
(e) [(η^5-cycloheptadienyl)(η^6-cycloheptatriene)ruthenium](1+)
(f) (η^6-benzene)(η^5-cyclopentadienyl)manganese
(g) bis(bis(trimethylsilyl)methyl)lead
(h) bis(μ-dimethylamino)tetramethyldialuminum
(i) bis(pentamethylcyclopentadienyl)tin
(j) bis[tetrakis(tetrahydrofuran)lithium] [tetraphenyldi(μ-phenyl)dimanganate](2–)
(k) lithium dimethylcuprate(1–)
(l) diallyldi-μ-chlorodirhodium

13.3 Problem: PhC≡CPh has a C≡C distance of 119.8 pm and $\nu_{C\equiv C}$ = 2223 cm^{-1}. Formulate the best VB structure for each of the following compounds in light of the structural and spectroscopic data given:
(a) CpRe(CO)$_2$(C$_2$Ph$_2$): d_{C-C} = 123.2 pm; < PhC–C–Ph = 151.6°; $\nu_{C\equiv C}$ = 1848 cm^{-1}.
(b) [CpMo(CO)(PPh$_3$)(C$_2$Ph$_2$)]$^+$: d_{C-C} = 129 pm; < Ph–C–C–Ph = 135° and 141°; $\nu_{C\equiv C}$ = 1645 cm^{-1}.

13.3 Solution:
(a) metallacyclopropene; C≡C bond length is considerably lengthened by back donation, and $\nu_{C\equiv C}$ lowered.
(b) metallacyclopropene; same reasoning as in **a**.

13.4 Problem: Give an MO analysis of the bonding between the allyl and $Co(CO)_3$ fragments in $(\eta^3\text{-}C_3H_5)Co(CO)_3$. Draw pictorial representations of interacting frontier orbitals. Give a qualitative MO diagram.

13.4 Solution: We consider the complex to contain $\eta^3\text{-}C_3H_5^-$ and $Co(CO)_3^+$. The metal fragment is isoelectronic with $Fe(CO)_3$ and so has the same frontier orbitals. Thus, the empty metal a_1 hybrid can overlap with filled Ψ_1 of $C_3H_5^-$. Also d_{yz} and Ψ_2 interact as do d_{xz} and Ψ_3, giving two bonding and two antibonding MOs; the available electrons occupy only the bonding levels.

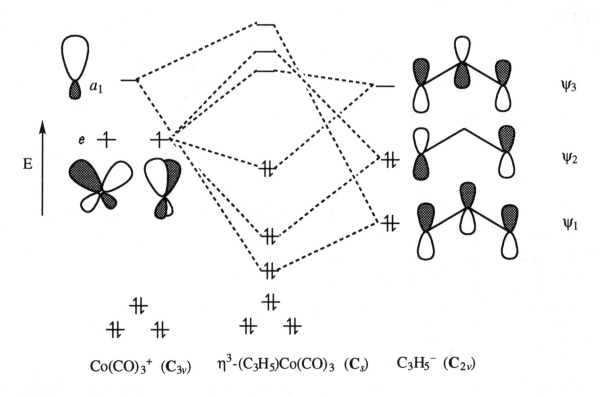

$Co(CO)_3^+$ (C_{3v}) $\eta^3\text{-}(C_3H_5)Co(CO)_3$ (C_s) $C_3H_5^-$ (C_{2v})

13.5 Problem: (a) The allylic H in propene is easily lost to give an allyl. A reaction thought to be important in catalysis is the so-called 1,3-hydride shift. An allylic H of an η^2-propene adds to the metal giving an η^3-allyl hydride product with the metal's oxidation number increased by two. Show how the reversibility of this process can afford a mechanism for shifting H from C_1 to C_3—a so-called 1,3-hydride shift.
(b) Account for the following observation by a mechanism involving the 1,3-hydride shift:

In the complex [structure with D]$Fe(CO)_4$, olefin is isomerized to give [structure with ($^1/_3$D) labels] $Fe(CO)_4$ and D is

scrambled to all methyl groups. (See C. P. Casey and C. R. Cyr, *J. Am. Chem. Soc.* **1973**, *95*, 2248.)

13.5 Solution:

(a) [scheme showing propene coordination to M^{n+}, insertion into M-H to form metallacyclic/σ-alkyl intermediates (all $M^{(n+2)+}$), equivalent to π-allyl MH, then β-H elimination giving $M^{(n+1)+}$ with CH₃-substituted vinyl]

(b) [scheme of allyl-$Fe(CO)_4$ ⇌ η³-allyl-$Fe(CO)_3$ + CO, with D-label migration through successive σ/π-allyl interconversions leading to scrambling]

The third reversible allylic shift could occur to the opposite direction as well, and these lead to the scrambling of D to the other two terminal methyls.

13.6 Problem:
$[(\eta^5\text{-}C_5Me_5)Ir(acetone)_3]^{2+}$ reacts with $CH_2\text{=}CH_2$ in acetone at room temperature to produce $[(\eta^5\text{-}C_5Me_5)IrC_5H_9]^+$, where C_5H_9 represents more than one organic ligand. Given the following NMR data, what is the structure? (See J. B. Wakefield and J. M. Stryker, *Organometallics* **1990**, *9*, 2428.)

¹H NMR δ (splitting, relative intensity): 4.11 (triplet of triplets, 1H), 3.79 (doublet of doublets, 2H), 3.11 (singlet, 2H), 2.60 (singlet, 2H), 2.04 (doublet of doublets, 2H), 1.91 (singlet, 15H).

¹³C NMR δ (splitting): 100.6 (quartet), 86.3 (doublet), 47.6 (triplet), 47.0 (broad triplet), 8.5 (quartet).

13.6 Solution: Consult Table 12.9 in DMA, 3rd ed. for the interpretation of ^{13}C NMR spectra. The ^{13}C signal at δ 100.6 is in the region for cyclopentadienyl C's and is split into a quartet by the 3 H on the attached methyl group. Similarly the quartet at 8.5 must be due to Cp* methyl C's, each split by 3 H.

In the ^1H NMR spectrum, the 15 H singlet at δ 1.91 is assigned to Cp* methyl protons. The triplet of triplets at δ 4.11 is due to a single H split by two different chemically equivalent sets of two H. The two doublets of doublets (intensity 2 H) must result from splitting of two different chemically equivalent sets of two H by a single H. This suggests an allylic arrangement:

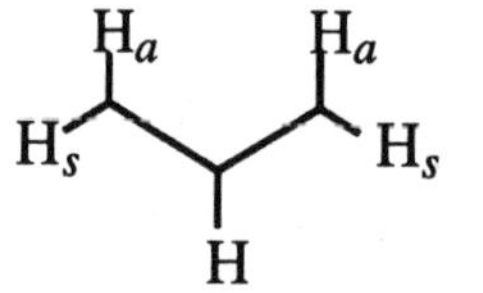

The ^{13}C doublet at δ 86.3 is assigned to the center allyl C split by one H. The broad triplet at 47.0 is due to the two end C's, each split by 2 H, broadening may result from unresolved splitting by H on C_2. Thus we have accounted for a $C_3H_5^-$ ligand.

One ^{13}C signal remains along with two proton signals; the ligand atoms unassigned are C_2H_4. The ^{13}C triplet can be assigned to two ethylene C's and the two ^1H singlets at δ 3.11 and 2.60 to two chemically nonequivalent sets of ethylene protons—one "close to" and one "away from" Cp*. Another structure which has the allyl ligand oriented with C_2 toward the Cp* ring is also consistent with the data.

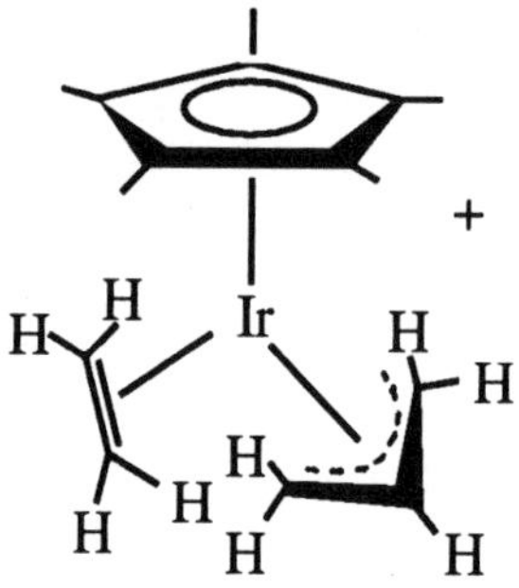

13.7 Problem: Explain how the facts that $(\eta^6\text{-}C_6H_5CO_2H)Cr(CO)_3$ is a stronger acid than benzoic acid and that $(\eta^6\text{-}C_6H_5NH_2)Cr(CO)_3$ is a weaker base than aniline show that the $Cr(CO)_3$ group withdraws electrons from the aromatic rings.

13.7 Solution: Any electron-withdrawing group strengthens a carboxylic acid by helping to delocalize negative charge on the carboxylate ion remaining after proton removal and so stabilizing it. Conversely, any group which supplies electrons (whether by inductive effect or conjugation) localizes negative charge and destabilizes the conjugate bases, thus reducing the acidity. Arguments along the opposite line can be applied to base strengths since their conjugate acids are positively charged relative to the base.

13.8 Problem: Write a mechanism for dimerization of propene to 2-methylpent-1-ene using Al, H_2 and a catalytic quantity of $AlEt_3$.

13.8 Solution:

$$2\,Al + 3\,H_2 + 2\,AlEt_3 \rightleftarrows 2\,HAlEt_2 + 2\,H_2AlEt$$

$$HAlEt_2 + CH_3CH=CH_2 \rightleftarrows CH_3CH_2CH_2AlEt_2 + CH_3CH(CH_3)AlEt_2$$

$$CH_3CH_2CH_2AlEt_2 + CH_3CH=CH_2 \rightleftarrows CH_3CH_2CH_2CH(CH_3)CH_2AlEt_2$$

$$CH_3CH_2CH_2CH(CH_3)CH_2AlEt_2 \rightarrow CH_3CH_2CH_2(CH_3)C=CH_2 + HAlEt_2$$

Since individual steps are reversible, the thermodynamic product is obtained.

13.9 Problem: (a) The butene product from the reaction

$$CpFe(CO)(PPh_3)(n\text{-}C_4H_9) \rightarrow CpFe(CO)(PPh_3)H + \text{butene}$$

consists of 1-butene as well as *cis*- and *trans*-2-butene. Keeping in mind the reversibility of the β-hydride elimination, write a mechanism to account for this.
(b) When the reaction in **a** is run with $CpFe(CO)(PPh_3)(CD_2CH_2Et)$, complete scrambling of D occurs. Write a mechanism to account for this observation.

13.9 Solution:

(a) $CpFe(CO)(PPh_3)CH_2CH_2Et \rightleftarrows CpFe(CO)CH_2CH_2Et + PPh_3$

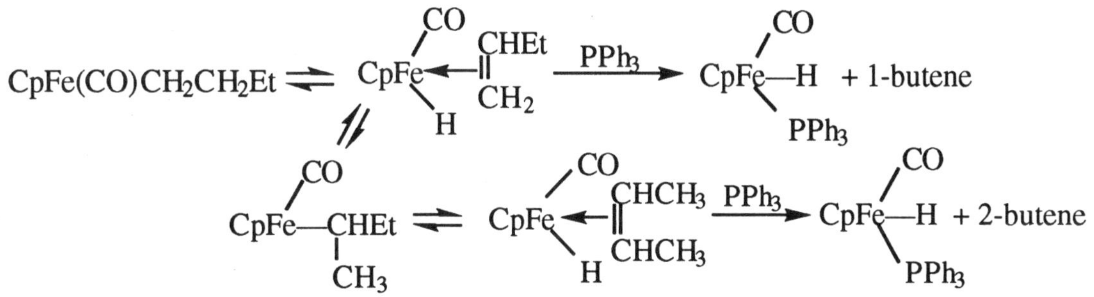

$$CpFe(CO)CH_2CH_2Et \rightleftarrows CpFe\begin{array}{c}CO\\\diagup\\\diagdown\\H\end{array}\!\!\!\overset{CHEt}{\underset{CH_2}{\|}} \xrightarrow{PPh_3} CpFe\begin{array}{c}CO\\\diagup\\\diagdown\\PPh_3\end{array}\!\!-H + \text{1-butene}$$

$$CpFe\!-\!\underset{CH_3}{\overset{CO}{\overset{|}{CHEt}}} \rightleftarrows CpFe\begin{array}{c}CO\\\diagup\\\diagdown\\H\end{array}\!\!\!\overset{CHCH_3}{\underset{CHCH_3}{\|}} \xrightarrow{PPh_3} CpFe\begin{array}{c}CO\\\diagup\\\diagdown\\PPh_3\end{array}\!\!-H + \text{2-butene}$$

The β-elimination from the *sec*-C_4H_9 species can produce *cis*- or *trans*-2-butene depending on the rotamer from which the elimination occurs.
(b) A reversible mechanism such as that in (a) provides a pathway for transferring D to Fe and transferring it back to different C's. C–H addition and β-elimination must be faster than final displacement of olefin.

13.10 Problem: Coordination of ligands to transition metals often makes them subject to nucleophilic attack. Make a list of reactions given in Chapter 13 in DMA, 3rd ed. involving nucleophilic attack on coordinated ligands.

13.10 Solution: Nucleophilic attack is involved in the following reaction numbers in DMA, 3rd ed.: (13.6), (13.14) if the alkyne is precoordinated to a Zr hydride, (13.15), (13.19), (13.25), (13.43), (13.44), the reverse of (13.61), (13.76) if the olefin is precoordinated to Pt, and (13.81).

13.11 Problem: Identify the following reactions by type and predict the products, e. g. $Mo(CO)_6 + PPh_3$ is CO displacement by a nucleophile giving $Mo(CO)_5(PPh_3) + CO$.

(a) $Fe(CO)_5 + EtCO_2CH=CHCO_2Et \xrightarrow{h\nu}$

(b) $[CpMo(CO)_3]^- + C_3H_5Br \rightarrow \mathbf{1} \xrightarrow{h\nu} \mathbf{2}$

(c) $[\{Os(\eta^4-C_8H_{12})Cl_2\}_x] + Sn(n-Bu_2)_2(C_5Me_5)_2$
(d) $[1-4\eta^4-C_6H_8]Fe(CO)_3 + Ph_3C^+$
(e) $EtMgI + SnCl_4$
(f) $[CpFe(CO)_2]^- + C_6H_5CH_2Cl$
(g) $MeLi(excess) + CuCl_2$
(h) $Cp_2Fe + n\text{-BuLi}$
(i) $(CO)_5CrC(OMe)Ph + BI_3$
(j) $Li + C_6H_5Cl \xrightarrow{ether}$
(k) $(CO)_5MoC(OMe)Ph + CH_2=CHCO_2Et$
(l) $(Me_3CCH_2)_3NbCHCMe_3 + PMe_3$
(m) $Cp(CO)_2WCMe + HI$
(n) $[(CO)_5WCNPh_2]^+ + CN^-$
(o) $[CpRe(CO)_2(\eta^3-C_3H_5)]^+ + CH(CO_2Et)_2^-$

13.11 Solution:

(a) CO substitution; $(\eta^2\text{-}EtCO_2CH=CHCO_2Et)Fe(CO)_4 + CO$
(b) S_{N2} displacement of Br^- followed by internal CO displacement. $\mathbf{1} = CpMo(CO)_3(\eta^1\text{-}C_3H_5)$; $\mathbf{2} = CpMo(CO)_2(\eta^3\text{-}C_3H_5) + CO$
(c) Displacement of μ-Cl to give monomer: $Cp^*Os(\eta^4\text{-}C_8H_{12})Cl$
(d) H^- abstraction; $[(\eta^5\text{-}C_6H_7)Fe(CO)_3]^+ + Ph_3CH$
(e) Alkylation; $Et_4Sn + MgClI$
(f) Halide displacement; $CpFe(CO)_2CH_2C_6H_5 + Cl^-$
(g) Reduction and alkylation; $Li[Me_2Cu]$
(h) Metallation; $(LiC_5H_4)FeCp + C_4H_{10}$
(i) Abstraction of MeO^- by Lewis acid followed by CO displacement; $\textit{trans}\text{-}I(CO)_4Cr\equiv CPh$ + "$B(OMe)I_2$"
(j) Direct synthesis of metal alkyl; $PhLi + LiCl$
(k) Carbene transfer;

$$\begin{array}{c} Ph\diagdown \ \diagup OMe \\ C \\ \triangle \\ EtCO_2C\text{———}CH_2 \end{array}$$

(l) Attack by PMe_3 followed by α-H abstraction; $(PMe_3)Nb(\equiv CCMe_3)(CH_2CMe_3)_2 + CMe_4$
(m) Addition of an electrophile to carbyne C; $[Cp(CO)_2W=C(H)(Me)]^+$
(n) Attack on electrophilic C of a cationic carbyne; $(CO)_5W=[C(CN)(NPh_2)]$
(o) Nucleophilic attack on coordinated allyl polarized by + charge; this involves formal reduction of Re^{III} to Re^I. $CpRe(CO)_2[\eta^2\text{-}CH_2=CHCH_2CH(CO_2Et)_2]$

13.12 Problem: Alkyl compounds of Zn and Cd inflame spontaneously in air, but $Hg(CH_3)_2$ is stable toward air and water. How is the stability of Hg alkyls related to the problem of Hg pollution?

13.12 Solution: Microorganisms can transform Hg compounds into CH_3Hg^+ which is stable and concentrates in the food chain. There have been serious problems with Hg contamination in fish and other seafood. Hg has a high affinity for S and this is presumed to be related to the inhibition of enzymes by Hg. Hg has a cumulative effect. It has no biological function and hence there is no mechanism to regulate the level of Hg.

13.13 Problem: Develop a fragment analysis of the bonding in $(\eta^3\text{-}C_3H_3)Co(CO)_3$. (See D. L. Lichtenberger, M. L. Hoppe, L. Subramanian, E. M. Kober, R. P. Hughes, J. L. Hubbard, and D. S. Tucker, *Organometallics* **1993**, *12*, 2025.)

13.13 Solution: We can view the compound as made up of the aromatic $[4(0) + 2] - e$ $C_3H_3^+$ cation and the $Co(CO)_3^-$ anion. We first develop the MO's for the π system of the cyclopopenium cation of $\mathbf{D}_{3h}$ symmetry. Since the overall symmetry of $(\eta^3\text{-}C_3H_3)Co(CO)_3$ is $\mathbf{C}_{3v}$, we work in $\mathbf{C}_{3v}$ and realize that the symmetry labels of all π orbitals will have double prime added in $\mathbf{D}_{3h}$ since they change sign on reflection in the molecular plane.

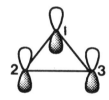

	E	$2C_3$	$3\sigma_v$
Γ_{red}	3	0	1

By inspection, this reduces to $A + E$. We now construct LCAO's of appropriate symmetry.

	E	$C_3(1)$	$C_3(2)$	$\sigma_v(1)$	$\sigma_v(2)$	$\sigma_v(3)$
π_1	π_1	π_3	π_2	π_1	π_3	π_2

$$\Psi a_1 = \pi_1 + \pi_3 + \pi_2 + \pi_1 + \pi_3 + \pi_2$$
$$= \frac{1}{\sqrt{3}}(\pi_1 + \pi_2 + \pi_3)$$
$$\Psi e(1) \sim 2\pi_1 - \pi_3 - \pi_2$$
$$= \frac{1}{\sqrt{6}}[2\pi_1 - \pi_2 - \pi_3]$$

By performing symmetry operations on π_2 we could generate $\Psi'e = \frac{1}{\sqrt{6}}[2\pi_2 - \pi_1 - \pi_3]$; this is not an independent orbital, but only corresponds to re-numbering atoms. However, by taking the linear combination $2\Psi'e - \Psi e(1)$, we can generate

$$\Psi_e(2) = \frac{1}{\sqrt{2}}[\pi_3 - \pi_2]$$

Not surprisingly, these resemble the LGO's for π-bonding in BF_3 (See DMA, 3rd ed., p. 183).

The Co(CO)$_3^-$ fragment orbitals are depicted in Problem 13.4. Below we have an energy-level diagram for (η^3-C$_3$H$_3$)Co(CO)$_3$ [Co–CO bonds are not shown].

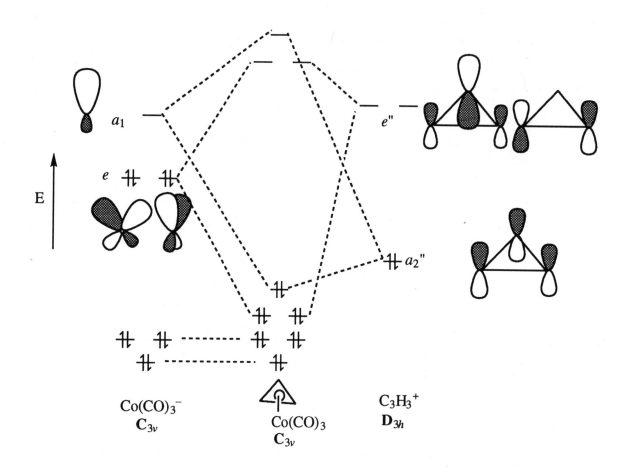

V. ORGANOMETALLIC CHEMISTRY
14
Reaction Mechanisms and Catalysis of Organometallic Compounds

14.1 Problem: Activation parameters for CO exchange by $M(CO)_6$ (M = Cr, Mo, W) given below are for the gas-phase reactions. In decalin solution, the $\Delta H^{\ddagger}$ values are: Cr, 163.2 kJ/mol; Mo, 132.2 kJ/mol and W, 169.9 kJ/mol. What is the significance of the agreement of gas-phase and solution activation enthalpies?

Activation parameters for some carbonyl substitutions		
Reaction	$\Delta H^{\ddagger}$ kJ/mol	$\Delta S^{\ddagger}$ J/mol K
$Cr(CO)_6 + {}^*CO \rightarrow Cr(CO)_5({}^*CO) + CO$	161.9	77.4
$Mo(CO)_6 + {}^*CO \rightarrow Mo(CO)_5({}^*CO) + CO$	126.4	−1.7
$W(CO)_6 + {}^*CO \rightarrow W(CO)_5({}^*CO) + CO$	166.5	46

14.1 Solution: The fact that the activation enthalpies are nearly the same suggests that the mechanisms in both the gas phase and in decalin must be the same. The large positive values imply a dissociative mechanism.

14.2 Problem: The substitution reaction $Mo(PPh_3)_2(CO)_4 + bipy \rightarrow Mo(CO)_4(bipy) + 2PPh_3$ obeys the usual two-term rate law:
$$\text{rate} = (k_1 + k_2[bipy])[Mo(PPh_3)(CO)_4]$$

Activation parameters are as follows: For the *cis* complex $\Delta H^{\ddagger}$ = 109.5 kJ/mol, and $\Delta S^{\ddagger}$ = 25 J/mol K; for the *trans* complex $\Delta H^{\ddagger}$ = 99.9, $\Delta S^{\ddagger}$ = −33. Comment on the significance of these numbers for the reaction mechanisms.

14.2 Solution: The large positive values for $\Delta H^{\ddagger}$ and $\Delta S^{\ddagger}$ are consistent with a dissociative mechanism for reaction of the *cis* complex. Since bipy is a didentate ligand, displacement of the second CO should be facilitated so that loss of the first CO is expected to be rate-determining. PPh_3 is a poorer π-acceptor than CO; thus, one might expect a weaker bond between Mo—PPh_3 *trans* to another PPh_3 accounting for a lower $\Delta H^{\ddagger}$ for dissociative replacement in the *trans* complex. In both cases the product contains bipy occupying *cis* positions because of the size of its "bite". So, the negative $\Delta S^{\ddagger}$ for the *trans* complex reflects the geometric requirements of a transition state which involves rearrangement to a *cis* product.

14.3 Problem: The following data refer to the reaction $(cod)Mo(CO)_4 + 2L \rightarrow Mo(CO)_4L_2$. What do these data suggest about the common mechanism for these reactions?

$$\text{rate} = k_1[(cod)Mo(CO)_4] + k_2[(cod)Mo(CO)_4][L]$$

Ligand	$\Delta H_1^{\ddagger}$ kJ/mol	$\Delta S_1^{\ddagger}$ J/mol K	$\Delta H_2^{\ddagger}$ kJ/mol	$\Delta S_2^{\ddagger}$ J/mol K
PPh_3	104.5	9	87.7	−13
$AsPh_3$	104.5	9	71.0	−71
$SbPh_3$	104.5	9	75.2	−4
py	100.3	13	83.6	−50
PCl_2Ph	100.3	13	36.8	−41

14.3 Solution: As in Problem 14.2, replacement of the first cod double bond is taken to be rate-determining. Otherwise, the rate law would involve [L] to a power higher than one. Activation parameters for the k_1 term involve large positive $\Delta H^{\ddagger}$ and rather small positive $\Delta S^{\ddagger}$, neither of which depends much on the identity of the entering ligand. This is consistent with a dissociative interchange mechanism (I_d). A **D** pathway would be expected to have a large positive $\Delta S^{\ddagger}$. Activation parameters for the k_2 term are quite dependent on the nature of the entering ligand consistent with an associative mechanism as are negative $\Delta S^{\ddagger}$.

14.4 Problem: The salt $[Cp_2Co^+][Co(CO)_4^-]$ is dark red even though both ions are separately colorless; this is due to an intense charge-transfer band at 532 nm. When a solution of the salt is irradiated in the region of this band in thf in the presence of PPh_3, a 65% yield of $Co_2(CO)_6(PPh_3)_2$ is isolated. Explain this observation. (See T. M. Bockman and J. K. Kochi, *J. Am. Chem. Soc.* **1989**, *111*, 4669.)

14.4 Solution: Light absorption in the region of the charge-transfer band produces short-lived caged radicals $Cp_2Co\cdot$ and 17-e $Co(CO)_4\cdot$ which undergo reverse transfer relaxing back the starting salt. On irradiation, energy is continuously supplied producing a bigger concentration of radicals. With PPh_3 present, rapid substitution occurs giving $Co(PPh_3)(CO)_3\cdot$ which diffuses away from $Cp_2Co\cdot$ and dimerizes to $Co_2(CO)_6(PPh_3)_2$.

14.5 Problem: Substitution of $(\eta^3\text{-}C_3H_5)Mn(CO)_4$ by PPh_3 could be envisioned as occurring either by associative or dissociative pathways.
(a) Write mechanisms for both pathways.
(b) The substitution reaction at 45° C is first-order in $[(\eta^3\text{-}C_3H_5)Mn(CO)_4]$ and $k_1 = 2.8 \times 10^{-4}$ s^{-1} at 45° C. The rate constant for decarbonylation of $(\eta^1\text{-}C_3H_5)Mn(CO)_5$ at 80° C was found to be 1.64×10^{-4} s^{-1}. How do these facts enable a choice between the two possible mechanisms?

14.5 Solution:
(a) Associative
$$(\eta^3\text{-}C_3H_5)Mn(CO)_4 + PPh_3 \rightarrow (\eta^1\text{-}C_3H_5)Mn(CO)_4(PPh_3)$$
$$(\eta^1\text{-}C_3H_5)Mn(CO)_4(PPh_3) \rightarrow (\eta^3\text{-}C_3H_5)Mn(CO)_3(PPh_3) + CO$$
Dissociative
$$(\eta^3\text{-}C_3H_5)Mn(CO)_4 \rightarrow (\eta^3\text{-}C_3H_5)Mn(CO)_3 + CO$$
$$(\eta^3\text{-}C_3H_5)Mn(CO)_3 + PPh_3 \rightarrow (\eta^3\text{-}C_3H_5)Mn(CO)_3(PPh_3)$$

(b) The dimensions of the rate constant k_1 indicate that attack by PPh_3 is not the rate-determining step, whichever of the above mechanisms is followed. Thus, the rate-determining step is either the second step of the associative mechanism or the first step of the dissociative one. We would expect CO dissociation from $(\eta^1\text{-}C_3H_5)Mn(CO)_4(PPh_3)$ to be slower than that from $(\eta^1\text{-}C_3H_5)Mn(CO)_5$ since the M-CO bond should be stronger in the former. The rate of CO loss from $(\eta^1\text{-}C_3H_5)Mn(CO)_5$ at 80° C is lower than the rate of the substitution reaction *at 45° C*. This rules out the associative mechanism as a possibility.

14.6 Problem: With poor nucleophiles, $V(CO)_5(NO)$ undergoes substitution via a dissociative pathway. *trans*-$V(CO)_4(PPh_3)(NO)$ undergoes carbonyl substitution exclusively via a dissociative pathway 10^6 times slower than in $V(CO)_5(NO)$ and 10^4 times faster than dissociative substitution in isoelectronic $Cr(CO)_6$. Explain these observations. (See Q.-Z. Shi, T. G. Richmond, W. C. Trogler and F. Basolo, *Inorg. Chem.* **1984**, *23*, 957.)

14.6 Solution: The associative mechanism for $V(CO)_5(NO)$ substitution with good nucleophiles discussed in DMA, 3rd ed., Section 14.1.1 apparently depends on the ability of the nitrosyl to withdraw electron density from V by bending. With poor nucleophiles, the electron density donated to V by the incoming ligand becomes less and its withdrawal by NO becomes energetically less significant. In these cases, a slower dissociative path becomes competitive and, with NEt_3, exclusive. In the case of *trans*-$V(CO)_4(PPh_3)(NO)$, the phosphine increases electron density around V and apparently saturates the ability of NO to drain off more, thus blocking an associative pathway and channeling the reaction via the slower dissociative route.

Two competing effects influence the substitution rate in $V(CO)_4(PPh_3)(NO)$ as compared to that in $Cr(CO)_6$. PPh_3 should labilize the CO *cis* to it while a CO *trans* to the excellent π-acceptor CO should be more difficult to dissociate than one *trans* to the even better π-acceptor NO^+ in $V(CO)_5(NO)$. Evidently, the former ground state feature determines the kinetics. A faster dissociative substitution than in $Cr(CO)_6$ reflects the kinetic *cis* labilization of PPh_3.

14.7 Problem: The Figure on the next page shows the cyclic voltammogram of Cp*W(CO)(p-tolyl)$_2$. Note that the voltage sweep is initially cathodic. What can you say about the electrochemical behavior of this complex? Why is such behavior reasonable?

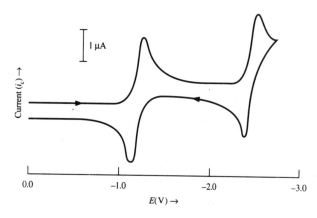

Cyclic voltammogram for Cp*W(NO)(p-tolyl)$_2$ in thf. Sweep is 400 mV/s. (Reprinted with permission from N. H. Dryden, P. Legzdins, S. J. Rettig, and J. E. Veltheer, *Organometallics*, **1992**, *11*, 2583. Copyright, 1992, American Chemical Society.)

14.7 Solution: The complex is reduced to a dianion in two steps which seem reversible since the cathodic and anodic peaks seem to be the same height. Cp*W(NO)(p-tolyl)$_2$ is an 18-e complex; however, a 20-e dianion could be avoided if NO assumed the NO$^-$ coordination mode with bending.

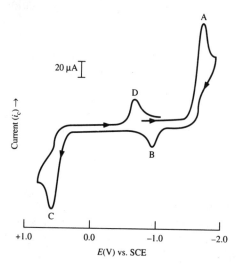

Cyclic voltammogram for [{(η^5-Me$_4$C$_5$)SiMe$_2$(MMe)}Fe(CO)$_2$]$_2$ in MeCN. Sweep is 200 mV/s. (Reprinted with permission from M. Moran, M. C, Pascual, I. Cuadrado, and J. Losada, *Organometallics*, **1993**, *12*, 811. Copyright, 1993, American Chemical Society.)

14.8 Problem: The Figure above shows the cyclic voltammogram of [{η^5-Me$_4$C$_5$)SiMe$_2$(NMe$_2$)}Fe(CO)$_2$]$_2$. Coulometric measurements show that all electrochemical reactions involve two electrons. Write the equation which corresponds to each of the peaks.

14.8 Solution: The initial sweep to negative potential results in a 2e reduction in this dimer which is isoelectronic with [CpFe(CO)$_2$]$_2$. Presumably the Fe—Fe bond is reduced:

A [{(η^5-Me$_4$C$_5$)SiMe$_2$(NMe$_2$)}Fe(CO)$_4$]$_2$ + 2e → 2[{(η^5-Me$_4$C$_5$)SiMe$_2$(NMe$_2$)}Fe(CO)$_2$]$^-$

On reversal of the scan direction reoxidation occurs:

B $2[\{(\eta^5\text{-}Me_4C_5)SiMe_2(NMe_2)\}Fe(CO)_2]^- \rightarrow [\{(\eta^5\text{-}Me_4C_5)SiMe_2(NMe_2)\}Fe(CO)_2]_2 + 2e$

The lower current of peak B as compared to peak A indicates depletion of the concentration of the anion produced at A over the time period required to sweep the voltage back to A. At yet more positive voltage, oxidation occurs, again with Fe—Fe bond rupture. Coordination of a solvent molecule gives an 18-e cation:

C $[\{(\eta^5\text{-}Me_4C_5)SiMe_2(NMe_2)\}Fe(CO)_2]_2 + 2\,MeCN \rightarrow$
$\qquad\qquad 2[\{(\eta^5\text{-}Me_4C_5)SiMe_2(NMe_2)\}Fe(CO)_2(NCMe)]^+ + 2e$

At D the cation is again reduced to the dimer. Decreased current at D as compared to C indicates reaction or decomposition of part of the cations before reduction.

D $2[\{(\eta^5\text{-}Me_4C_5)SiMe_2(NMe_2)\}Fe(CO)_2(NCMe)]^+ + 2e \rightarrow$
$\qquad\qquad [\{(\eta^5\text{-}Me_4C_5)SiMe_2(NMe_2)\}Fe(CO)_2]_2 + 2\,MeCN$

14.9 Problem: Show that the product distribution for carbonylation depicted below would be different if the mechanism involved movement of coordinated CO which inserted itself into the Mn—C bond. (This is so-called "CO insertion" mechanism for carbonylation.)

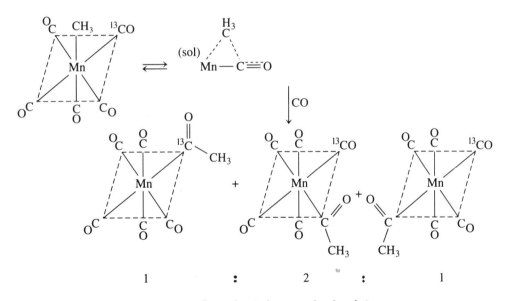

Stereochemical course of carbonylation.

14.9 Solution: If CO moved to accomplish the insertion, we would have:

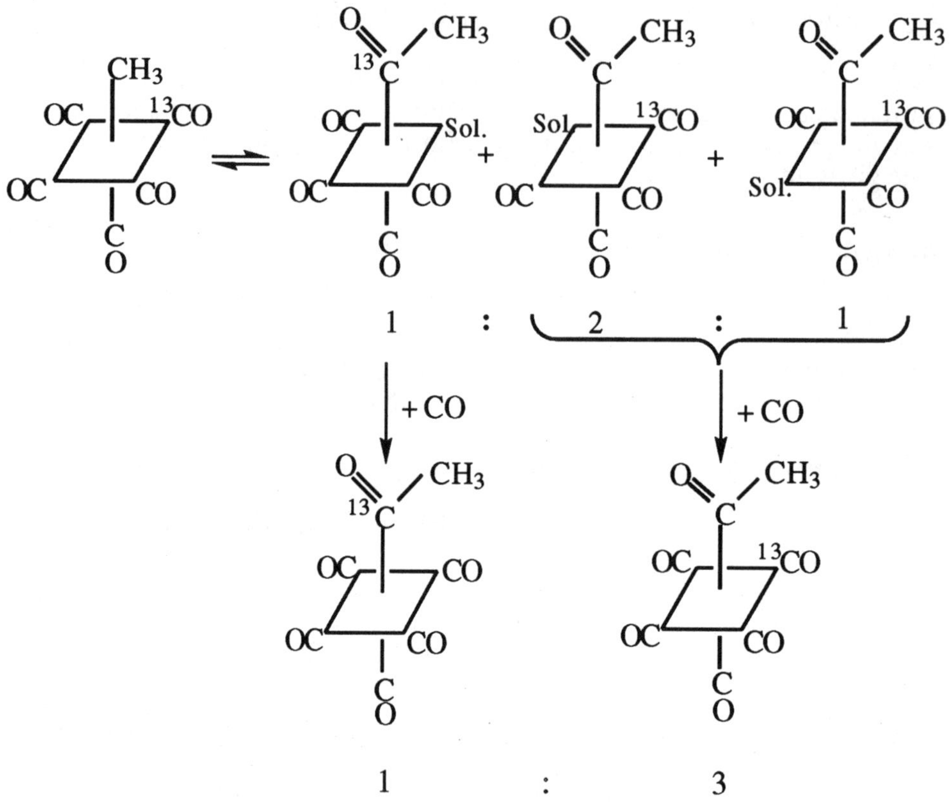

[Structural diagrams showing equilibrium of octahedral metal complexes with CH₃, CO, and ¹³CO ligands, with solvent (Sol.) replacing CO in three positions in ratios 1 : 2 : 1, followed by +CO addition yielding acetyl products in ratio 1 : 3]

No acetyl product having ^{13}CO *trans* to acetyl would be produced.

14.10 Problem: Some data for the insertion reaction $Cp^*_2Nb(H)(\eta^2\text{-}CH_2=CHR) + CNMe \rightarrow Cp^*_2Nb(CH_2CH_2R)(CNMe)$ are given below. Rate = $k_2[Cp^*_2Nb(H)(\eta^2\text{-}CH_2=CHR)][CNMe]$.

R	k_1, s^{-1} (50° C in C_6D_6)
H	2.62
Me	890
Ph	3.18
$p\text{-}NMe_2C_6H_4$	6.80
$p\text{-}MeOC_6H_4$	4.81
$p\text{-}MeC_6H_4$	3.47
$p\text{-}CF_3C_6H_4$	0.91

(a) Recognizing that k is determined by the energy difference between the ground state and the transition state, what assumption needs to be made in comparing activation parameters as a function of R?

(b) If this assumption is made, what can be said about the nature of the transition state from these data? (See. N.M. Doherty and J. E. Bercaw, *J. Am. Chem. Soc.* **1985**, *107*, 2670.)

14.10 Solution:
(a) The geometries of the ground and transition states must all be comparable.
(b) R groups which donate electrons by induction or by resonance (to a lesser extent) stabilize the transition state. This is consistent with a developing positive charge on the olefinic C to which H⁻ migrates.

14.11 Problem: Show how a reversible olefin insertion ⇌ β-elimination can lead to equilibration of D as shown below (dmpe = $Me_2PCH_2CH_2PMe_2$):

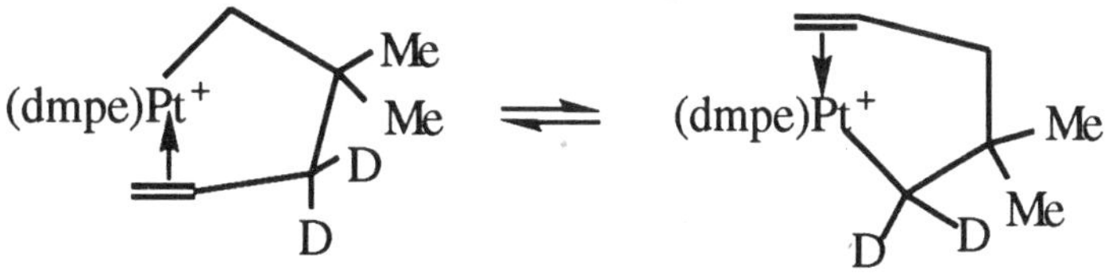

(See T. C. Flood and S. P. Bitler, *J. Am. Chem. Soc.* **1984**, *106*, 6076.)

14.11 Solution:

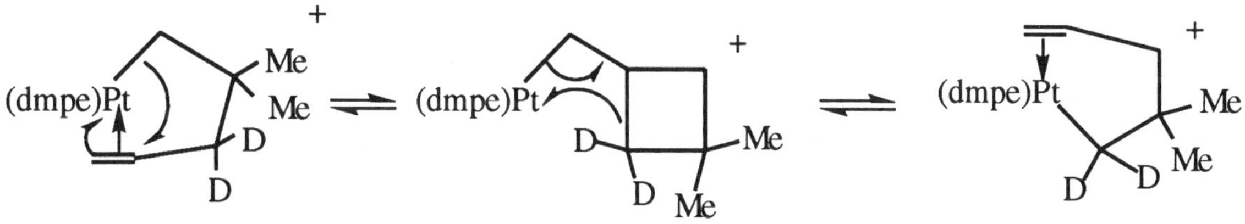

14.12 Problem: The ease of oxidative addition and the tendency toward five-coordination for d^8-metals increase as shown:

⟵ *Tendency to give five-coordination*

| Fe⁰ | Co¹ | Ni^II |
| Ru⁰ | Rh¹ | Pd^II | ↓ *Ease of oxidative addition*
| Os⁰ | Ir¹ | Pt^II |

Explain these trends.

14.12 Solution: Because additions result in formal oxidation state increase for the metal, their ease should parallel ease of removal of two metal electrons. Ionization energies decrease on descending each group. Earlier members of Group VIII (8–10) have lower effective nuclear charges. Hence, their electrons are more available for back-donation to π-acid ligands, thereby strengthening the fifth metal–ligand bond.

14.13 Problem: The rate of reaction of O_2 with *trans*-IrX(PPh$_3$)$_2$(CO) in benzene decreases in the order X = NO_2 > I > ONO_2 > Br > Cl > N_3 > F. Explain this observation. (See L. Vaska, L. S. Chen and C. V. Senoff, *Science* **1971**, *174*, 587.)

14.13 Solution: The rates decrease as the electron-withdrawing ability of X increases. This reduces the electron density on Ir and increases the enthalpy of activation since the transition state must involve transfer of electron density from Ir to O_2.

14.14 Problem: Account for the following observations. (*Hint:* β-elimination and/or reductive elimination are involved.)
(a) (dppe)Pt(CH$_2$CH$_3$)$_2$ affords C$_2$H$_6$ and C$_2$H$_4$ on heating in xylene at 150°C. (dppe = Ph$_2$PCH$_2$CH$_2$PPh$_2$)
(b) (dppe)Pt(CH$_2$CH$_3$)(OMe) affords C$_2$H$_6$ and C$_2$H$_4$ in a 40:60 ratio as well as MeOH and H$_2$C=O on heating in toluene at 100°C.
(c) (dppe)Pt(OMe)$_2$ affords MeOH and H$_2$C=O on stirring in CH$_2$Cl$_2$ at 25°C. (See H. E. Bryndza, J. C. Calabrese, M. Marsi, D. C. Roe, W. Tam and J. E. Bercaw, *J. Am. Chem. Soc.* **1986**, *108*, 4805.)

14.14 Solution:
(a) β-elimination gives C$_2$H$_4$ and (dppe)Pt(H)(C$_2$H$_5$) which can reductively eliminate C$_2$H$_6$ leaving dppe and Pt^0.
(b) β-elimination gives C$_2$H$_4$ and (dppe)Pt(H)(OMe) or H$_2$C=O (if β-elimination occurs from OMe) and (dppe)Pt(H)(C$_2$H$_5$). Reductive elimination from the two Pt complexes affords MeOH and C$_2$H$_6$, respectively. The 40:60 ratio of hydrocarbons is determined by the relative rates of β-elimination from methoxide and ethyl ligands, respectively.
(c) β-elimination gives (dppe)Pt(H)(OMe) and H$_2$C=O. Reductive elimination affords MeOH and Pt^0 products.

14.15 Problem: Propose a mechanism for the stoichiometric decarbonylation of C$_6$H$_5$CH$_2$C(O)Cl by Rh(PPh$_3$)$_3$Cl giving benzyl chloride. Keep in mind the 16- and 18-electron rule.

14.15 Solution:

Rh(PPh$_3$)$_3$Cl + C$_6$H$_5$CH$_2$C(O)Cl → C$_6$H$_5$CH$_2$C(O)Rh(PPh$_3$)$_3$Cl$_2$	oxidative addition
C$_6$H$_5$CH$_2$C(O)Rh(PPh$_3$)$_3$Cl$_2$ ⇌ C$_6$H$_5$CH$_2$C(O)Rh(PPh$_3$)$_2$Cl$_2$ + PPh$_3$	ligand dissociation
C$_6$H$_5$CH$_2$C(O)Rh(PPh$_3$)$_2$Cl$_2$ → C$_6$H$_5$CH$_2$Rh(CO)(PPh$_3$)$_2$Cl$_2$	alkyl migration
C$_6$H$_5$CH$_2$Rh(CO)(PPh$_3$)$_2$Cl$_2$ → C$_6$H$_5$CH$_2$Cl + Rh(CO)(PPh$_3$)$_2$Cl	reductive elimination

14.16 Problem: Generation of Cp*Rh(PMe$_3$) in a mixed benzene/propane solvent at –55°C results in a 4:1 ratio of Cp*Rh(PMe$_3$)(C$_6$H$_5$)(H): Cp*Rh(PMe$_3$)(C$_3$H$_7$)(H).

For the elimination reaction

Cp*Rh(PMe$_3$)(C$_6$H$_5$)(H) → Cp*Rh(PMe$_3$) + C$_6$H$_6$

$\Delta H^{\ddagger}$ = 128 kJ/mol; $\Delta S^{\ddagger}$ = 62.3 J/molK; and $\Delta G^{\ddagger}$ (–17°C) = 111.7 kJ/mol

Cp*Rh(PMe$_3$)(C$_3$H$_7$)(H) → Cp*Rh(PMe$_3$) + C$_3$H$_8$

$\Delta G^{\ddagger}$ (–17°C) = 77.8 kJ/mol

(a) Show that the difference in activation energy for oxidative addition of benzene and propane to Cp*Rh(PMe)$_3$ is 2.5 kJ as depicted in the Figure.
(b) Verify the value for K_{eq} at –17° C as 2.7 x 10^7.

14.16 Solution: **(a)** At the low temperature (–55°C) at which Cp*Rh(PMe$_3$) is generated by irradiation, the relative amounts of products are determined by the relative rate constants for each separate reaction. Each rate constant is proportional to exp(–$\Delta G^\ddagger$/RT). So, we have

$$\frac{[Cp*Rh(PMe_3)(Ph)H]}{[Cp*Rh(PMe_3)(C_3H_7)H]} = \frac{e^{-\Delta G^\ddagger_{benzene}/RT}}{e^{-\Delta G^\ddagger_{propane}/RT}} = e^{(\Delta G^\ddagger_{propane} - \Delta G^\ddagger_{benzene})/RT} = 4$$

and ($\Delta G^\ddagger$ propane – $\Delta G^\ddagger$ benzene)/RT = ln 4. Thus,
$\Delta G^\ddagger$ propane – $\Delta G^\ddagger$ benzene = 2.5 kJ/mol.

(b) For the reaction
Cp*Rh(PMe$_3$)(C$_3$H$_7$)(H) + benzene ⇌ Cp*Rh(PMe$_3$)(C$_6$H$_5$)(H) + propane

$$K_{eq} = \frac{[Cp*Rh(PMe_3)(C_6H_5)(H)][propane]}{[Cp*Rh(PMe_3)(C_3H_7)(H)][benzene]}$$

From the data for the reductive elimination on the Figure,
$$-\Delta G^0 = -[\Delta G^\ddagger_{benzene} + 2.5kJ - \Delta G^\ddagger_{propane}]$$
$$= -[111.7 \text{ kJ} + 2.5 \text{ kJ} - 77.8 \text{ kJ}]/mol$$
$$= -36.4 \text{ kJ/mol} = -RT \ln K_{eq}$$

$$\ln K_{eq} = \frac{36.4 \text{ kJ/mol}}{(98.31 \text{ J/molK})(256 \text{ K})} = 17.1$$

$$K_{eq} = 2.7 \times 10^7$$

A more refined treatment would correct the difference in free energies for addition to unsaturated Cp*Rh(PMe$_3$) to a temperature of –17°C giving $\Delta\Delta G^* = 2.9$ kJ/mol and $\Delta G^0 = -36.8$ kJ/mol with $K_{eq} = 3.3 \times 10^7$.

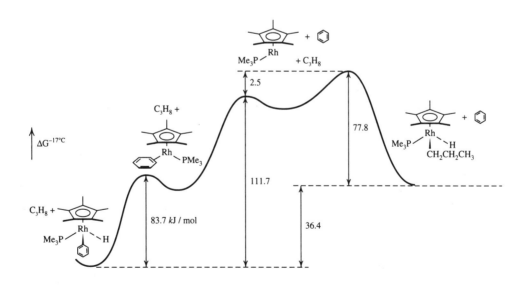

Energy profile for oxidative addition to Cp*Rh(PMe$_3$) and reductive elimination from Cp*Rh(PMe$_3$)(R)(H).
(Reprinted with permission from W. D. Jones and F. J. Feher, *Acc. Chem. Res.* **1989**, *22*, 91.
Copyright American Chemical Society, 1989.)

14.17 Problem: Write a cycle for the reaction of $(CH_3)_2C=CHCH_2O_2CCH_3$ with $Na^+CH(CO_2Et)_2^-$ catalyzed by $Pd(PPh_3)_4$. (*Hint:* The mechanism involves oxidative addition of the acetate to generate an η^3-allyl complex.)

14.17 Solution:
$Pd(PPh_3)_4 \rightarrow Pd(PPh_3)_3 + PPh_3$
$Pd(PPh_3)_3 \rightarrow Pd(PPh_3)_2 + PPh_3$
$Pd(PPh_3)_2 + (CH_3)_2C=CHCH_2O_2CCH_3 \rightarrow$
$\qquad\qquad\qquad [\eta^3\text{-}(CH_3)_2CCHCH_2]Pd(PPh_3)(O_2CCH_3) + PPh_3$
$[\eta^3\text{-}(CH_3)_2CCHCH_2]Pd(PPh_3)(O_2CCH_3) + Na^+CH(CO_2Et)_2^- + PPh_3 \rightarrow$
$\qquad Na^+\{[\eta^2\text{-}(CH_3)_2C=CHCH_2(CH(CO_2Et)_2)]Pd(PPh_3)_2(O_2CCH_3)\}^-$
$Na^+\{[\eta^2\text{-}(CH_3)_2C=CHCH_2(CH(CO_2Et)_2)]Pd(PPh_3)_2(O_2CCH_3)\} \rightarrow$
$(CH_3)_2C=CHCH_2(CH(CO_2Et)_2) + Na^+(O_2CCH_3)^- + Pd^0$

14.18 Problem: Predict the products of the following reactions:
(a) *cis*-$PtCl_2(PPh_3)(\eta^2\text{-}C_2H_4) + xs\ NMe_2H$
(b) $[CpMo(CO)(NO)(\eta^3\text{-}C_8H_{13})]^+ + PhS^-$
(c) $(\eta^4\text{-}C_4H_6)Fe(CO)_3 + HBF_4$
(d) $(\eta^4\text{-}C_6H_8)Fe(CO)_3 + Ph_3C^+BF_4^-$
(e) $(\eta^4\text{-}C_6H_8)Fe(CO)_3 + Li^+CH_2CN^-$
(f) $[\eta^5\text{-}(6\text{-phenylhexadienyl})Mn(CO)_2(NO)]^+ + PPh_3$
(g) $[(\eta^5\text{-}C_6H_7)Fe(CO)_3]^+ + I^-$
(h) $[(\eta^6\text{-}C_6H_6)Mn(CO)_3]^+ + NaBH_4$
(i) $[(\eta^7\text{-}C_7H_7)W(CO)_3]^+ + MeO^-$

14.18 Solution:
(a) *cis* -$PtCl_2(PPh_3)(CH_2CH_2NMe_2H)$; nucleophilic attack on coordinated ethylene.
(b) $CpMo(CO)(NO)[1,2\text{-}\eta^2\text{-}3\text{-}(SPh)C_8H_{13}]$; attack on allyl $\rightarrow$ diene
(c) $[(\eta^3\text{-}CH_2CHCHCH_3)Fe^{II}(CO)_3]^+BF_4^-$; electrophilic attack on Fe^0 with oxidation to Fe^{II} and reduction to H^-; hydride transfer to allyl$^-$.
(d) $[(\eta^5\text{-}C_6H_7)Fe(CO)_3]^+BF_4^- + Ph_3CH$; H^- abstraction; an additional C is placed in conjugation.
(e) $Li^+[(\eta^3\text{-}1\text{-}(CH_2CN)C_6H_8)Fe(CO)_3]^-$; nucleophilic attack on coordinated diene $\rightarrow$ allyl.
(f) $[(\eta^5\text{-}PhC_6H_6)Mn(CO)(PPh_3)(NO)]^+ + CO$; nucleophilic displacement of CO.
(g) $[(\eta^5\text{-}C_6H_7)Fe(CO)_2I + CO$; nucleophilic displacement of CO.
(h) $(\eta^5\text{-}C_6H_7)Mn(CO)_3 + Na^+$; nucleophilic addition to coordinated "triene" $\rightarrow$ dienyl.
(i) $[\eta^6\text{-}C_7H_7(OMe)]W(CO)_3$; nucleophilic attack on trienyl$^+$ $\rightarrow$ triene.

14.19 Problem: Exchange of ^{17}O in the reaction $+ H_2^{17}O \rightarrow [Re(CO)_5(C^{17}O)]^+ + H_2O$ obeys the rate law, rate = $k [Re(CO)_6^+][H_2O]^2$. The experiments were conducted by adding measured amounts of ^{17}O-enriched water to a MeCN solution of $[Re(CO)_6]PF_6$. Addition of OH^- does not appreciably increase the rate. (See R. L. Kump and L. J. Todd, *Inorg. Chem.* **1981**, *20*, 3715.)

14.19 Solution: Ligands coordinated to positively charged metals are subject to nucleophilic attack. The results on addition of OH^- indicates that hydroxide ion is not the principal attacking species. A possible mechanism is:

$$Re(CO)_6^+ + H_2^{17}O \rightleftharpoons (CO)_5ReC(=O)-^{17}OH_2 \quad K$$

$$(CO)_5ReC(=O^+)(^{17}OH_2) + H_2O \rightarrow (CO)_5ReC(=O)(^{17}OH) + H_3O^+ \quad k$$

$$(CO)_5ReC(=O)(^{17}OH) \rightarrow (CO)_5ReC(OH)(=^{17}O) \quad \text{fast}$$

$$(CO)_5ReC(OH)(=^{17}O) + H_3O^+ \rightarrow (CO)_5Re(C^{17}O)^+ + 2 H_2O \quad \text{fast}$$

rate = $k[(CO)_5ReC(=O^+)(^{17}OH_2)][H_2O]$

$K = \dfrac{[Re(CO)_5C(=O^+)-^{17}OH_2]}{[Re(CO)_6^+][H_2^{17}O]}$

rate = $kK [Re(CO)_6^+][H_2^{17}O][H_2O]$

But $[H_2^{17}O] = f[H_2O]$ where f is the fraction containing ^{17}O, so

rate = $kKf [Re(CO)_6^+][H_2O]^2 = k_{obs}[Re(CO)_6^+][H_2O]^2$

14.20 Problem: Give electron counts for all the species postulated to be involved in the catalytic cycle for hydroformylation (see the Figure on the next page).

14.20 Solution:

Sixteen Electrons	Eighteen Electrons
$HCo(CO)_3$	$Co_2(CO)_8$
$RCH_2CH_2Co(CO)_3$	$HCo(CO)_4$
$CH_3\overset{R}{C}HCo(CO)_3$	$HCo(CO)_3(\eta^2\text{-}CH_2=CHR)$

$$\text{RCH}_2\text{CH}_2\overset{\overset{O}{\|}}{\text{C}}\text{Co(CO)}_3 \qquad \underset{\underset{}{}}{\overset{R}{|}}\text{CH}_3\text{CHCo(CO)}_4$$

$$\text{CH}_3\overset{}{\underset{\underset{R}{|}}{\text{CH}}}\overset{\overset{O}{\|}}{\text{C}}\text{Co(CO)}_3 \qquad \text{RCH}_2\text{CH}_2\overset{\overset{O}{\|}}{\text{C}}\text{Co(CO)}_4$$

$$\text{CH}_3\underset{\underset{R}{|}}{\text{CH}}\overset{\overset{O}{\|}}{\text{C}}\text{Co(CO)}_4$$

$$\text{RCH}_2\text{CH}_2\overset{\overset{O}{\|}}{\text{C}}\text{Co(H)}_2(\text{CO})$$

$$\text{CH}_3\underset{\underset{R}{|}}{\text{CH}}\overset{\overset{O}{\|}}{\text{C}}\text{Co(H)}_2(\text{CO})_3$$

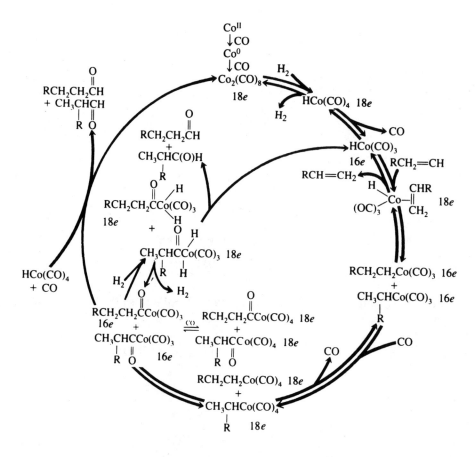

Catalytic cycle for hydroformylation of olefins.

14.21 Problem: See the cycle on the previous page. Kinetic studies indicate that hydroformylation reaction rate is enhanced by an increase in H_2 pressure and retarded by an increase in CO pressure. How is the mechanism proposed in Chapter 14 DMA, 3rd ed. consistent with these observations?

14.21 Solution: H_2 pressure leads to conversion of $Co_2(CO)_8$ to the active catalyst $HCo(CO)_4$. CO pressure leads to conversion of 16-e tricarbonyl species to stable 18-e tetracarbonyl species thus preventing ß-elimination and/or H_2 oxidative addition steps. However, CO is required for the reaction stoichiometry.

14.22 Problem: What products would you expect from hydroformylation of $C_3H_7C(CH_3)DCH=CH_2$? Show how each is obtained.

14.22 Solution: Of importance here is the reversibility of the ß-elimination.

$$HCo(CO)_4 \rightleftharpoons HCo(CO)_3 + CO$$

$$C_3H_7CD(CH_3)CH=CH_2 + HCo(CO)_3 \rightleftharpoons H(CO)_3Co \leftarrow \| \begin{array}{c} CH_2 \\ C \\ H \quad C(D)(CH_3)(C_3H_7) \end{array}$$

$$H(CO)_3Co \leftarrow \| \begin{array}{c} CH_2 \\ C \\ H \quad C(D)(CH_3)(C_3H_7) \end{array} \rightleftharpoons C_3H_7CD(H_3C\text{-}Co(CO)_3)CHCH_3 + C_3H_7CD(CH_3)CH_2CH_2Co(CO)_3$$

$$C_3H_7CD(H_3C\text{-}Co(CO)_3)CHCH_3 + CO \rightleftharpoons C_3H_7CD(H_3C\text{-}Co(CO)_4)CHCH_3 \rightleftharpoons C_3H_7CD(H_3C)CHCCo(CO)_3\,(=O)$$

$$C_3H_7CD(CH_3)CH_2CH_2Co(CO)_3 + CO \rightleftharpoons C_3H_7CD(CH_3)CH_2CH_2Co(CO)_4 \rightleftharpoons C_3H_7CD(CH_3)CH_2CH_2CCo(CO)_3\,(=O)$$

$$C_3H_7CD(H_3C)CHCCo(CO)_3\,(=O) + H_2 \rightleftharpoons C_3H_7CD(H_3C)CHCC(OH)(H)Co(CO)_3 \rightarrow \boxed{C_3H_7CD(H_3C)CHCH(=O)(CH_3)} + HCo(CO)_3$$

$$C_3H_7CD(CH_3)CH_2CH_2CCo(CO)_3\,(=O) + H_2 \rightleftharpoons C_3H_7CD(CH_3)CH_2CH_2C(OH)(H)Co(CO)_3 \rightarrow \boxed{C_3H_7CD(CH_3)CH_2CH_2CH(=O)} + HCo(CO)_3$$

β-elimination and olefin insertion can occur in the opposite sense to move the D to the next C:

$$C_3H_7CDCHCH_3 \rightleftharpoons D(CO)_3Co \leftarrow \| \rightleftharpoons C_3H_7CCDHCH_3$$
$$\quad | \qquad\qquad\qquad\qquad\qquad\qquad | $$
$$H_3CCo(CO)_3 \qquad\qquad\qquad\qquad H_3C$$

(with intermediate showing H₃C, H on one C and H₃C, C₃H₇ on the other C of the olefin coordinated to D(CO)₃Co; product has Co(CO)₃ group)

$$\underset{\underset{CDHCH_3}{|}}{\overset{\overset{CH_3}{|}}{C_3H_7C-Co(CO)_3}} + CO \rightleftharpoons \underset{\underset{CDHCH_3}{|}}{\overset{\overset{CH_3}{|}}{C_3H_7C-Co(CO)_4}} \rightleftharpoons \underset{\underset{CDHCH_3}{|}}{\overset{\overset{H_3C\ O}{|\ \|}}{C_3H_7C-CCo(CO)_3}}$$

$$\underset{\underset{CDHCH_3}{|}}{\overset{\overset{H_3C\ O}{|\ \|}}{C_3H_7C-CCo(CO)_3}} + H_2 \rightleftharpoons \underset{\underset{CH_3HDC}{|}}{\overset{\overset{H_3C\ O\ H}{|\ \|\ |}}{C_3H_7C-CCo(CO)_3}} \rightarrow \boxed{\underset{\underset{CDHCH_3}{|}}{\overset{\overset{H_3C\ O}{|\ \|}}{C_3H_7C-CH}}} + HCo(CO)_3$$

This product, which has the aldehyde function coordinated to a tertiary C, is not actually observed. (R. Heck and D. S. Breslow, *J. Am. Chem. Soc.* **1961**, *88*, 4023.) The likely reason is the instability of the tertiary alkyl precursor which can undergo ß-elimination and reinsertion leading to a primary alkyl:

$$\underset{\underset{CDHCH_3}{|}}{\overset{\overset{CH_3}{|}}{C_3H_7C-Co(CO)_3}} \rightleftharpoons HCo(CO)_3 \leftarrow \| \rightleftharpoons \underset{\underset{CHDCH_3}{|}}{C_3H_7CHCH_2Co(CO)_3}$$

(intermediate olefin CH₂=C with H₇C₃ and CHDCH₃ substituents)

$$\underset{\underset{CHDCH_3}{|}}{C_3H_7CHCH_2Co(CO)_3} + CO \rightleftharpoons \underset{\underset{CHDCH_3}{|}}{C_3H_7CHCH_2Co(CO)_4} \rightleftharpoons \underset{\underset{CHDCH_3}{|}}{\overset{\overset{O}{\|}}{C_3H_7CHCH_2CCo(CO)_3}}$$

$$\underset{\underset{CHDCH_3}{|}}{\overset{\overset{O}{\|}}{C_3H_7CHCH_2CCo(CO)_3}} + H_2 \rightleftharpoons \underset{\underset{CHDCH_3}{|}}{\overset{\overset{O}{\|}}{C_3H_7CHCH_2CCo(CO)_3}} \rightarrow \boxed{\underset{\underset{CHDCH_3}{|}}{\overset{\overset{O}{\|}}{C_3H_7CHCH_2CH}}} + HCo(CO)_3$$

14.23 Problem: $Ni[P(OEt)_3]_4$ reacts with H^+ affording $HNi[P(OEt)_3]_4^+$.
(a) Formulate the electronic structure of this cation.
(b) The cation is known to catalyze olefin isomerization. Write a cycle for the isomerization of 1-butene catalyzed by $HNi[P(OEt)_3]_4^+$.

14.23 Solution:
(a) $Ni^{II} + H^- + 4\,P(OEt)_3$
    $8e^- + 2e + 4 \times 2e = 18e$

(b) The cation {HNi[P(OEt)$_3$]$_4$}$^+$ could be regarded as containing protonated Ni^0 or as a hydride complex of Ni^{II} (d^8). The latter is probably the better formulation since the insertion of olefin into the Ni–H bond postulated in the cycle below is known with other d^8 hydrides.

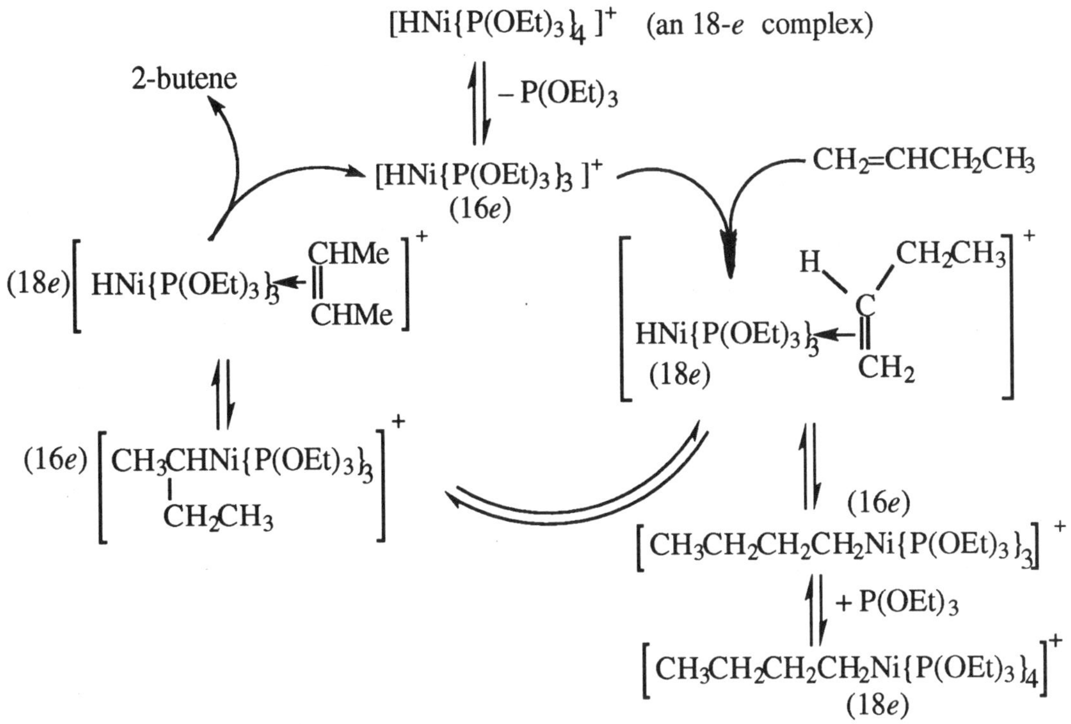

[HNi{P(OEt)$_3$}$_4$]$^+$ (an 18-e complex)

$\rightleftharpoons$ –P(OEt)$_3$

2-butene

[HNi{P(OEt)$_3$}$_3$]$^+$ (16e)

CH$_2$=CHCH$_2$CH$_3$

(18e) [HNi{P(OEt)$_3$}$_3$(CHMe=CHMe)]$^+$

$\left[\text{HNi\{P(OEt)}_3\}_3 \leftarrow \| \begin{matrix} \text{H} & \text{CH}_2\text{CH}_3 \\ \text{C} \\ \text{CH}_2 \end{matrix} \right]^+$ (18e)

(16e) [CH$_3$CHNi{P(OEt)$_3$}$_3$ | CH$_2$CH$_3$]$^+$

[CH$_3$CH$_2$CH$_2$CH$_2$Ni{P(OEt)$_3$}$_3$]$^+$ (16e)

$\rightleftharpoons$ +P(OEt)$_3$

[CH$_3$CH$_2$CH$_2$CH$_2$Ni{P(OEt)$_3$}$_4$]$^+$ (18e)

14.24 Problem: Cyclopentene undergoes *R*ing *O*pening *M*etathesis *P*olymerization (ROMP) with a catalyst of R$_3$Al/WCl$_6$, producing $\left(\begin{matrix} \text{H} \\ \text{C}=\text{C} \\ \text{H} \quad \text{(CH}_2\text{)}_3 \end{matrix} \right)_n$ Write the mechanism for the reaction.

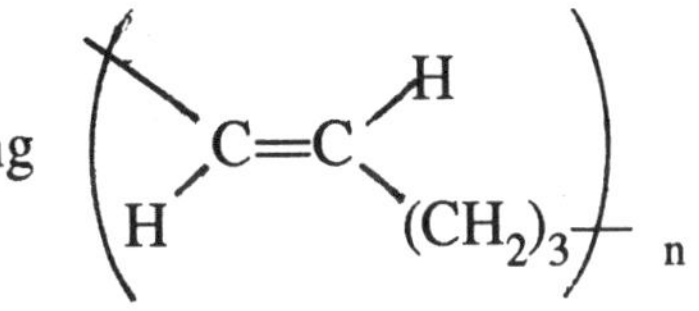

14.24 Solution:

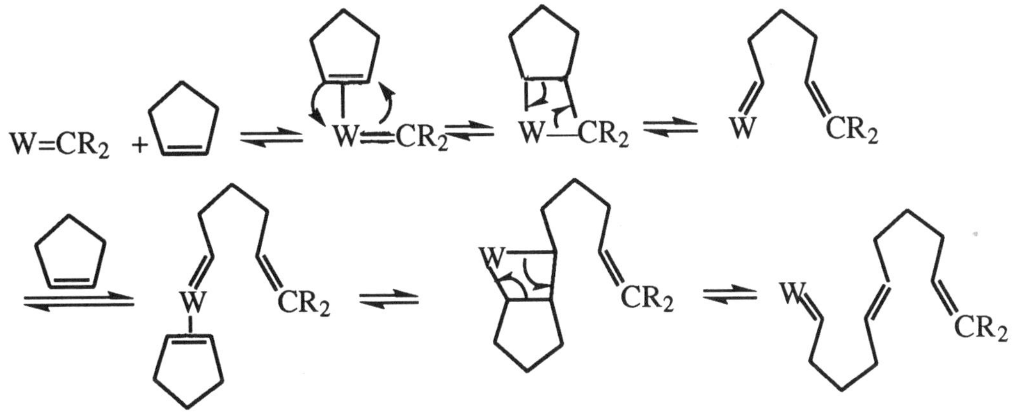

W=CR$_2$ + ⬠ ⇌ W=CR$_2$(⬠) ⇌ W–CR$_2$(metallacycle) ⇌ W=CH–(CH$_2$)$_3$–CH=CR$_2$

⇌ W=CH–(CH$_2$)$_3$–CH=CR$_2$ (with cyclopentene coordinated) ⇌ (metallacycle intermediate) CR$_2$ ⇌ W=CH–(CH$_2$)$_3$–CH=CH–(CH$_2$)$_3$–CH=CR$_2$

, etc.

14.25 Problem: Write a catalytic cycle for the production of acetone from propylene via the Wacker process.

14.25 Solution:

$$Pd^{2+} + CH_3CH=CH_2 \rightleftharpoons \left[Pd-\parallel\overset{H\diagdown\underset{C}{}\diagup CH_3}{\underset{CH_2}{}} \right]^{2+}$$

$$\left[Pd-\parallel\overset{H\diagdown\underset{C}{}\diagup CH_3}{\underset{CH_2}{}} \right]^{2+} + OH^- \rightleftharpoons \left[Pd-\underset{HO}{\overset{H}{\underset{|}{C}}}-H \atop CHCH_3 \right]^+ \rightleftharpoons \left[Pd-\underset{HO}{\overset{H}{\underset{|}{}}}\parallel\overset{CH_2}{\underset{C}{\diagdown CH_3}} \right]^+$$

$$\left[Pd-\underset{HO}{\overset{H}{\underset{|}{}}}\parallel\overset{CH_2}{\underset{C}{\diagdown CH_3}} \right]^+ \rightarrow CH_3\overset{O}{\overset{\parallel}{C}}CH_3 + H^+ + Pd^0$$

$$Pd^0 + 2Cu^{2+} \rightarrow Pd^{2+} + 2Cu^+$$
$$2Cu^+ + \tfrac{1}{2}O_2 + 2H^+ \rightarrow 2Cu^{2+} + H_2O$$

14.26 Problem: Write a catalytic cycle for the production of methyl acetate via the Monsanto acetic acid synthesis.

14.26 Solution: The reaction is run in a medium with high MeOH:H$_2$O ratio. The cycle will be the same as that shown on the next page for acetic acid except that the elimination step involves the reaction

$$CH_3\overset{O}{\overset{\parallel}{C}}OH + CH_3OH \rightarrow CH_3\overset{O}{\overset{\parallel}{C}}OCH_3 + H_2O$$

This reaction tends to occur as water is depleted in the reaction mixture, and so methyl acetate becomes a byproduct.

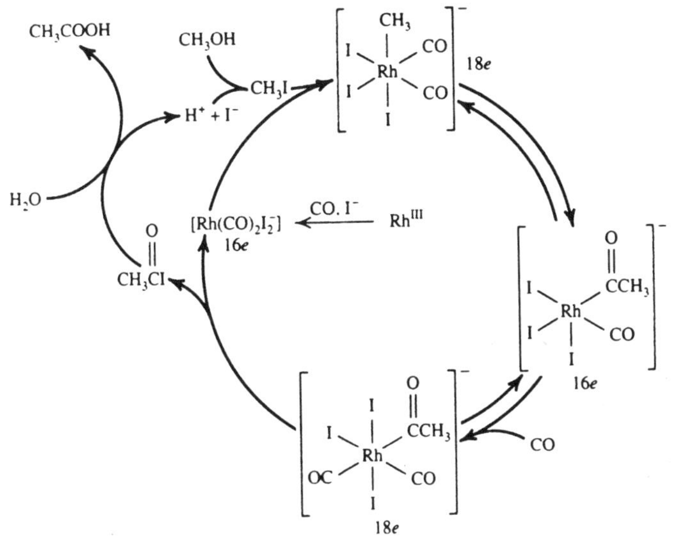

Catalytic cycle for the Monsanto acetic acid synthesis

14.27 Problem: Propose a mechanism for the following reaction:
$IrCl_3(PEt_3)_3 + C_2H_5OH + KOH \rightarrow HIrCl_2(PEt_3)_3 + CH_3CHO + KCl + H_2O$. (*Hint:* The mechanism involves ß-elimination.) (See J. Chatt and B. L. Shaw, *Chem. Ind. (London)* **1960**, 931.)

14.27 Solution:

$C_2H_5OH + KOH \leftrightarrows K^+(OC_2H_5)^- + H_2O$	deprotonation
$IrCl_3(PEt_3)_3 + K^+(OC_2H_5)^- \rightarrow Ir(OC_2H_5)Cl_2(PEt_3)_3 + KCl$	ligand displacement
$Ir(OC_2H_5)Cl_2(PEt_3)_3 \leftrightarrows Ir(OC_2H_5)Cl_2(PEt_3)_2 + PEt_3$	dissociation

$$Ir(OC_2H_5)Cl_2(PEt_3)_2 \rightarrow Cl_2Ir(H)(PEt_3)_2 \leftarrow \| \atop CH-CH_3 \atop O \quad\quad \text{ß-elimination}$$

$$Cl_2Ir(H)(PEt_3)_2 \leftarrow \| \atop CH-CH_3 \atop O \; + PEt_3 \rightarrow CH_3CHO + Ir(H)Cl_2(PEt_3)_3 \quad \text{substitution}$$

13.28 Problem: Discuss the addition of H_2 to Vaska's compound from the viewpoint of orbital symmetry.

13.28 Solution: The oxidative additions involve transfer of electron density from the highest occupied molecular orbital (HOMO) of Vaska's compound to the lowest unoccupied MO (LUMO) of the adding molecule. The LUMO of H_2 is σ_u^* having two lobes of opposite sign. The activated complex for H_2 addition must involve overlap of orbitals having like sign. If we imagine H_2 to approach $Ir(CO)(PPh_3)_2Cl$ along the z-axis, in-phase overlap could occur between lobes of the filled d_{xz} or d_{yz} orbital of Ir and the σ_u^* of H_2. This arrangement would lead to *cis* addition of H_2. On the other hand approach along the x or y axis of H_2 could lead to favorable overlap of σ_u^* of H_2 with Ir d_{xz} or d_{yz}, resulting in addition of one H above and one below the molecular plane (*trans* addition), but this pathway is not as attractive from steric considerations. Reversible *cis* addition of H_2 to vaska's compound is observed.

A third possibility would be electron transfer via overlap of the Ir d_{z^2} orbital with the + lobe of the σ_u^* orbital of H_2. This would lead to breaking the H_2 bond with bonding of H^+ to Ir and departure of H^- which could then attack the cation. This is energetically very unfavorable for H_2, but possible in solution for alkyl halides, which are found to undergo *trans* addition.

14.29 Problem: The complex *trans*-$[(PPh_3)_2PtBr(\eta^1-C_7H_7)]$ has been prepared recently.
(a) Draw its structure and give the electron count.
(b) When this compound is allowed to react with $Ph_3C^+BF_4^-$, a product of formula $[(PPh_3)_2Pt(C_7H_6)]^+BF_4^-$ is obtained. What are reasonable possible structures for this product?
(c) Spectroscopic data for the C_7H_6 ligand are given below. (Coupling to ^{195}Pt is ignored.)

^1H NMR δ (relative intensity) 8.89 (2H), 8.09 (2H), 7.42 (2H).
^{13}C NMR δ 210.5, 162.5, 143.5, 146.5.

What is the structure of the compound? (See Z. Lu, W. M. Jones, and W. R. Winchester, *Organometallics* **1993**, *12*, 1344.)

14.29 Solution:
(a)

Br—Pt(PPh₃)(PPh₃)—(C₇H₇) or Br—Pt(PPh₃)(PPh₃)—(C₇H₇) or Br—Pt(PPh₃)(PPh₃)—(C₇H₇)

On the basis of the information provided, it is not possible to distinguish which C in the ring is saturated. For the electron count, we have:

Pt^{II} + $2PPh_3$ + Br^- + $C_7H_7^-$
8e + 2 x 2e + 2e + 2e = 16e

(b) Ph_3C^+ abstracts a hydride from the $C_7H_7^-$ ring giving neutral C_7H_6 and Ph_3CH. We first consider what VB structures can be drawn for C_7H_6. Hydride abstraction places an additional π orbital on C into conjugation, but there are no additional π electrons since both C—H bonding electrons are removed with H^-. Possible structures are

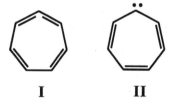

I   **II**

I is a bent allene structure while **II** is a carbene. Each could act as a 2-*e* donor given a 16-*e* cation. Resonance structures are possible for both **I** and **II**, but bonding to Pt will select out just one canonical structure. Possibilities are

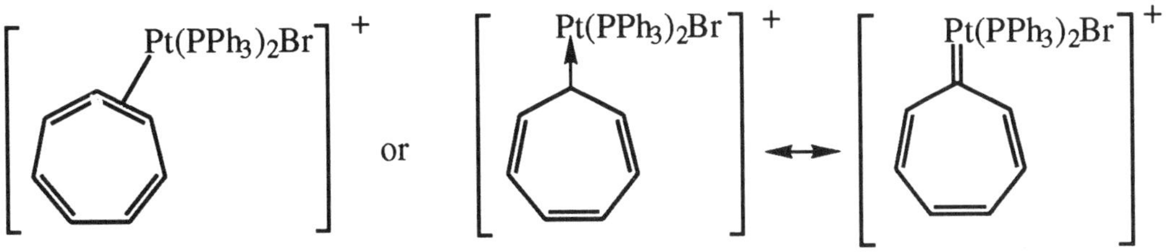

(c) There is no reason to asssume that PPh$_3$ ligands are not still *trans*. The ^{1}H NMR data indicate three types of olefinic H. This is not compatible with structure **III** unless the Pt flips back and forth to the other allene double bond faster than the NMR measurement time scale. The ^{1}H NMR data are compatible with the two canonical forms of structure **IV**. Definitive evidence in favor of **IV** is provided by the ^{13}C signal at 210.5 which is indicative of a carbene C.

14.30 Problem: When *cis*-[(CO)$_4$(CN*p*-C$_6$H$_4$Me)ReCH$_2$C$_6$H$_4$*p*-Cl] is refluxed in toluene, the product has the structure shown below.

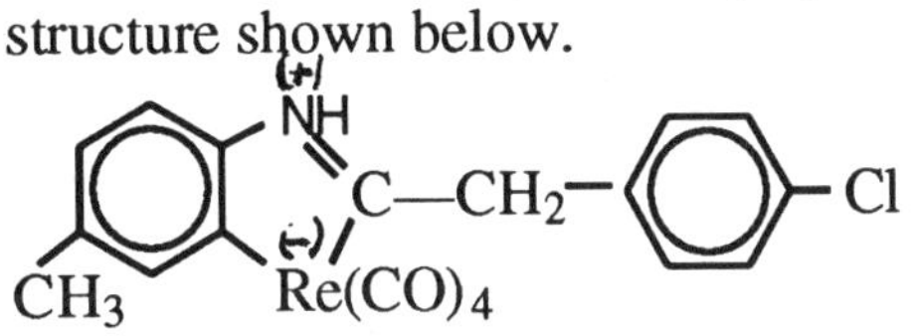

Propose a mechanism for formation of this product. (L. L. Padolik, J. J. Alexander, and D. M. Ho, unpublished observations.)

14.30 Solution: The imino group could be produced by migratory insertion of isocyanide (RNC) which is isoelectronic with CO.

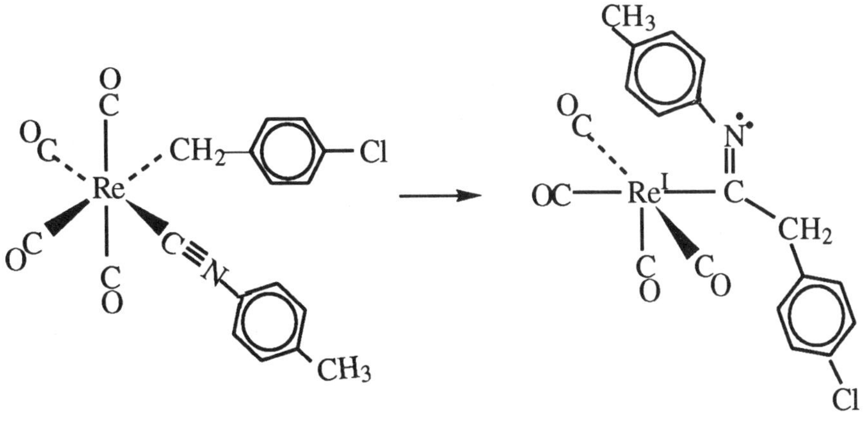

Oxidative addition (*ortho*-metallation) could then add a C—H bond to Re giving

[Structure: aryl-Re(CO)$_4^{III}$(H) complex with C=N-aryl(4-CH$_3$) and CH$_2$-C$_6$H$_4$-Cl groups]

H$^+$ could then be transferred to $\ce{>N:}$ leaving the two electrons behind to give the product.

[Structure: product with N(+)-H, C-CH$_2$-C$_6$H$_4$-Cl, and Re^I(CO)$_4$(-) showing metallacycle]

VI. SELECTED TOPICS
15
Metals
Chemistry and Periodic Trends

15.1 Problem: Write balanced equations for the preparation of
(a) Na_2CO_3 (b) MgF   (c) $TiBr_3$ (d) $ZrCl_3$ (e) $VOCl_3$ (f) WCl_6

15.1 Solution: (a) Na_2CO_3 is the cheapest strong base. It is made efficiently from inexpensive starting materials (limestone and NaCl). Limestone ($CaCO_3$) provides CaO and CO_2. The reaction of CO_2, NH_3, and NaCl depends on the low solubility of $NaHCO_3$.

$$CO_2 + NH_3 + H_2O + NaCl \rightarrow NaHCO_3(s) + NH_4Cl(aq)$$

$$2NaHCO_3(s) \xrightarrow{\Delta} Na_2CO_3 + H_2O + CO_2 \text{ (recycle)}$$

$$2NH_4Cl(aq) + Ca(OH)_2 \rightarrow CaCl_2 + 2H_2O + 2NH_3 \text{ (recycle)}$$

(b) The direct reaction of Mg and a limited supply of F_2 should produce some MgF in the gas phase at high temperature. It disproportionates to form $Mg + MgF_2(s)$ on cooling.

$$Mg + \tfrac{1}{2} F_2 \rightarrow MgF(g) \text{ stable at high T}$$

(c) Ti^{IV} compounds are hydrolyzed in aqueous solution. $TiBr_4$ can be reduced to $TiBr_3$ by H_2 at 500 to 1000°C.

$$TiBr_4 + \tfrac{1}{2}H_2 \xrightarrow{\Delta} TiBr_3 + HBr$$

(d) $ZrCl_4$ can be reduced to $ZrCl_3$ by Zr at elevated temperature.

$$3ZrCl_4)(s) + Zr \xrightarrow{\Delta} 4ZrCl_3(s)$$

(e) Vanadium is oxidized by a strong oxidizing agent to VO^{3+} and, in the presence of HCl, $VOCl_3$ is formed.

$$3V + 5HNO_3(dil) + 9HCl \rightarrow 3VOCl_3 + 5NO + 7H_2O$$

(f) The free elements combine to form WCl_6 at moderate temperature. At higher temperatures

WCl_5, WCl_4, and WCl_3 are formed since their ΔH_f^0's are less negative.
$$W + 3Cl_2 \xrightarrow{\Delta} WCl_6$$

15.2 Problem: The bond dissociation energies for the alkali metal M_2 molecules decrease regularly from 100.9 kJ/mol for Li_2 to 38.0 kJ/mol for Cs_2. The bond dissociation energies are *greater* for the M_2^+ ions, decreasing from 138.9 kJ/mol for Li_2^+ to 59 kJ/mol for Cs_2^+. Explain why the dissociation energies are greater for the M_2^+ ions.

15.2 Solution: The bond dissociation energies are greater for M_2^+ ions, even though they have a bond order of 0.5, because of polarization. M^+ polarizes M, with the effect increasing from Li → Cs, so that the percentage increase (M_2^+ compared to M_2) is greatest for Cs_2^+. This effect is not important for H_2^+ because the H atom has very low polarizability.

15.3 Problem: Predict the following for Fr.
(a) The product of the burning of Fr in air.
(b) An insoluble compound of Fr.
(c) The structure of FrCl.
(d) The relative heats of formation of FrF and FrI.

15.3 Solution: (a) FrO_2—The larger alkali metals stabilize O_2^-.
(b) $FrClO_4$ and Fr_2PtCl_6—These are the salts of low solubility for the larger alkali metal ions.
(c) CsCl structure—C.N. 8 is favored by the large cation size.
(d) ΔH_f is more negative for FrF—the heats of formation of all alkali metal fluorides are more negative than those of the other halides.

15.4 Problem: How could you remove unreacted Na (metal) in liquid ammonia safely?

15.4 Solution: Excess Na in liquid ammonia can be removed safely by adding NH_4Cl (a strong acid in NH_3) gradually to form NaCl and H_2.

15.5 Problem: Write equations for the preparation of the following metals.
(a) Na   (b) K   (c) Cs   (d) Mg

15.5 Solution: (a) Electrolysis is used to obtain Na since ordinary reducing agents do not reduce alkali metal ions.
$$2NaCl(l) \xrightarrow{electr.} 2Na(l) + Cl_2(g)$$
(b) CaC_2 reduces KF at high temperature because CaC_2 decomposes and CaF_2 has very high lattice energy.
$$2KF + CaC_2 \xrightarrow{\Delta} CaF_2 + 2K + 2C$$
(c) Al reduces Cs_2O because Cs is volatile and Al_2O_3 has very high lattice energy.
$$3Cs_2O + 2Al \rightarrow Al_2O_3 + 6Cs(g)$$
(d) The electrolysis of fused $MgCl_2$ is made more energy efficient by adding KCl to lower the

melting point of $MgCl_2$.

$$KCl-MgCl_2(fused) \xrightarrow{electr.} Mg + Cl_2(g)$$

15.6 Problem: How can you stabilize solutions containing Na^-?

15.6 Solution: Na^- can be stabilized by dissolving Na in ethylenediamine and adding a cryptand ligand (see DMA, 3rd ed., p. 715). This increases the solubility of Na by coordinating Na^+, shifting the equilibrium to the right:

$$2Na + cryptate \leftrightarrows Na(crypt.)^+ + Na^-$$

15.7 Problem: What are laboratory and/or everyday uses of Li, K, and Cs?

15.7 Solution: Li is used in batteries and alloys. K analytical reagents are used because K^+ salts are not so highly solvated as those of Na^+, making it easier to obtain anhydrous reagents. Cs is used in photocells because of its low ionization energy.

15.8 Problem: Cite several properties that show the diagonal relationships between Li^+ and Mg^{2+}, and between Be^{2+} and Al^{3+}.

15.8 Solution:
The ionic charge densities (charge/radius) are very similar for each pair. The solubilities of the fluorides, carbonates, and phosphates of Li and Mg are relatively low, unlike these salts of other members of the families. Li^+ and Mg^{2+} hydrolyze appreciably. Li alkyls and aryls are used in organic syntheses as are Grignard reagents (RMgX). Li and Mg form nitrides by reaction with N_2.

Fused $BeCl_2$ and $AlCl_3$ are poor electrolytes. BeO and Al_2O_3 are hard and refractory. $BeCO_3$ and $Al_2(CO_3)_3$ are unstable except in the presence of CO_2. Be_2C and Al_4C are unusual among metal carbides, hydrolyzing to form methane. Be^{2+} and Al^{3+} hydrolyze extensively. Be^{2+} and Al^{3+} have a great tendency to form complexes with F^- and oxoanions.

15.9 Problem: The element pairs Li–Mg, Na–Ca, and Be–Al are closely related because of the diagonal relationship. Would you expect Mg^{2+} to be more closely related to Sc^{3+} or to Ga^{3+}? Why?

15.9 Solution: Mg^{2+} is expected to be much more similar to Sc^{3+} because of the similar radii and similar electron configurations. Ga^{3+} has a much smaller radius than Mg^{2+} and Sc^{3+} because of the transition series contraction. Ga^{3+} has an 18-electron configuration while the others have 8-electron configurations. Ga^{3+} is more polarizing and more polarizable.

15.10 Problem: Why is Au expected to form Au^-?

15.10 Solution: The electron affinity of Au is higher than that of *any* element other than the halogens. It should form an anion if any metal does.

15.11 Problem: Write equations for the reduction of Cu^{2+} with a limited amount of CN^- and with an excess of CN^-.

15.11 Solution: Cyanide ion reduces Cu^{2+} to precipitate CuCN which forms a soluble complex ion in the presence of excess CN^-.

$$2Cu^{2+} + 4CN^- \rightarrow 2CuCN(s) + (CN)_2$$

$$CuCN(s) + CN^-(xs.) \rightarrow Cu(CN)_2^-$$

15.12 Problem: Why is *aqua regia* (HCl-HNO_3) effective in oxidizing noble metals when neither HCl nor HNO_3 is effective alone?

15.12 Solution: Formation of chloro complexes stabilizes the positive oxidation states and lowers the emf required for oxidation to the useful range for HNO_3.

15.13 Problem: Why are large anions effective in stabilizing unusually low oxidation states, such as Cd^I?

15.13 Solution: Unusual low oxidation states, such as Cd^I, are often stabilized using $AlCl_4^-$ or some other large anion that has little tendency to form stable complexes with the metal in its higher oxidation state. This avoids stabilization of the the disproportionation products. (See DMA, 3rd ed., p. 234.)

15.14 Problem: Sketch a linear combination of s, p_z, and d_{z^2} orbitals that would give very favorable overlap for bonding in a linear MX_2 molecule.

15.14 Solution:

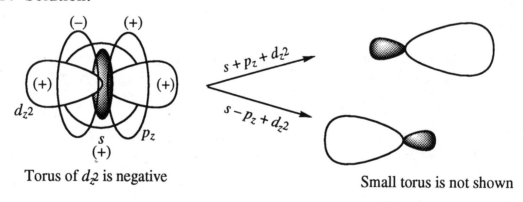

Torus of d_{z^2} is negative

Small torus is not shown

A third linear combination has a large donut in the xy plane (nonbonding).

15.15 Problem: The structures of ZnO, CdO, and HgO are quite different. Describe the structures.

15.15 Solution: ZnO has the zinc blende (*ccp* O^{2-}) or wurtzite (*hcp* O^{2-}) structures with C.N. 4. CdO has the NaCl structure with C.N. 6. HgO (orthorhombic) has zigzag chains of essentially linear O—Hg—O, giving C.N. 2. (See DMA, 3rd ed, p. 731).

15.16 Problem: The structures of HgF_2, $HgCl_2$, $HgBr_2$, and HgI_2 show interesting variations. Describe the structures.

15.16 Solution: HgF_2 has the fluorite structure with C.N. 8 for Hg^{2+}. $HgCl_2$ contains discrete linear $HgCl_2$ molecules—a molecular structure. Red HgI_2 has a layer structure and, at 126°C it goes to a yellow form containing linear HgI_2 molecules. $HgBr_2$ has a layer structure, but the presence of two Br closer than the other two indicates a transition between layer and molecular structures.

15.17 Problem: Sketch the structure of the silicate anion in beryl.

15.17 Solution: Beryl, $Al_2Be_3[Si_6O_{18}]$, contains a cyclic "discrete" silicate anion.

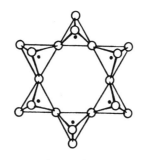

$[Si_6O_{18}]^{12-}$

15.18 Problem: Give equations for reactions that could be used to separate a mixture of Zn^{2+}, Cd^{2+}, and Hg^{2+} present in solution.

15.18 Solution: Zn^{2+} can be separated easily since it is the only amphoteric ion of the group.
$$Zn^{2+} + Cd^{2+} + Hg^{2+} + \text{excess } OH^- \rightarrow [Zn(OH)_4]^{2-} + CdO(s) + HgO(s)$$
$$CdO(s) + HgO(s) + H^+ \rightarrow Cd^{2+} + Hg^{2+} \xrightarrow{H_2S} CdS(s) + HgS(s)$$
$$3CdS(s) + 3HNO_3 \rightarrow 3Cd(NO_3)_2 + 3S + 2NO + 4H_2O$$
HgS does not dissolve in HNO_3.

15.19 Problem: Can one obtain
(a) Hg^{2+} salts free of Hg_2^{2+}? (b) Hg_2^{2+} salts free of Hg^{2+}?
Explain.

15.19 Solution: (a) Yes, Hg_2^{2+} can be oxidized completely to Hg^{2+}.
(b) Hg_2^{2+} salts cannot be obtained free of Hg^{2+} because Hg_2^{2+} is in equilibrium with Hg^{2+} and Hg ($K = 166$).

15.20 Problem: Discuss the factors involved in determining the following solubility patterns: LiF is much less soluble than LiCl, but AgF is much more soluble than AgCl.

15.20 Solution: LiF has such a high U_0 that the high solvation energy of F^- is not sufficient to make it more soluble than LiCl. The difference in U_0 is not so great for AgF and AgCl and the high solvation energy of F^- determines the relative solubilities.

15.21 Problem: The metal perchlorates have been referred to as *universal solutes*. What properties are important in causing most metal perchlorates to be quite soluble in water and several other solvents?

15.21 Solution: U_0 is fairly low for salts of the large ClO_4^- and the solvation energy is fairly high in a wide range of solvents. There is little tendency for formation of perchlorate complexes in most solvents.

15.22 Problem: For which elements in the rare earth series are M^{II} and M^{IV} oxidation numbers expected?

15.22 Solution: M^{IV} is expected for f^1 (Ce) and for f^8 (Tb) since removal of one more electron gives empty (Ce^{IV}) or half-filled (Tb^{IV}) f orbitals. M^{II} is expected just before Gd (Eu) and Lu (Yb) since these ions achieve half-filled (Eu^{II}) and filled (Yb^{II}) f orbitals.

15.23 Problem: Why should it have been expected that there would be more uncertainty concerning the identity of the first member of the second inner transition series compared to the lanthanide series?

15.23 Solution: The energy levels of the valence electrons are close together for the larger actinides and greater variation of oxidation states is expected.

15.24 Problem: One of the common M_2O_3 structures is that of α-alumina (Al_2O_3). Another is the La_2O_3 structure (C.N. 7). Check the La_2O_3 structure and describe the coordination about La^{3+}. (See A. F. Wells, *Structural Inorganic Chemistry*, 5th edition, Oxford Press, Oxford, 1984.)

15.24 Solution: In the La_2O_3 structure (A-M_2O_3 type) La^{3+} has C.N. 7, with 3 O^{2-} at 238 pm, 1 at 245 pm, and 3 at 272 pm. Approximately octahedral LaO_6 unit share edges, with the oxygen of another octahedron also shared through one of the octahedral faces.

15.25 Problem: Ti is the ninth most abundant element in the earth's crust, and its minerals are reasonably concentrated in nature. Why is Ti less commonly used than much rarer metals?

15.25 Solution: Ti is a rather active metal with a great affinity for oxygen. Its compounds are not easily reduced. Advanced technology is required for obtaining and fabricating the metal.

15.26 Problem: Some elements are known as dispersed elements, forming no independent minerals, even though their abundances are not exceptionally low; whereas others of comparable or lesser abundance are highly concentrated in nature. Explain the following cases:
(a) *Dispersed:* Rb, Ga, Ge, Hf. (b) *Concentrated:* Li, Be, Au.

15.26 Solution:(a) Rb^+ has about the same size and follows the more abundant K^+; it does not form independent minerals. Similarly, Ga^{3+} follows the very abundant Al^{3+} (almost same radii). Ge^{4+} follows the very abundant Si^{4+}. Hf^{4+} follows the more abundant Zr^{4+} (same radii).

(b) Li^+ and Be^{2+} differ so much from other members of their families that they concentrate in nature in spite of their low abundances. Au is much more noble than other metals and concentrates as the dense free element.

15.27 Problem: The occurrence of stable well-characterized oxidation states of the lanthanides other than +III can be explained in terms of empty, half-filled, and filled f orbitals. Attempt to apply a similar approach to the series Sc through Zn. Where does it work well? Why is it much less useful? Is this approach effective for the actinide elements?

15.27 Solution: The "characteristic" oxidation number for the first transition series is M^{2+} since the "regular" configuration is $3d^x 4s^2$. Expected "other" oxidation states are Sc^{3+} (d^0), Fe^{3+}(d^5), and Cu^+(d^{10}). Mn^{2+}(d^5) and Zn^{2+}(d^{10}) are expected to be the stable oxidation states for these metals. The situation is complicated by the importance of ligand-field effects, so the half-filled t_{2g} orbitals (Cr^{3+}) and filled t_{2g} orbitals (low-spin Fe^{2+} and Co^{3+}) need to be considered also. It does not work as well for the transition metals because the d electrons are valence electrons (low ionization energies) and the f electrons are not. It does not work as well for the actinides as for the lanthanides because the energy levels of the valence electrons are closer together for the larger actinides.

15.28 Problem: Compare the syntheses of the highest and lowest stable oxidation states of Mn and Re. Which halide ions can be oxidized by MnO_4^- and by ReO_4^-?

15.28 Solution: Mn^{II} is obtained by dissolution of Mn in acid or by reduction of higher oxidations states by most strong reducing agents in acidic solution. Re^{IV} is obtained by strong reduction of Re_2O_7 with H_2 to form ReO_2. MnO_4^- is produced by electrolytic oxidation of lower oxidation states or by Cl_2 oxidation of K_2MnO_4. ReO_4^- is obtained by mild oxidation of lower oxidation states using O_2 or milder oxidizing agents.

MnO_4^- in acidic solution will oxidize Cl^- (slowly), Br^-, and I^-. ReO_4^- will not oxidize any halide.

15.29 Problem: Would the removal of Hf from Zr be important in most applications of zirconium compounds? In the use of Zr metal in flash bulbs?

15.29 Solution: For most applications the removal of Hf from Zr compounds would be unimportant because they are so similar. This is certainly true for use in flash bulbs where only the ease of oxidation of the metal is important.

15.30 Problem: What properties of tungsten make it so suitable for filaments for light bulbs?

15.30 Solution: The very low volatility of W and its suitable resistance (higher than that of many metals) are most important.

15.31 Problem: Give an example of
(a) An acidic oxide of a metal.

(b) An amphoteric oxide of a transition metal.
(c) A diamagnetic rare earth metal ion.
(d) A compound of a metal in the +VIII oxidation state.
(e) A liquid metal chloride.
(f) A compound of a transition metal in a negative oxidation state.

15.31 Solution:
(a) CrO_3, WO_3, V_2O_5—Oxides of metals in very high oxidation states are acidic.
(b) Cr_2O_3—Oxides of transition metals in intermediate oxidation states are amphoteric.
(c) Ce^{4+}, La^{3+}, Lu^{3+}, Yb^{2+}—Rare earth d^0f^0 and d^0f^{14} ions are diamagnetic.
(d) OsO_4—The +8 oxidation state is expected for larger Group 8 metals in combination with O or possibly F.
(e) $TiCl_4$—Only chlorides of metals in oxidation states of +4 or higher are expected to be liquids.
(f) $Na_2Fe(CO)_4$, $NaCo(CO)_4$.

15.32 Problem: Describe the quadruple bonding in $[Re_2Cl_8]^{2-}$ in terms of the bond types (σ, π, etc.) and the atomic orbitals involved. What is significant about the eclipsed configuration?

15.32 Solution: The planar $ReCl_4$ units use dsp^2 (σ) hybrids involving $d_{x^2-y^2}$ and p_x, p_y. The M–M bond involves a p_z-d_{z^2} hybrid. Two π bonds can be formed using the d_{xz} and d_{yz} orbitals on each Re. The δ bond uses the d_{xy} orbitals on each Re. The eclipsed configuration is required for overlap of the d_{xy} orbitals to form the δ bond.

15.33 Problem: Determine the bond order and oxidation number for $Mo_2(O_2CCH_3)_4$ (bridging acetate ions), $W_2Cl_4(PR_3)_4$ (no bridging ligands), and $[Mo_2(O\text{-}t\text{-}Bu)_6\text{-}\mu\text{-}CO]$.

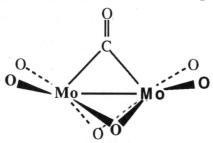

O-*t*-Bu =*t*-butoxide (only the oxygens coordinated of O-*t*-Bu are shown)

15.33 Solution:

	B.O.	Oxidation Number
$Mo_2(O_2CCH_3)_4$	4	II
$W_2Cl_4(PR_3)_4$	4	II
$[Mo_2(O\text{-}t\text{-}Bu)_6CO]$	2	III

For $Mo_2(O_2CCH_3)_4$ and $W_2Cl_4(PR_3)_4$ there are two square-planar MX_4 units joined by quadruple bonds, as in $Re_2Cl_9^{2-}$. For each Mo^{III} of $[Mo_2(O\text{-}t\text{-}Bu)_6CO]$, one electron is used for bonding to the bridging CO, and four electron pairs are provided by the butoxide ions. This leaves two electrons per Mo for forming an Mo–Mo double bond in the diamagnetic complex.

15.34 Problem: Give an example of each type of Co complex.
- (a) Co^{II} tetrahedral
- (b) Co^{II} square planar
- (c) Co^{III} octahedral, high-spin
- (d) Co^{III} optically active

15.34 Solution:

(a) $[CoCl_4]^{2-}$—Co^{II} with anionic ligands, where ligand-ligand repulsion is strong.

(b) $Co(dimethylglyoximate)_2$—Co^{II} (d^7) with a ligand favoring a square-planar arrangement.

(c) $[CoF_6]^{3-}$—only very-weak-field ligands give high-spin for Co^{III} (d^6).

(d) $[Co(en)_3]^{3+}$—a complex with chelate ligands and no S_n axis.

15.35 Problem: Give one example of a nickel complex illustrating square-planar, tetrahedral, and octahedral coordination. What type of ligands favor each of these cases?

15.35 Solution: Square-planar Ni^{II} complexes are formed by very strong-field ligands, especially π acceptor ligands—$[Ni(CN)_4]^{2-}$. Tetrahedral complexes of Ni^{II} are formed by anionic weak-field ligands—$[NiCl_4]^{2-}$. Those of Ni^0 are formed by π acceptor ligands—$Ni(CO)_4$. Octahedral complexes are formed in all cases not favoring C.N. 4, e.g., $[Ni(NH_3)_6]^{2+}$.

15.36 Problem: The structures of complexes with high coordination number involving didentate groups can be described in terms of the "average" positions of the didentate groups. Describe $[Ce(NO_3)_5]^{2-}$ and $[Ce(NO_3)_6]^{2-}$ in this way.

15.36 Solution: The structure of $[Ce(NO_3)_5]^{2-}$ (C.N. 10) is a bicapped trigonal antiprism. Three of the didentate NO_3^- are coordinated along the edges of the trigonal antiprism (a trigonal prism with one triangular face twisted relative to the other—the octahedron is a trigonal antiprism with the twist angle 60° and all edges of equal length). One of the other two didentate NO_3^- is coordinated above the top triangular face and one below the bottom triangular face. The "average" positions of the NO_3^- ligands correspond to the positions of the N atoms—three in a trigonal plane, with one above and one below—describing a trigonal bipyramid.

The structure of $[Ce(NO_3)_6]^{2-}$ (C.N. 12) is an icosahedron. With the icosahedron oriented so that edges at the top and bottom are horizontal, the N of the NO_3^- ions spanning these edges are along the z axis of an octahedron. For the two NO_3^- spanning opposite vertical edges, the N atoms are along the x axis of an octahedron. The remaining pair of NO_3^- span opposite horizontal edges with the N atoms along the y axis of an octahedron. A model of an icosahedron (constructed in Problem 3.24) is helpful. (See DMA, 3rd ed,, p. 431 for simplified shapes of complexes of didentate ligands giving high C.N. and M. G. B. Drew, *Coord. Chem. Rev.* **1977**, *24*, 179.)

15.37 Problem: How can one account for the color of the following?
(a) Fe_3O_4
(b) Ag_2S
(c) $KFeFe(CN)_6$
(d) $KMnO_4$
(e) $[Ti(H_2O)_6]^{3+}$
(f) $[Cu(NH_3)_4]^{2+}$

15.37 Solution:
(a) Black, Fe is present in two oxidation states. This is a defect structure involving charge delocalization between Fe^{II} and Fe^{III} (see DMA, 3rd ed., p. 267).
(b) Black, there is an intense charge-transfer absorption band involving $S^{2-} \rightarrow Ag^+$ donation (see DMA, 3rd ed., p. 463)
(c) Deep blue, Fe is in two oxidation states.
(d) Deep magenta (red-purple), the charge-transfer band O→Mn, is more intense than the usual $d \rightarrow d$ bands observed for transition metal complexes.
(e) Red-purple and (f) Blue-purple, these involve $d \rightarrow d$ transitions. The number of bands depends on the number of d electrons and the symmetry of the complex. The band energies depend on the ligand field strength (see DMA, 3rd ed., p. 442f).

15.38 Problem: Give the bond order and show the occupancy of bonding and antibonding orbitals for M—M bonds in the following. (Ligands bridge metals except for **c** and **d**; there are axial ligands for **e** and **f**.)
(a) $[Mo_2(SO_4)_4]^{4-}$
(b) $[Mo_2(SO_4)_4]^{3-}$
(c) $[Mo_2(NMe_2)_6]$
(d) $[Tc_2Cl_8]^{3-}$
(e) $[Tc_2(O_2CCMe_3)_4Cl_2]$
(f) $[Ru_2(O_2CC_2H_5)_4(OCMe_2)_2]$

15.38 Solution: (a) $2Mo^{II}$ (d^4) give $\sigma^2\pi^4\delta^2$, B.O. 4
(b) Mo^{II} (d^4) and Mo^{III} (d^3) give $\sigma^2\pi^4\delta^1$, B.O. 3.5
(c) $2Mo^{III}$ (d^3) give $\sigma^2\pi^4$, B.O. 3
(d) Tc^{II} (d^5) and Tc^{III} (d^4) give $\sigma^2\pi^4\delta^2\delta^{*1}$, B.O. 3.5
(e) $2Tc^{III}$ (d^4) give $\sigma^2\pi^4\delta^2$, B.O. 4
(f) $2Ru^{II}$ (d^6) give $\sigma^2\pi^4\delta^2\delta^{*2}\pi^{*2}$, B.O. 2

15.39 Problem: Count the valence shell electrons for the compounds in Problem 15.38.

15.39 Solution: (a) Mo^{II} $4e$ (d) + $8e$ (4 O) + $4e$ (M—M) = 16 e.
(b) Count the odd bonding electron for Mo^{III}
Mo^{II} $4e$ (d) + $8e$ (4 O) + $3e$ (M—M) = $15e$
Mo^{III} $3e$ (d) + $8e$ (4 O) + $4e$ (M—M) = $15e$

(c) Mo^{III} $3e$ (d) + $6e$ (3 N) + $3e$ (M—M) = $12e$

(d) Tc^{II} $5e$ (d) + $8e$ (4 Cl) + $3e$ (M—M) = $16e$
Tc^{III} $4e$ (d) + $8e$ (4 Cl) + $4e$ (M—M) = $16e$

(e) Tc^{III} $4e$ (d) + $8e$ (4 O) + $2e$ (Cl) + $4e$ (M—M) = $18e$

(f) Ru^{II} $6e$ (d) + $8e$ (4 O) + $2e$ (O, axial) + $2e$ (M—M) = $18e$

15.40 Problem: The isopoly anions containing MO_6 can be considered as fragments of oxide structures. Why are the arrangements usually *ccp*?

15.40 Solution: The *ccp* structure involves sharing octahedral edges only, while *hcp* requires sharing faces also. Sharing faces brings the metal ions at the center of the octahedra closer together, increasing cation-cation repulsion.

15.41 Problem: The simple anions CrO_4^{2-}, MoO_4^{2-}, and WO_4^{2-} are tetrahedral. Why do the polyacids and polyanions of Cr differ structurally from those of Mo and W?

15.41 Solution: The small Cr^{VI} cannot expand its C.N., while Mo^{VI} and W^{VI} become 6-coordinate in polyacids and polyanions. The electroneutrality principle (see Problem 2.14) leads us to expect higher C.N. when the oxygens are shared.

15.42 Problem: Give syntheses for *cis-* and *trans*-$[PtCl_2(NH_3)_2]$, starting with $[Pt(NH_3)_4]^{2+}$ and $[PtCl_4]^{2-}$. (see DMA, 3rd ed., p. 502f for the *trans* effect.)

15.42 Solution:

$$[Pt(NH_3)_4]^{2+} + HCl \rightarrow [PtCl(NH_3)_3]^+ + NH_4^+$$
$$[PtCl(NH_3)_3] + HCl \rightarrow trans\text{-}[PtCl_2(NH_3)_2] + NH_4^+$$
$$[PtCl_4]^{2-} + NH_3 \rightarrow [PtCl_3(NH_3)]^- + Cl^-$$
$$[PtCl_3(NH_3)]^- + NH_3 \rightarrow cis\text{-}[PtCl_2(NH_3)_2] + Cl^-$$

In each case a ligand *trans* to Cl^- is displaced more readily than one *trans* to NH_3.

15.43 Problem: The colors of lanthanide ions arise primarily from *f*-electron transitions. Which of the lanthanide ions might be expected to be colorless? Which of the following might be expected to be the same color: Ce^{3+}, Pr^{3+}, Pm^{3+}, Tm^{3+}, Tb^{3+}? Explain.

15.43 Solution: Ions with f^0 (La^{3+}), f^{14} (Lu^{3+}), and half-filled f^7 (Gd^{3+}) configurations are expected to be colorless. One expects from the hole formalism that ions of the same charge for which there are n electrons or n holes will have similar absorption bands (all are expected to be weak field complexes since the *f* orbitals are rather well shielded from their environment). The number of *f* electrons of the ions are Ce^{3+} 1, Pr^{3+} 2, Pm^{3+} 4, Tm^{3+} 12, and Tb^{3+} 8. Praseodymium and thulium salts are both green, cerium(III) colorless, terbium(III) pale pink, and promethium(III) pink, thus Pr^{3+} and Tm^{3+} fulfill our expectations.

15.44 Problem: Metals such as Ti and Al are useful structural materials even though they are easily oxidized. Na and K (among other metals) have emf values comparable in magnitude to those of Ti and Al, yet are not useful for applications involving contact with air. Explain.

15.44 Solution: The molar volumes of TiO_2 and Al_2O_3 are larger than the molar volumes of Ti and Al, respectively. Hence, the oxides occupy larger volumes than the metals from which they were formed and thus cover the metal surface with an impervious oxide layer. Sometimes oxide layers are deposited electrochemically, a process known as anodizing the metal, for protection. Actually these oxide layers represent defect structures varying in composition from the stoichiometric oxide on the surface (in contact with O_2) to the metal structure below. Because of this there is no sharp phase boundary that would result in flaking. For Na and K, the oxides have *smaller* volumes than the metals. Hence, O_2 can continue to penetrate the structure and oxidize the underlying metal at a rate governed by the O_2 diffusion rate. The alkali metals react vigorously with H_2O.

15.45 Problem: The structure of ReO_3 is similar to that of perovskite, $CaTiO_3$, without the Ca. Describe the structure in terms of the roles of the ions in a close-packed structure.

15.45 Solution: The oxides are *ccp* with 1/4 of the *ccp* packing sites (those for Ca^{2+} in perovskite) vacant and with Re in 1/4 of the octahedral holes (there are as many octahedral holes as packing positions). A beautiful model can be built using octahedra (ReO_6) centered at each corner of a cube and sharing the octahedral apices (each O^{2-} is shared by two Re, see the structure in the figure after Problem 5.5). The *PTOT* notation is $6P_{3/4}O_{1/4}$, the 6 refers to the 6 layers (3*P* and 3*O*) in the repeating *ccp* arrangement.

15.46 Problem: Generally abundances of elements decrease with increasing atomic number. However, the terrestial and cosmic abundances of the light elements Li, Be, and B are very low. Explain the low abundances of these elements.

15.46 Solution: The elements Li, Be, and B have low atomic numbers and undergo thermonuclear reactions readily because of low barriers for proton or alpha capture. These elements did not accumulate since they were used up as formed.

VI. SELECTED TOPICS
16
Chemistry of Some Nonmetals

16.1 Problem: The electron affinity of N (–7 kJ) seems anomalous. Explain the order of electronegativities for Group 15 and why one can conclude that N is really more "regular" than O.

Electron affinities (kJ) for Group 15 and 16 elements

N	P	As	Sb	Bi
–7	72	78	103	91

O	S	Se	Te	Po
141	200	195	190	180

16.1 Solution: The ionization energies (IE's) change by small amounts from one element to the next for P→Bi, but the IE of N is much higher than that of P. The electron affinities (EA's) in general for the Group 15 elements are lower than those of the neighboring elements in the same period because an electron must be added to the half-filled p^3 configuration. Unlike the trend for most main group families, the EA's for Group 15 *increase* with increasing atomic number, except for Bi. The increase might be expected because of the decrease in pairing energy with increasing atomic radius. The EA of N is slightly negative. This results from the combination of three trends: (1) It is expected to be lower than that of P because of the high pairing energy of N, (2) It is expected to be lower than the values of C or O because of the half-filled configuration of N, and (3) The first members of Groups 15–17 are anomalous with respect to further increase in electron density for the small compact atoms, as seen in their low EA's and single bond energies. The EA of N appears to deviate more from the family trend than in the case of O or F. However, because of the opposite trend in EA's for groups 15 and 16, plots of IE vs. EA reveal that N comes closer to the straight line determined by the rest of the family than is the case for O. (See the plot on the next page.)

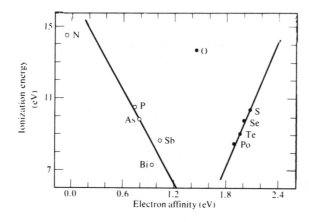

Ionization energy versus electron affinity plots for Groups 15 and 16. (Adapted from P. Politzer, in *Homoatomic Rings, Chains and Macromolecules of Main Group Elements*, A. L. Rheingold, Ed., Elsevier, Amsterdam, 1977, p. 95.)

16.2 Problem: What compound is decomposed for inflating airbags in cars?

16.2 Solution: NaN_3, which decomposes thermally to give N_2.

16.3 Problem: Write balanced equations for the preparation of HNO_3, starting with N_2.

16.3 Solution:

$$N_2 + 3H_2 \underset{500°C,\ 500\ atm}{\overset{Fe\ catalyst}{\rightleftharpoons}} 2NH_3 \qquad \text{Haber process}$$

$$4NH_3 + 5O_2 \xrightarrow{Pt\ catalyst} 4NO + 6H_2O \qquad \text{Ostwald process}$$

$$2NO + O_2 \rightarrow 2NO_2$$

$$2NO_2 + H_2O \rightarrow HNO_3 + HNO_2$$

HNO_2 disproportionates to give HNO_3 and NO, which is recycled.

16.4 Problem: When dilute nitric acid reacts with Cu turnings in a test tube, a colorless gas is formed that turns brown near the mouth of the tube. Explain the observations and write equations for the reactions involved.

16.4 Solution:

$$3Cu + 8HNO_3 \rightarrow 3Cu(NO_3)_2 + 2NO + 4H_2O$$
colorless

$$2NO + O_2 \rightarrow 2NO_2$$
brown

The brown color appears as NO mixes with O_2 (in air) and is oxidized to NO_2.

16.5 Problem: Write equations for the preparation of five nitrogen oxides.

16.5 Solution:

$$NH_4NO_3\ (aq) \xrightarrow{heat} N_2O + 2H_2O$$ Heat the solution containing some Cl^-.

NO is obtained in the Ostwald process for the catalytic oxidation of NH_3 (See Problem 16.3). Other oxides can be obtained from NO as follows:

$$\text{ONNO} \underset{\text{cooling}}{\rightleftharpoons} \text{NO} \xrightarrow{O_2} \text{NO}_2 \underset{\text{cooling}}{\rightleftharpoons} \text{N}_2\text{O}_4$$

with ONON (acid) above NO, NO$_2$ branching to O$_2$NNO, O$_3 \downarrow$ giving NO$_3 \xrightarrow{NO_2}$ N$_2$O$_5$, and N$_2$O$_4 \downarrow O_3$

(See DMA, 3rd ed., p. 768)

HNO$_3$ may be used for the laboratory preparation of the common nitrogen oxides other than N$_2$O. At low temperatures, 100% HNO$_3$ may be dehydrated with P$_4$O$_{10}$ to give N$_2$O$_5$ (at elevated temperatures the N$_2$O$_5$ dissociates to give NO$_2$ and O$_2$). The reduction products with glassy As$_2$O$_3$ are a function of the acid concentration as indicated by the density of the acid.

density (g/cm^3)	1.2	1.35	1.45
products	NO	N$_2$O$_3$	NO$_2$ + 10% N$_2$O$_3$

Dry NO$_2$ may be prepared by thermal decomposition of predried Pb(NO$_3$)$_2$.

$$\text{Pb(NO}_3)_2 \rightarrow \text{PbO} + \tfrac{1}{2}\text{O}_2 + 2\text{NO}_2$$

16.6 Problem: Sketch the *cis*- and *trans*-isomers of hyponitrous acid.

16.6 Solution:

trans: H—O—N=N—O—H (trans arrangement)   cis: H—O, O—H on same side of N=N

16.7 Problem: What properties of polymers of (SN)$_x$ and (N-PR$_2$)$_x$ make them of interest for practical uses?

16.7 Solution: (SN)$_x$ forms chains with high electrical conductivity along the chain, sometimes referred to as a "one-dimensional metal" (See DMA, 3rd ed., p. 280). (N-PR$_2$)$_x$ polymers have high thermal stability, oil resistance, etc. (See DMA, 3rd ed., p. 772-773).

16.8 Problem: Give the formula of a biodegradable polyphosphazene polymer that could be used for medical devices.

16.8 Solution:

$$[-\text{NP}-]_x \text{ with substituents NHCH}_2\text{CO}_2\text{C}_2\text{H}_5 \text{ on P}$$

16.9 Problem: PI$_5$ is not known. What would be its likely structure in the solid?

16.9 Solution: [PI$_4$]$^+$I$^-$. P is not expected to show C.N. 5 with atoms as large as I. In the solid PCl$_5$ exists as [PCl$_4$]$^+$[PCl$_6$]$^-$ and PBr$_5$ exists as [PBr$_4$]$^+$Br$^-$.

16.10 Problem: (a) What N species are compatible with the H_3PO_4/PO_4^{3-} species from the Pourbaix diagrams given by J. A. Campbell and R. A. Whiteker (*J. Chem. Educ.* **1969**, *46*, 90)? (b) What P species are compatible with $NH_3(aq)$?

16.10 Solution:
(a) Phosphoric acid represents the highest oxidation state of P, so the stable N species are those existing in the same high E^0 region, N_2 and NO_3^-. If we superimpose Pourbaix diagrams on one another, species in the same region are compatible.
(b) Ammonia is the reduced form of N existing at high pH. The P species compatible (existing in the same pH–E^0 region) are PH_3 and $H_2PO_3^-$.

16.11 Problem: How can white P be separated from red P?

16.11 Solution: White P (P_4) dissolves in diethyl ether or benzene, polymeric red P does not.

16.12 Problem: How are P_4, P_4O_6, and P_4O_{10} related structurally?

16.12 Solution: All have tetrahedral arrangements of four P—six O are inserted in the edges of the tetrahedron to form P_4O_6 and four more are added at the apices to form P_4O_{10} (See DMA, 3rd ed., p. 770.)

16.13 Problem: In what way are phosphine ligands in metal complexes similar to CO?

16.13 Solution: PR_3 ligands have vacant d orbitals (or possibly P–OR σ^* orbitals for phosphites) which make them π acceptors (back bonding) as in the case of CO (using π^* orbitals).

16.14 Problem: Write the formulas for the diethylester of phosphorous acid and the monoethylester of hypophosphorous acid. Would these be protonic acids?

16.14 Solution: In both compounds P has C.N. 4. Neither is protonic, the hydrogens bonded to P are not acidic.

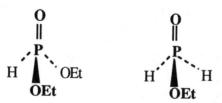

16.15 Problem: Sketch the $p\pi d$ bonding in cyclic hexachlorotriphosphazene $\{(PCl_2N)_3\}$ and in cyclic $(PCl_2N)_4$, assuming it to be planar.

16.15 Solution: The signs are mismatched for any odd n in $(PN)_n$. Only lobes of the P $3d$ orbitals and N $2p$ orbitals above the plane of the paper are shown.

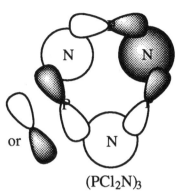
or
(PCl$_2$N)$_3$
Signs mismatched

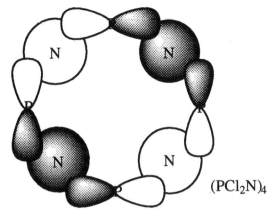
(PCl$_2$N)$_4$
Signs are matched

16.16 Problem: Write balanced equations and calculate E^0_{cell} for the reaction in acidic solution of (a) H$_2$O$_2$ with NO$_2^-$ and (b) H$_2$O$_2$ with MnO$_4^-$.

16.16 Solution: H$_2$O$_2$ can be oxidized or reduced.

$$O_2 \xrightarrow{0.68} H_2O_2 \xrightarrow{1.77} H_2O$$

(a)
HNO$_2$ + H$^+$ + e → NO + H$_2$O	$E^0 = 1.00$V
H$_2$O$_2$ → O$_2$ + 2H$^+$ + 2e	$E^0 = -0.68$V
HNO$_2$ + H$_2$O → NO$_3^-$ + 3H$^+$ + 2e	$E^0 = -0.94$V
H$_2$O$_2$ + 2H$^+$ + 2e → 2H$_2$O	$E^0 = 1.78$V

The two reactions

2HNO$_2$ + H$_2$O$_2$ → 2NO + 2H$_2$O + O$_2$	$E^0_{cell} = +0.32$V
HNO$_2$ + H$_2$O$_2$ → NO$_3^-$ + H$_2$O + H$^+$	$E^0_{cell} = +0.84$V

are favorable, with the latter having the greater force. Oxidation of HNO$_2$ to NO$_3^-$ should be the major reaction, but, in effect, both reactants disproportionate.

(b) H$_2$O$_2$ cannot oxidize Mn^{2+} in acidic solution, so MnO$_4^-$ oxidizes H$_2$O$_2$.

MnO$_4^-$ + 8H$^+$ + 5e → Mn^{2+} + 4H$_2$O	$E^0 = 1.51$V
H$_2$O$_2$ → O$_2$ + 2H$^+$ + 2e	$E^0 = -0.68$V
2MnO$_4^-$ + 5H$_2$O$_2$ + 6H$^+$ → 2Mn^{2+} + 5O$_2$ + 8H$_2$O	$E^0_{cell} = 0.83$V

16.17 Problem: What thermodynamic factors are important in determining the relative stabilities of solid metal oxides, peroxides, and superoxides? What cation characteristics (size and charge) would you choose to prepare oxides, peroxides, superoxides, and ozonides?

16.17 Solution: The oxides are favored by high lattice energies and this is particularly favorable for small cations. Peroxides are favored over oxides because the electron repulsion is lower for O_2^{2-} than for O^{2-} and the fact that it is not necessary to dissociate O_2 to form O_2^{2-}. The lattice energies are lower for the larger O_2^{2-}. Peroxides are favored for cations of intermediate size. Superoxides are favored over peroxides by the favorable electron affinity. Superoxides are favored by large cations with low charge (actually +1). Ozonides (O_3^-) are expected only for very large +1 ions where the more favorable lattice energies of MO_2, M_2O_2, and M_2O are less important.

16.18 Problem: Give a description of bonding in O_2F_2 to account for the very short O—O bond and very long O—F bonds.

16.18 Solution: The structure of O_2F_2 is similar to that of H_2O_2, but with O—O⟨F bond angles of 109.5° and the angle between the planes containing the O—F bonds is 87.5°. The O—O bond distance corresponds to a bond order of 2 and the O—F bond to a fractional bond order. This suggests that the O—O bond is much like that of O_2 with a singly occupied σ orbital of each F atom overlapping with one of the singly occupied π* orbitals of oxygen to form a weak 3-center electron pair bond. Because of the high electronegativity of F, little electron density is transferred into the π* orbital to weaken the O—O bond.

Bond Distances for Some Oxygen and Fluorine Compounds

Molecule	O—O distance	O—F distance
OF_2		142 pm
H_2O_2	148 pm	
O_2F_2	122	158
O_2	121	
Sum of covalent radii	148	145

16.19 Problem: Write the equations for the oxidation of $S_2O_3^{2-}$: (a) by I_2 (b) by H_2O_2.

16.19 Solution: (a) $2 S_2O_3^{2-} + I_2 \rightarrow S_4O_6^{2-} + 2I^-$
   tetrathionate

(b) $2 S_2O_3^{2-} + 4 H_2O_2 \rightarrow S_3O_6^{2-} + SO_4^{2-} + 4 H_2O$
   trithionate

16.20 Problem: In the Pourbaix diagram for oxygen, H_2O_2 does not appear. Explain. (If you care to check the diagram, see reference in Problem 16.10.)

16.20 Solution: H_2O_2 is thermodynamically unstable (disproportionates) over the entire pH/E^0 range. The only species shown in a Pourbaix diagram are the thermodynamically stable ones.

16.21 Problem: From Pourbaix diagrams for S and Se on the next two pages:
(a) What S and Se species are stable in contact with O_2?
(b) What S and Se species are stable in contact with H_2?

(c) What S species are stable in contact with Se?

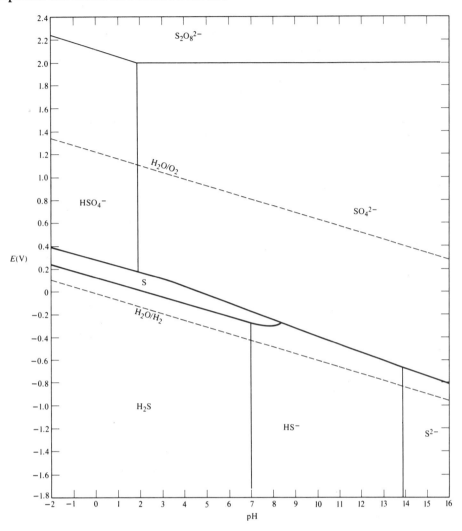

Pourbaix diagram for sulfur. [Adapted with permission from M. J. N. Pourbaix, in *Atlas of Electrochemical Equilibria in Aqueous Solution* (English translation, 2nd ed., by J. A. Franklin, National Association of Corrosion Engineering, Houston, TX, 1974, p. 551).]

16.21 Solution:

(a) HSO_4^-/SO_4^{2-} and $S_2O_8^{2-}$; $HSeO_4^-/SeO_4^{2-}$. These are the species not oxidized by O_2.

(b) H_2S/S^{2-}, and H_2Se/Se^{2-}. these are the species not reduced by H_2.

(c) H_2S/S^{2-}, S, and HSO_4^-/SO_4^{2-}.

Compatible species exist in the same pH/E^0 area of Pourbaix diagrams.

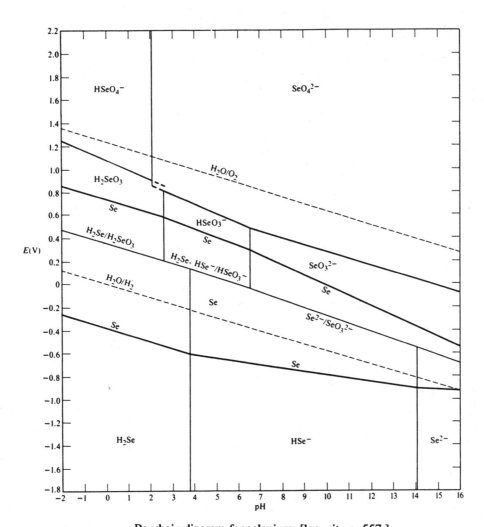

Pourbaix diagram for selenium. [loc. cit., p. 557.]

16.22 Problem: How are dithionates and polythionates alike, and in what respects do they differ?

16.22 Solution: Dithionates are related to polythionates $(O_3S-S_n-SO_3)^{2-}$ only in the sense that $n = 0$ for dithionate ion. Dithionates are stable with respect to reactions with H_2O and resist oxidation or reduction. They are obtained by oxidation of SO_3^{2-}. Polythionates decompose readily to form S and S oxides or oxoacids. They are formed by a variety of reactions involving $S_2O_3^{2-}$.

16.23 Problem: What are the products of reaction of SF_4 with alkylcarbonyls, carboxylic acids, and phosphonic acids [R(RO)PO(OH)]?

16.23 Solution: SF_4 reacts with $RR'C=O$ to give $RR'CF_2$, with $R-CO_2H$ to give $R-CF_3$, and with R(RO)PO(OH) to give $R(RO)PF_3$.

16.24 Problem: Give valence bond descriptions for SF_3 and SF_5 and indicate why they have low stability.

16.24 Solution: SF_2 has two ordinary 2-center bonds. As in the case of the interhalogen compounds, the VB treatment leads us to expect SF_4 (promotion of 1 e) and SF_6 (promotion of 2 e). These correspond in bonding descriptions to ICl_2^- and IF_6^-. SF_3 and SF_5 are radicals, with only 1e for bonding to the third or fifth F. Such radicals, with weak bonding, are expected to be very reactive.

16.25 Problem: Write balanced equations for the following preparations:
(a) Cl_2 (from NaCl)   (d) Br_2 (recovery from seawater)
(b) I_2 (from $NaIO_3$)   (e) HF (from CaF_2)
(c) HCl (from NaCl)

16.25 Solution: (a) Cl_2 can be obtained by oxidation of Cl^- with an inexpensive oxidizing agent, such as MnO_2 in acidic solution. On a commercial scale electrolysis of aqueous NaCl is used.

$$2\,NaCl + MnO_2 + 2\,H_2SO_4 \rightarrow MnSO_4 + Cl_2 + Na_2SO_4 + 2\,H_2O \text{ or}$$

$$2\,Na^+ + 2\,Cl^- + 2\,H_2O \xrightarrow{\text{electr.}} 2\,Na^+ + 2\,OH^- + Cl_2 + H_2$$

(b) $NaIO_3$ can be reduced to I_2 using $NaHSO_3$ as an inexpensive reducing agent.

$$2\,NaIO_3 + 5\,NaHSO_3(aq) \rightarrow 3\,NaHSO_4 + 2\,Na_2SO_4 + H_2O + I_2$$

(c) NaCl is the important Cl^- ore. It is converted to HCl using the cheapest strong acid of low volatility, H_2SO_4.

$$NaCl + H_2SO_4 \rightarrow NaHSO_4 + HCl$$

(d) Br_2 is recovered by first evaporating seawater partially to concentrate the Br^- and the Br^- is oxidized by Cl_2 in an air stream blown through the solution. The Br_2 in the air stream is absorbed in a solution of a strong base (Na_2CO_3 is used because it is cheap) since the Br_2 disproportionates. The Br_2 is recovered by acidifying the solution since BrO_3^- oxidizes Br^- in acidic solution.

$$Cl_2 \text{ (in air stream)} + 2\,Br^- \rightarrow Br_2 + 2\,Cl^-$$

$$3\,Br_2 + 3\,CO_3^{2-} \text{ (solution)} \rightarrow 5\,Br^- + BrO_3^- + 3\,CO_2$$

Add acid for recovery of Br_2 from the Na_2CO_3 solution.

$$5\,Br^- + BrO_3^- + 6\,H^+ \rightarrow 3\,Br_2 + 3\,H_2O$$

(e) CaF_2 is the important F^- ore. As noted for the production of HCl, H_2SO_4 is the best choice for displacing the volatile HF.

$$CaF_2 + H_2SO_4 \rightarrow CaSO_4 + 2\,HF$$

16.26 Problem: Draw the structures of the following species, indicating the approximate bond angles: IF_2^+, IF_4^+, and IF_6^+.

16.26 Solution: The bond angle for IF_2^+ is smaller than the tetrahedral angle because of lone pair-lone pair repulsion. IF_4^+ has four bonding electron pairs and one lone pair. The lone pair occupies an equatorial position. The F atoms bend away from the lone pair because the lone pair-bonding pair repulsion is most important. IF_6^+ (isoelectronic with SF_6) has six bonding electron pairs and no lone pairs. The structure is a regular octahedron.

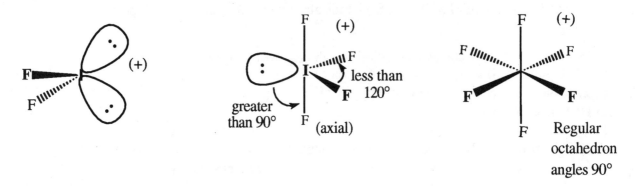

16.27 Problem: Indicate reactions which might be suitable for the preparation of:
(a) Anhydrous tetramethylammonium fluoride
(b) Aluminum bromide
(c) Barium iodide

16.27 Solution: (a) Tetraalkylammonium chlorides are generally available. The fluoride can be obtained by precipitating AgCl using AgF in CH_3OH.

$$AgF + Me_4NCl \xrightarrow{CH_3OH} Me_4NF + AgCl$$

(b) $AlBr_3$ cannot be obtained readily by metathesis from $AlCl_3$. Direct reaction of Al and Br_2 is preferable to obtain an anhydrous product.

$$2Al + 3Br_2 \rightarrow 2AlBr_3$$

(c) BaI_2, free of other salts, can be obtained in a nonprotonic solvent (pyridine here) by reaction of BaH_2 and with NH_4I, since NH_4^+ and H^- give H_2 and NH_3.

$$BaH_2(xs.) + 2NH_4I \xrightarrow{py} BaI_2 + 2NH_3 + H_2$$

16.28 Problem: The ΔH_f^0 of $BrF(g)$, $BrF_3(l)$, and $BrF_5(l)$ are –61.5, –314, and –533 kJ/mol, respectively. Which of these species should be predominant on reacting Br_2 and F_2 under standard conditions?

16.28 Solution: Since ΔH_f^0 of BrF_3 is more negative than that of BrF and that of BrF_5 is still more negative, the reaction will proceed to the extent of the availability of F_2. BrF_5 will be formed if sufficient F_2 is available.

16.29 Problem: Bromites and perbromates have been obtained only recently. Give the preparation of each.

16.29 Solution: Anodic oxidation in basic solution of Br⁻ produces BrO_2^- and anodic oxidation of BrO_3^- produces BrO_4^-. Perbromates can be obtained also by XeF_2 or F_2 in basic solution. Oxoanions in general are obtained more easily in basic solution.

$$Br^- + 4OH^- \xrightarrow{\text{anodic oxid.}} BrO_2^- + 2H_2O + 4e$$

$$BrO_3^- + 2OH^- \xrightarrow{\text{anodic oxid.}} BrO_4^- + H_2O + 2e$$

16.30 Problem: What are the expected structures of ClO_3^+, $F_2ClO_3^-$, and $F_4ClO_2^-$? Sketch the regular figure and indicate deviations from idealized bond angles.

16.30 Solution: ClO_3^+ with no unshared electrons on Cl is planar with 120° bond angles (SO_3 structure). Pi bonding, using the unhybridized p orbitals on Cl, does not alter the molecular geometry. There are no unshared electrons on Cl in $ClO_3F_2^-$, so a trigonal bipyramid with normal bond angles results. The electroneutrality principle (See Problem 2.14) favors partial Cl–O π bonding, placing the O atoms in equatorial positions, but causing no distortion. In the octahedral arrangement of $F_4ClO_2^-$ (no unshared electrons on Cl) the partially π bonded oxygens would be expected to occupy *trans* positions with all 90° angles. There is some expectation of preference for the *cis*-configuration (See K. O. Christe and C. J. Schack, *Adv Inorg. Chem. Radiochem.* **1976**, *18*, 319).

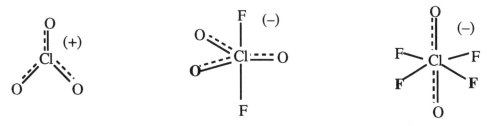

16.31 Problem: Use the emf data to predict the results of mixing the following:
(a) Cl⁻ and BrO_3^- in 1M acid
(b) Cl_2 and IO_3^- in 1M base
(c) At_2 and Cl_2 in 1M base

$$E^0_{acid} \quad ClO_4^- \xrightarrow{1.20} ClO_3^- \xrightarrow{1.18} HClO_2 \xrightarrow{1.70} HClO \xrightarrow{1.63} Cl_2 \xrightarrow{1.36} Cl^-$$

$$BrO_4^- \xrightarrow{1.85} BrO_3^- \xrightarrow{1.45} HBrO \xrightarrow{1.60} Br_2(l) \xrightarrow{1.07} Br^-$$

$$E^0_{base} \quad ClO_4^- \xrightarrow{0.37} ClO_3^- \xrightarrow{0.30} ClO_2^- \xrightarrow{0.68} ClO^- \xrightarrow{0.42} Cl_2 \xrightarrow{1.36} Cl^-$$

$$H_3IO_6^{2-} \xrightarrow{0.65} IO_3^- \xrightarrow{0.15} IO^- \xrightarrow{0.42} I_2 \xrightarrow{0.535} I^-$$

$$AtO_3^- \xrightarrow{0.5} AtO^- \xrightarrow{0.0} At_2 \xrightarrow{0.2} At^-$$

(emf data from DMA, 3rd ed., Appendix E)

16.31 Solution: (a) $Cl_2 + 2e \rightarrow 2Cl^-$ $E^0 = 1.36V$

$2BrO_3^- + 12H^+ + 10e \rightarrow Br_2 + 6H_2O$ 1.50

$\overline{2BrO_3^- + 12H^+ + 10Cl^- \rightarrow 5Cl_2 + Br_2 + 6H_2O}$ $\overline{0.14V}$

Br_2 does not oxidize Cl^-, so the reaction stops with Br_2. BrO_3^- cannot oxidize Cl_2 to $HClO$.

(b) $Cl_2 + 2e \rightarrow 2Cl^-$ $E^0 = 1.36V$

$H_3IO_6^{2-} + 2e \rightarrow IO_3^- + 3OH^-$ 0.65

$\overline{IO_3^- + 3OH^- + Cl_2 \rightarrow 2Cl^- + H_3IO_6^{2-}}$ $\overline{0.71V}$

Cl_2 easily oxidizes IO_3^- to $H_3IO_6^{2-}$ and Cl_2 cam be reduced only to Cl^-.

(c) $Cl_2 + 2e \rightarrow 2Cl^-$ $E^0 = 1.36V$

$2AtO_3^- + 6H_2O + 10e \rightarrow At_2 + 12OH^-$ 0.3

$\overline{5Cl_2 + At_2 + 12OH^- \rightarrow 10Cl^- + 2AtO_3^- + 6H_2O}$ $\overline{1.1V}$

Cl_2 is the stronger oxidizing agent, oxidizing At_2 to AtO_3^- in basic solution.

16.32 Problem: Compare the expected *rate* of reaction of AtO_3^- and IO_3^- as oxidizing agents. What formula is expected for perastatinate? Why?

16.32 Solution: AtO_3^- should be a faster oxidizing agent than IO_3^-, since the rate order is

$$IO_3^- > BrO_3^- > ClO_3^-$$

Perastinate ion is expected to be $H_3AtO_6^{2-}$. C.N. 6 is expected since At has a larger radius than I.

16.33 Problem: Give practical uses of He, Ne, and Ar, and give the sources of each.

16.33 Solution: Helium is used for cryogenic work because of its very low boiling point, and for lighter-than-air craft because of its low density and non-combustability. He is obtained from deposits of natural gas such as those in the southwestern U. S. Ne is used in Ne signs and some lamps. It is recovered from air. Ar is used to provide an inert atmosphere and in light bulbs. It is recovered from fractional distillation of liquid air.

16.34 Problem: What known compounds related in bonding and structure to the Xe halides should have prompted a search for the Xe halides earlier?

16.34 Solution: The interhalogen halides, for example, ICl_2^- and ICl_4^-, are isoelectronic (considering valence electrons) with XeF_2 and XeF_4. The existence of such compounds of halogens with positive oxidation states should have prompted the search for fluorides, at least, of Xe. The ionization of Xe is about the same as that of O_2 and the discovery of compounds of O_2^+ led to the discovery of Xe compounds.

16.35 Problem: Give the expected shape and approximate bond angles for ClF$_3$O, considering the effects of the lone pair on Cl and the directional effects of the Cl=O π bond. (See K. O. Christe and H. Oberhammer, *Inorg. Chem.* **1981**, *20*, 297.)

16.35 Solution: Both the lone pair and the Cl—O bond with double bond character (see Problem 16.30) cause the axial F to bend away, but the effect of the π bonding to O is slightly greater. Both the lone pair and double bonding to O decrease the OClF$_{eq}$ angle to much less than 120°. The OClF$_{ax}$ angle is influenced primarily by Cl–O double bonding. The rather large increase (above 90°) suggests that the π bond is perpendicular to the equatorial plane, using d_{xz} (or d_{yz}). See K. O. Christe and H. Oberhammer for discussion of directional effects of double bonds.

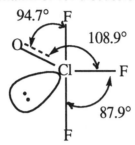

16.36 Problem: Draw the structure expected for XeOF$_4$ and indicate the approximate bond lengths (the Xe—F bond length in XeF$_4$ is 195 pm).

16.36 Solution: A square pyramid is expected to accomodate one lone pair. The Xe—F bond lengths should be about the same as for XeF$_4$. The Xe—O bond should be shorter because of some double bond character, perhaps about 170–180 pm. The bond lengths are 190 pm for Xe—F and 170 pm for Xe—O.

16.37 Problem: XeF$_6$ gives solutions in HF which conduct electricity. How might one distinguish between the following possible modes of dissociation:

$$XeF_6 + HF \rightarrow XeF_5^+ + HF_2^-$$

and

$$XeF_6 + 2\,HF \rightarrow XeF_7^- + H_2F^+$$

16.37 Solution: One might attempt to obtain an IR spectrum of the solution and look for the bands associated with the known FHF$^-$ species. Alternatively one might attempt to assign bands to the new XeF$_5^+$ or XeF$_7^-$ species based on their symmetry. Addition of a strong fluoride ion acceptor, such as SbF$_5$, would increase the concentration of XeF$_5^+$ (if this were present) or decrease the concentration of XeF$_7^-$ (if this were present) and thus increase or decrease the intensity of the IR absorption. If fluoride exchange were not too rapid, the fluorine NMR spectrum might be used to distinguish between these possibilities.

16.38 Problem: How might one distinguish between Xe^+ ions and Xe_2^{2+} ions in the compound $XePtF_6$?

16.38 Solution: A distinction between Xe^+ and Xe_2^{2+} in $XePtF_6$ could be made by a determination of the Xe–Xe separation or possibly by a magnetic moment determination, Xe^+ would be paramagnetic and Xe_2^{2+} should be diamagnetic.

16.39 Problem: Heating B_2S_3 in a glass tube gave the compound B_8S_{16} with a planar porphine-like structure. On the basis of the chemistry of B and S, sketch the formula for B_6S_{16}.

16.39 Solution: B tends to form three bonds and S forms two bonds. Thus B is best suited for the bridge-heads of the five-membered rings and these are joined by S. (See B. Krebs and H.-U. Hürter, *Angew. Chem. Int. Ed. Engl.* **1980**, *19*, 482.)

B_8S_{16}

16.40 Problem: Ozone is used in place of Cl_2 for water purification in many European countries. Explain the advantages and disadvantages of the use of ozone for water treatment.

16.40 Solution: Ozone, like chlorine, is a strong oxidizing agent that will oxidize bacteria. Unlike Cl_2, ozone decomposes fairly rapidly, leaving no residual taste, and no potentially carcinogenic chlorocarbons. It must be generated on site by a process much more expensive than for chlorination (i.e., by an electric discharge in O_2).

16.41 Problem: Explain why the ozone molecule has a dipole moment.

16.41 Solution: In the contributing structures to the resonance hybrid for O_3 the "single bond" must be a dative bond giving rise to a positive formal charge on the central oxygen atom and a formal negative charge for the terminal oxygen.

$$\left\{ \begin{array}{c} (+) \\ :O: \end{array} \begin{array}{c} :\ddot{O}: \, (-) \\ \\ :\ddot{O}: \end{array} \longleftrightarrow \begin{array}{c} (+) \\ :O: \end{array} \begin{array}{c} :\ddot{O}: \\ \\ :\ddot{O}: \, (-) \end{array} \right\}$$

16.42 Problem: Explain the function of SiO_2 and C in the production of elemental phosphorus from phosphates. What other substances could be substituted for these? Why are these other substances not used?

16.42 Solution: The SiO_2 functions as a nonvolatile acid in the Lux-Flood sense—that is, as an oxide ion acceptor— and thus displaces the more volatile P_4O_{10}, which is reduced to P_2 by the C. The $P_2(g)$ condenses to white P_4. Other nonvolatile Lux-Flood acids are TiO_2, V_2O_5, MoO_3, etc.; other reducing agents should be easily separable from the product, and further the reduction of the Lux-Flood acid should not occur readily. A ferrosilicon alloy or hydrogen might function satisfactorily. The expense of the substitute reagents rules them out for commercial production of phosphorus.

16.43 Problem: Although the pentahalides of phosphorus are known for all but the iodide, only recently has a "pentahalide" been found for nitrogen, NF_5. Do you expect NCl_5 will soon be found? Explain.

16.43 Solution: Both PF_5 and PCl_5 exist in the vapor as discrete molecules, for which sp^3d hybridization may be assumed. The NF_4^+ ion has been prepared through the reaction of NF_3 with F_2 in the presence of a good F^- ion acceptor (AsF_5, SbF_5, BF_3). Metathesis reactions with a suitable metal fluoride (e.g., CsF) then give NF_4F. Since N has no low lying d orbitals, a molecular NF_5 should not exist. NF_3 is the only stable nitrogen trihalide known, the others being shock-sensitive compounds undergoing exothermic, explosive, decomposition. If NCl_4Cl could be put together, coordination of the Cl^- ion with an empty d orbital of a chlorine attached to N would provide a low energy path for decomposition to NCl_3 and Cl_2. The greater bond energy of Cl_2 than that of F_2, coupled with the lower lattice energy expected for NCl_4Cl than that of NF_4F, and the unimportance of resonance stabilization of NCl_4^+ compared to NF_4^+ all militate against the possible existence of NCl_4Cl.

16.44 Problem: Explain the trends in the melting and boiling points of the fluorides of the third period:

	NaF	MgF$_2$	AlF$_3$	SiF$_4$	PF$_5$	SF$_6$
Mp (°C)	992	1263	1270	−90	−83	−51
Bp (°C)	1704	−	1270 (subl.)	−86	−75	−64 (subl.)

16.44 Solution: Boiling points and melting points depend on forces between constituent particles. NaF and MgF$_2$ are both ionic solids. Their increasing melting points parallel increasing lattice energies. Boiling points are expected to follow the same trend because of the greater coulombic attraction for 2+ ions. For AlF$_3$ the lattice energy energy and melting points are still higher. In the solid, Al^{3+} has six F$^-$ neighbors. AlF$_3$ sublimes near the boiling point, indicating that once the AlF$_6^{3-}$ units are broken down on melting, AlF$_3$ vaporizes. We know that AlCl$_3$ vaporizes as Al$_2$Cl$_6$, two tetrahedra sharing an edge. The sublimation of AlF$_3$ suggests the formation of a molecular species in the vapor.

The small Si^{IV} is shielded by four F$^-$ to form discrete, polar covalent SiF$_4$ molecules with T_d symmetry. Only van der Waals forces bind the molecules together so the melting point and boiling point are very low. The trigonal bipyramidal PF$_5$ molecule has two long P–F bonds in the vapor state. Insofar as this persists in the solid, the very small change in melting point of PF$_5$ (compared to SiF$_4$) could reflect weak van der Waals interaction resulting from greater distances between molecular centers and less efficient packing. The normal increase in melting point for molecules with more electrons is seen for the highly symmetrical octahedral SF$_6$ molecule. The trend is often described as paralleling the change in molar mass, but it is not dependent on mass. The boiling points show the expected trend, increasing in the series from SiF$_4$ to PF$_5$ to SF$_6$. The boiling points being near the melting points indicate weak attraction between the compact, nonpolarizable fluoride molecules once the solid structure breaks down. No great significance can be associated with the fact that the boiling point of SF$_6$ is below the melting point. We choose arbitrarily to define the boiling point at 1 atmosphere. Variations could occur at another pressure.

VI. SELECTED TOPICS
17
Cluster and Cage Compounds

17.1 Problem: Classify the following species as *closo, nido, arachno* or *hypho*.
(a) C_8H_8 (cyclooctatetraene)   (d) B_4H_{10} (g) B_9H_9NH
(b) B_6H_{12} (e) B_4H_8 (h) $B_9H_{12}^-$
(c) B_9H_{15} (f) $B_6H_{11}^+$ (i) $B_9H_{13} \cdot CH_3CN$

17.1 Solution: Clusters having p vertices are classified by skeletal electron count as follows:

Skeletal e pairs	Classification
$p + 1$	*closo*
$p + 2$	*nido*
$p + 3$	*arachno*
$p + 4$	*hypho*

Each BH group is considered to contribute two electrons ($2e$, one pair) to skeletal bonding. Groups isoelectronic with BH (*e.g.*, CH^+, BeH^-) also contribute $2e$. Others contribute a number of e given by the formula $(v + x - 2)$ where v is the number of valence e of the atom in the cluster and x is the number contributed by *exo* bonded groups (*e.g.*, for CH, $v + x - 2 = 4 + 1 - 2 = 3$). "Extra" H's each contribute one e. Account must also be taken of the charge. The numbers of e pairs for each compound and the classifications are:

(a) C_8H_8 $(8 \times 3) = 24e = p + 4$ pairs; *hypho*
(b) B_6H_{12} $(6 \times 2) + 6 = 18e = p + 3$ pairs; *arachno*
(c) B_9H_{15} $(9 \times 2) + 6 = p + 3$ pairs; *arachno*
(d) B_4H_{10} $(4 \times 2) + 6 = p + 3$ pairs; *arachno*
(e) B_4H_8 $(4 \times 2) + 4 = p + 2$ pairs; *nido*
(f) $B_6H_{11}^+$ $(6 \times 2) + 5 - 1 = p + 2$ pairs; *nido*
(g) B_9H_9NH $(9 \times 2) + 4 = 22e = p + 1$ pairs; *closo*
(h) $B_9H_{12}^-$ $(9 \times 2) + 3 + 1 = 22e = p + 2$ pairs; *nido*
(i) $B_9H_{13} \cdot CH_3CN$ $(9 \times 2) + 4 + 2 = 24e = p + 3$ pairs; *arachno*

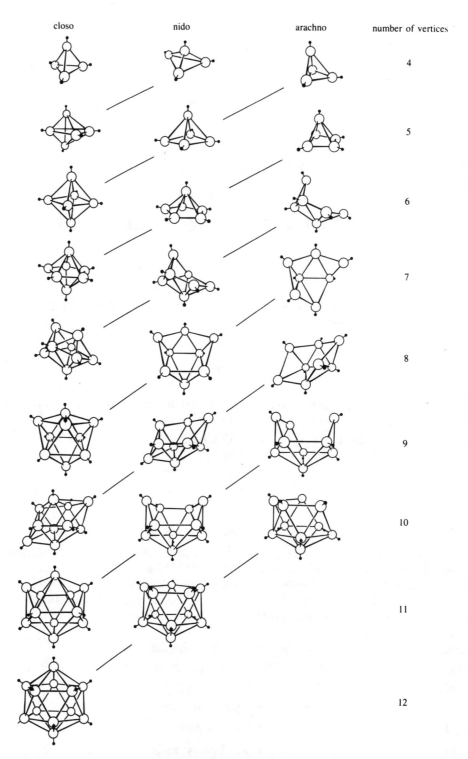

Idealized deltahedra and deltahedral fragments for boranes. The diagonal progressions represent excision of successive BH groups while retaining their electrons, thus generating *nido* from *closo* and *arachno* from *nido* species. (Reprinted with permission from R. W. Rudolph, *Acc. Chem.. Res.* **1976**, *9*, 446. Copyright 1976, American Chemical Society.)

17.2 Problem: What structure do you expect for each species in Problem 17.1?

17.2 Solution: (See DMA, 3rd ed. Sec. 17.3)
The first column of Rudolph's figure shown on p. 274 contains drawings of the *closo* frameworks for various numbers (p) of vertices. Excision of the most highly connected group at a vertex leads to a *nido* structure given in the second column. The p-vertex *nido* structure is related to the ($p + 1$)-vertex *closo* structure by removal of one group. Similarly, excision of a highly connected group from the ($p + 1$)-vertex *nido* structure structure produces the p-vertex *arachno* structure. Hence, each of p-vertex *arachno* framework is related to the ($p + 2$)-vertex *closo* framework by successive excision of two groups. *Hypho* structures contain one vertex fewer than *arachno* structures. Bridging H's are not shown.

(a) For C_8H_8 we expect the structure derived from excision of three BH vertices from the *closo* $B_{11}H_{11}^{2-}$ structure (octadecahedron). This provides a distorted version of the actual structure which is the boat form of a C_8H_8 ring.

(b) For B_6H_{12} the *arachno* structure for $p = 6$ is shown in the third column of the figure on p. 274.

(c) B_9H_{15} shown in the figure on p. 274, third column.

(d) B_4H_{10} shown in the figure on p. 274, third column.

(e) B_4H_8 shown in the figure on p. 274, second column.

(f) $B_6H_{11}^+$ is the protonated form of B_6H_{10}. All basal B—B bonds have μ-H. Shown in the second column, p. 274.

(g) B_9H_9NH has the structure of $B_{10}H_{10}^{2-}$ with one B replaced N. (See in the second column, p. 274.) A bicapped Archimedean antiprism is expected and this structure having NH in the 1-position has been found. (See A. Arafat, J. Baer, J. C. Huffman and L. J. Todd, *Inorg. Chem.* **1986**, *25*, 3757.)

(h) An Archimedean antiprism lacking one five-coordinate vertex is expected and this structure has been found for the PPN$^+$ salt. (See G. B. Jacobsen, D. G. Meina, J. H. Morris, C. Thomson, S. J. Andrews, D. Reed, A. J. Welch and D. F. Gaines, *J. Chem. Soc., Dalton Trans.* **1985**, 1645.)

(i) The 9-vertex *arachno* structure in the figure is expected and observed. However, the placement of H's around the open face differs from that in isoelectronic $B_9H_{14}^-$. (See G. B. Jacobsen, J. H. Morris and D. Reed, *J. Chem. Soc., Dalton Trans.* **1984**, 415.)

17.3 Problem: Calculate *styx* numbers and draw valence structures for the following:
(a) B_5H_{11}
(b) B_6H_{10}
(c) $B_5H_5^{2-}$
(d) B_8H_{12}

17.3 Solution: Lipscomb and co-workers developed the Equations of Balance as a way of counting available orbitals and electrons in the boron hydrides and distributing them among various possible types of bonds to give each B an octet of electrons. Formulas are written as
$[(BH)_p H_{(q+c)}]^{c+}$ where p is the number of vertices containing BH groups, ($q + c$) is the number of "extra" H, and c is the charge on the species. Of course, $c = 0$ for neutral species and is negative for anions. If we let

s = number of B–H–B bonds
t = number of B–B–B three-center bonds
y = number of B–B bonds
x = number of BH_2 groups

the Equations of Balance are (See DMA, 3rd ed., Section 17.2.2)
$$s + x = q + c$$
$$s + t = p + c$$
$$p = t + y + (q/2) + c$$

Sets of *styx* numbers may be calculated for any formula and used to draw valence bond (VB) structures. (See DMA, 3rd ed., p. 822.)

(a) $B_5H_{11} = (BH)_5H_6$: $p = 5, q = 6, c = 0$. The equations are

$$s + x = 6$$
$$s + t = 5$$
$$5 = t + y + 3$$

These are diophantine equations; that is, the solutions are whole, positive, numbers. Since the sum of t and y is 2, the possible solutions for t and y, respectively, are 0,2; 1,1; and 2,0. Since the sum of s and t is 5, and of s and x is 6, the complete set of possible *styx* numbers would be 5021, 4112, and 3203. This is an *arachno* species (see Problem 17.1). *Arachno* structures are derived from *closo* structures by successive excision of B–H groups. *Closo* structures have all BH groups at vertices of triangular polyhedral faces. Hence, each BH group is attached to at least four others by B–B bonds. If we start from the seven-vertex *closo* structure and excise two B–H groups, we must still have at least one B bonded to four other B's by B–B bonds. Only the 3203 structure incorporates this feature and so is preferred.

3203

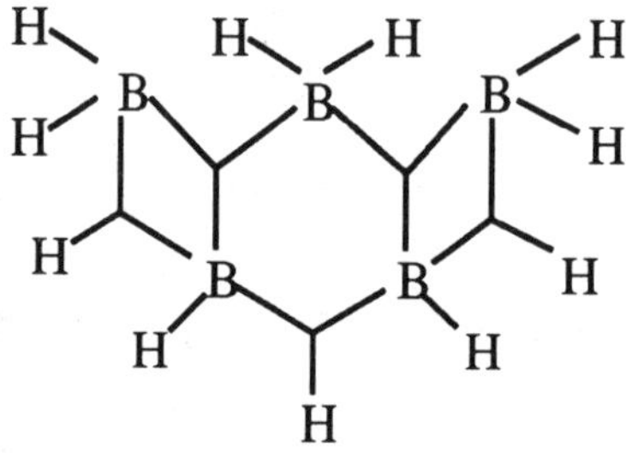

(b) B_6H_{10}: Possible *styx* numbers are 4220, 3311, and 2402. In this *nido* structure, the apical BH must be joined to the five basal ones. This will require at least two three-center bonds. Hence, we must have $t \geq 2$. However, t cannot be > 3. Otherwise, at least one pair of basal B's would be joined by two three-center B–B–B bonds. Hence, 4220 is preferred.

4220

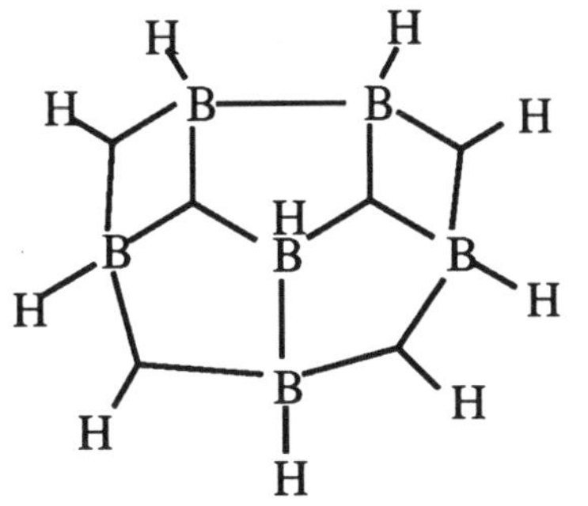

(c) $B_5H_5^{2-}$:

$s + x = q + c = $ # of "extra" H $= 0$ Hence, $q = 2$ since $c = -2$.
$s + t = p + c = 5 - 2 = 3$
$p = 2 = t + y + 1 - 2 = 5;$ $\therefore 6 = t + y$

The only solution is 0330, leading to a structure with open B–B–B bonds or the two structures with closed B–B–B bonds.

0330

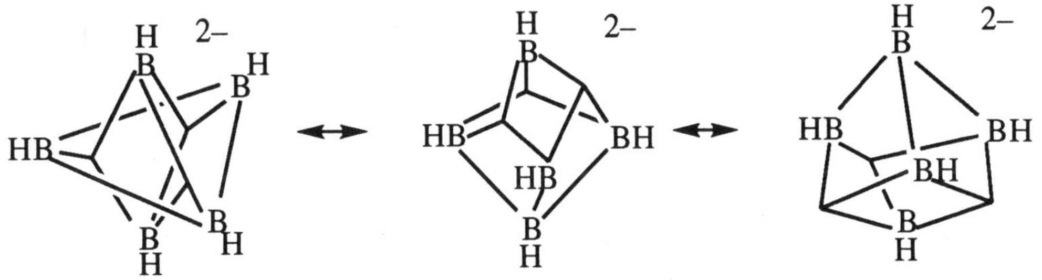

(d) B_8H_{12}: Solutions to the equations of balance are 4420, 3511, and 2602. Both 3511 and 2602 involve open B–B–B bonds. One 4420 structure need not have open B–B–B bonds and is preferred.

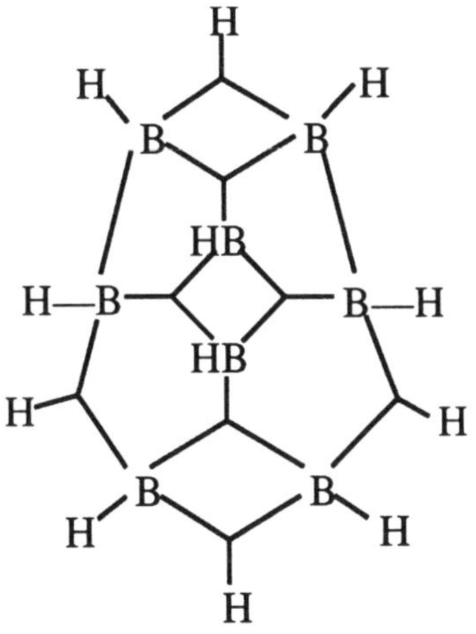

4420

Other 4420 strutures with open B–B–B bonds can be drawn.

17.4 Problem: Solve the Equations of Balance for $B_3H_6^+$ (not known) and write a "reasonable" structure for such a hydride.

17.4 Solution: See Problem 17.3 for the procedure used in solving the Equations of Balance. We formulate $B_3H_6^+$ as $(BH)_3H_3^+$ for which

$$p = 3 \qquad s + x = q + c = 3$$
$$q + c = 3 \qquad s + t = p + c = 4$$
$$c = +1 \qquad p = t + y + \frac{q}{2} + c = 3 = t + y + 1 + 1; \qquad \therefore t + y = 1$$
$$q = 2$$

3100 is the solution. The structure is:

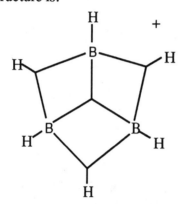

17.5 Problem: Show that the number of framework electrons contributed by the main group species NH^-, Si, Be^+, and BNR_3 is given by the formula $(v + x - 2)$. v is the number of valence electrons in the vertex atom and x is the number contributed by *exo* ligands. For example, for BH, $v = 3$ and $x = 1$.

17.5 Solution: We want to reserve one of the four valence orbitals of a main group atom to form a bond to an *exo* substituent or to contain a lone pair. The two e in this orbital may be supplied one each by the main group atom and *exo* ligand, both by the *exo* ligand (coordinate covalent bond) or both by the main group atom (lone pair). The remaining three orbitals and any remaining e are contributed to framework bonding.

NH^-: N^- has six valence e; one of these plus one from H form the NH bond leaving five framework e. $(v + x - 2) = 6 + 1 - 2 = 5$.

Si: Si has four valence e; two are used for a lone pair leaving two for framework bonding. $(v + x - 2) = 4 + 0 - 2 = 2$.

Be^+: Be^+ has one valence e. In order to have a lone pair, it must extract one e from the framework. $(v + x - 2) = 1 + 0 - 2 = -1$.

BNR_3: Both e in the B–N bond are supplied by N of the amine leaving the three B e for donation to the framework. $(v + x - 2) = 3 + 2 - 2 = 3$.

17.6 Problem: (1) Classify the following species as *closo, nido, arachno* or *hypho*.
(a) $2\text{-}CB_5H_9$ (f) $B_{11}SH_{10}Ph$ (Ph attached to B)
(b) $2\text{-}CH_3\text{-}2,3\text{-}C_2B_4H_7$ (g) $B_9H_{12}NH^-$

(c) 1,2-$C_2B_9H_{11}$ (h) $C_2B_6H_{10}$
(d) $B_9H_{11}S$ (i) $B_2H_4(PF_3)_2$
(e) 1,7-$B_{10}CPH_{11}$

(2) Name and sketch the structure of the above species.

17.6 Solution: See Problems 17.1 and 17.5 regarding structure and electron count.

(a) 2-CB_5H_9 $(CH)(BH)_5(H)_3$: $3e + 5 \times 2e + 3e = 16e = 8$ pr; *nido*; 2-carbahexaborane(9). All frameworks are depicted in the Figure of Problem 17.2. It is the $p = 6$ *nido* framework having a CH group in one of the basal positions. The three μ-H bridge the basal BH pairs.

(b) 2-CH_3-2,3-$C_2B_4H_7$ $(C-CH_3)(CH)(BH)_4(H)_2$: $3e + 3e + 4 \times 2e + 2e = 16e = 8$ pr; *nido*; 2,3-dicarba-2-methylhexaborane(7)). Same framework as above, except two CH groups are adjacent in the basal plane and a methyl group instead of H is bonded to C(2).

(c) 1,2-$C_2B_9H_{11}$ $(CH)_2(BH)_9$: $2 \times 3e + 9 \times 2e = 24e = 12$ pr; *closo*; 1,2-dicarbaundecaborane(11). This compound has the eleven-vertex *closo* framework (octadecahedron) with CH groups at two adjacent vertices in the top and next lower plane. The numbering of vertices (DMA, 3rd ed., p. 833) is described more fully in R. M. Adams and K. A. Jensen, *Pure Appl. Chem.* **1972**, *30*, 681.

(d) $B_9H_{11}S$ $(BH)_9(H)_2S$: $9 \times 2e + 2e + 4e = 24e = 12$ pr; *nido*; thiadecaborane(11). The framework is that of $B_{10}H_{14}$. The vertex occupied by S is not specified here, but is probably one of the two in the uppermost plane.

(e) 1,7-$B_{10}CPH_{11}$ $(CH)_1(BH)_{10}P$: $3e + 10 \times 2e + 3e = 26e = 13$ pr; *closo*; 1-carba-7-phosphadodecaborane(11). The twelve-vertex *closo* structure is the icosahedron. Here, the CH group is at the apex and P is at one of the five vertices in the lower pentagonal plane.

(f) $B_{11}SH_{10}Ph$ $(BPh)(BH)_{10}S$: $2e + 10 \times 2e + 4e = 26e = 13$ pr; *closo*; thiaphenyldodecaborane(10). The icosahedral framework has two unspecified vertices which contain BPh and S instead of BH.

(g) $B_9H_{12}NH^-$ $(BH)_9(H)_3(NH^-)$: $9 \times 2e + 3e + 5e = 26e = 13$ pr; *arachno*; imidododecahydrodecaborate(1–). The framework is the *arachno* $p = 10$ one derived by excision of two BH groups from an icosahedron. An NH^- group occupies an unspecified vertex, probably on the open face. The bridging H is between two BH groups. Two BH_2 groups are on the open face.

(h) $C_2B_6H_{10}$ $(CH)_2(BH)_6(H)_2$: $2 \times 3e + 6 \times 2e + 2e = 20e = 10$ pr; *nido*; dicarbaoctaborane(10). The *nido* eight-vertex framework is derived from the tricapped trigonal prism by removal of a capping vertex. The two CH groups are in non-adjacent positions in the most stable isomer and μ-H bridge BH groups.

(i) $B_2H_4(PF_3)_2$ $(BH)_2(H)_2(PF_3)_2$: $2 \times 2e + 2e + 2 \times 2e = 10e = 5$ pr; *arachno*; bis(trifluorophosphine)diborane(4). The structure contains two more e than B_2H_6 since PF_3 is a 2-e donor. Enough e are available to write an electron-precise structure.

$$F_3P \rightarrow \underset{H}{\overset{H}{B}} - \underset{H}{\overset{H}{B}} \leftarrow PF_3$$

17.7 Problem: Use the Equations of Balance to obtain a reasonable bonding picture of 1,5-dicarba-*closo*-pentaborane(5).

17.7 Solution: See Problem 17.3 for Equations of Balance. $1,5\text{-}C_2B_3H_5$ is isoelectronic with $B_5H_5^{2-}$. See Solution 17.3.

17.8 Problem: Predict the products of the following reactions.
(a) $B_5H_{11} + KH$
(b) $B_5H_9 + NMe_3$
(c) $B_{10}H_{14} + SMe_2$
(d) $B_5H_9 + HCl(l)$
(e) $B_6H_{10} + Br_2$
(f) $Li_2[o\text{-}C_2B_{10}H_{10}] + MeI$
(g) $Li_2[o\text{-}C_2B_{10}H_{10}] + R_3SiCl$
(h) $B_6H_9^- + Me_2SiCl_2$
(i) $2,3\text{-}C_2B_4H_8 + NaH$
(j) $ZrCl_4 + 4\,LiBH_4$
(k) $Ph_2PCl + LiAlH_4$

17.8 Solution:
(a) $K[B_5H_{10}] + 1/2\,H_2$; μ-H is acidic.
(b) $B_5H_9(NMe_3)_2$; B_5H_9 contains no BH_2 groups making fragmentation less likely. $B_5H_9(NMe_3)$ is an acceptable possibility but does not form in fact.
(c) $B_{10}H_{12}(SMe_2)_2 + H_2$
(d) $[B_5H_{10}]^+Cl^-$
(e) $B_6H_9Br + HBr$
(f) $o\text{-}(CMe)_2B_{10}H_{10} + 2\,LiI$
(g) $o\text{-}[C(SiR_3)]_2B_{10}H_{10} + 2\,LiCl$
(h) $μ,μ'\text{-}Me_2Si[B_6H_9]_2 + 2Cl^-$
(i) $Na^+[2,3\text{-}C_2B_4H_7]^- + H_2$
(j) $Zr(BH_4)_4 + 4\,LiCl$
(k) $Ph_2PH + LiCl + AlH_3$

17.9 Problem: Give a reasonable method for preparing and purifying B_2H_6. How might the purity of the sample be determined? How could one dispose of the diborane?

17.9 Solution: Reduction of boron halides by $LiAlH_4$ in ether, or displacement from BH_4^- salts by a stronger, nonoxidizing acid such as H_3PO_4 will produce B_2H_6. Either reaction should be carried out on a vacuum line, where the purity could be checked by the tensiometric homogeneity, *i.e.*, the constancy of the vapor pressure of different fractions. For disposal, produce an amine borane which could then be exposed to air and allowed to react with 2-propanol or *n*-butanol.

17.10 Problem: (a) How may $(CH_3)_2B_2H_4$ be prepared?
(b) Draw structural formulas for all isomers expected for $(CH_3)_2B_2H_4$.
(c) How might one identify these isomers if they were all separated?

17.10 Solution: Mixing $B(CH_3)_3$ and B_2H_6 would lead to a statistical redistribution of the methyl groups through H-bridged intermediates. Other possible preparative routes are:

$$4\,Li[CH_3AlH_3] + 4\,BCl_3 \xrightarrow{Et_2O} 2\,(CH_3)_2B_2H_4 + 4\,Li[AlCl_3H]$$

$$2[n\text{-}Bu_4N][CH_3BH_3] + 2\,BCl_3 \xrightarrow{25°C} (CH_3)_2B_2H_4 + 2[n\text{-}Bu_4N][HBCl_3]$$

Possible isomers are:

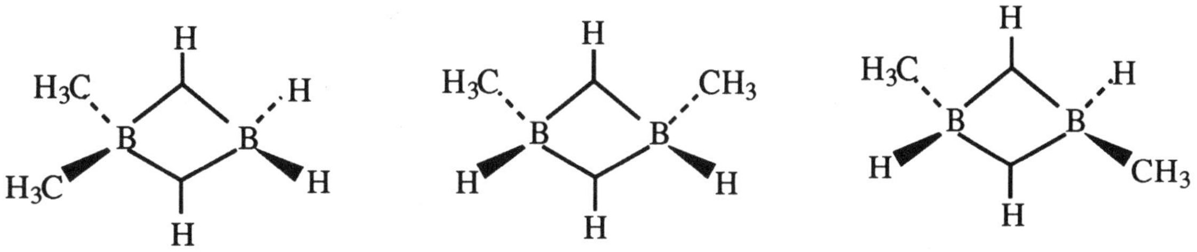

(a) **C**$_{2v}$ (b) **C**$_{2v}$ (c) **C**$_{2h}$

Bridging methyl groups are not found in boron compounds. The isomer (a) could be distinguished from (b) and (c) by ^{11}B and ^{1}H NMR. Isomer (a) would have two ^{11}B signals. Both (b) and (c) would have one ^{11}B signal, one ^{1}H signal each for methyl H, μ-H and terminal B–H. In all cases B signals will split by coupling to H and *vice versa*. (A good reference for ^{11}B is G. R. Eaton and W. N. Lipscomb, *NMR Studies of Boron Hydrides and Related Compounds*, W. A. Benjamin, New York, 1969). (b) would have a (small) dipole moment whereas (c) would have a zero dipole moment. (c) would have no coincident IR and Raman bands since it has an inversion center.

17.11 Problem: (a) Show that the diamond-square-diamond (DSD) mechanism shown in the figure below cannot account for the known thermal rearrangement of *o*-carborane to *p*-carborane.
(b) What experiments might be helpful in shedding light on the *o*- to *p*-rearrangement?
(c) Show how the ETR mechanism (Figure c, d) could lead to the *o*- to *p*-rearrangement.

17.11 Solution: (a) The DSD path allows a particular C to re-connect only to a B in the plane of atoms immediately above or below it in the starting isomer giving an *m* arrangement.
(b) One possibility might be rotation of half the icosahedron giving a bicapped pentagonal prism. Another might be rotation of triangular icosahedral faces. Labeling (*e.g.*, starting with a monobromo *o*-carborane) could distinguish via isomer distribution in the product.
(c) The figure shows how 120° rotations of the two highlighted triangular faces leads from *o*- to *p*-carborane with *m*-carborane a probable intermediate.

17.12 Problem: Write an essay on the possible relevance of transition metal cluster compounds to heterogeneous catalysis. (You will want to consult some of the references.)

17.12 Solution: Appropriate references to consult would be:
E. L. Muetterties, *Bull. Soc. Chim. Belg.* **1975**, *84*, 959.
E. L. Muetterties and J. Stein, *Chem. Rev.* **1979**, *79*, 479.
B. F. G. Johnson, Ed., *Transition Metal Clusters*, Wiley, New York, 1980.

17.13 Problem: Classify the following as *closo*, *nido* or *arachno*:
(a) [Co$_4$Ni$_2$(CO)$_{14}$]$^{2-}$ (e) [(Et$_3$P)$_2$Pt(H)]B$_9$H$_{10}$S
(b) [Fe(CO)$_3$]B$_4$H$_8$ (f) Rh$_6$(CO)$_{16}$
(c) (CpCo)C$_2$B$_7$H$_{11}$ (g) [Rh$_9$P(CO)$_{21}$]$^{2-}$ (P is at cluster center)
(d) Os$_5$(CO)$_{16}$

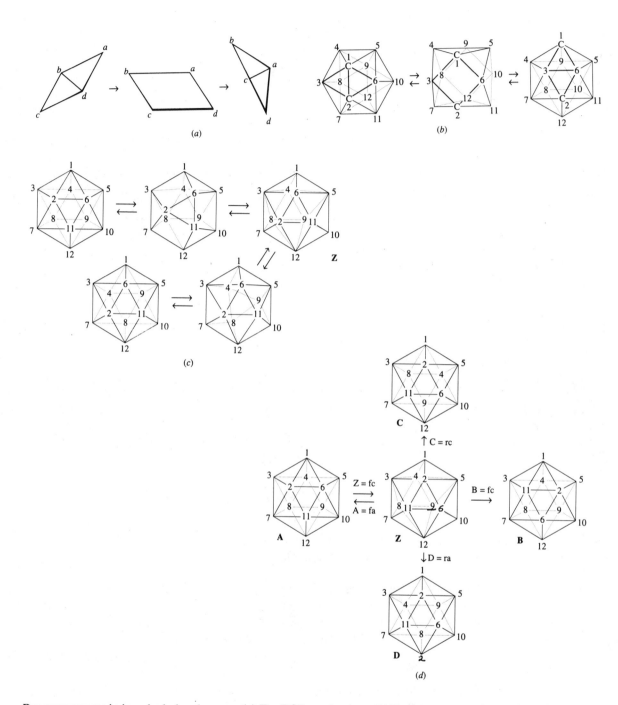

Rearrangements in icosahedral carboranes. (*a*) The DSD mechanism. (*b*) Rearrangement of *o*- to *m*-carborane via a cuboctahedral intermediate. (*c*) The ETR equivalent of DSD. (*d*) Possible fates of a Z-type intermediate showing possible rearrangement products. (Part *b* is from H. Beall in *Boron Hydride Chemistry*, E. L. Muetterties, Ed., Academic Press, New York, 1975. Parts *c* and *d* are reprinted with permission from S.-h. Wu and M. P. Jones, Jr., *J. Am. Chem. Soc.* **1989**, *111*, 5373. Copyright 1989, American Chemical Society.)

17.13 Solution: Rules formulated by Wade suggest that the relation between framework electron count and molecular geometry should be similar for transition metal clusters and boranes. The number of framework electrons contributed by transition-metal-containing groups is given by $(v + x - 12)$ where v is the number of valence electrons of the metal and x is the number of electrons contributed by all the attached ligands. Transition metals have nine valence orbitals ($5d$'s + $3p$'s + $1s$). By analogy with BH, we wish to allow three orbitals plus whatever electrons they contain to contribute to framework bonding. Hence, six orbitals must be employed in bonding ligands or containing lone pairs. This requires a total of twelve e which may come from the pool of metal valence electrons or from those contributed by the ligands. If the number of ligands attached to each metal is not given, the total number of metal valence electrons may be added to the total of electrons donated by ligands (*e.g.*, two for each CO) and the charge subtracted. Twelve e for each metal are then subtracted giving the final framework count. (See DMA, 3rd ed., Section 17.10.4 or K. Wade, *Chem. in Britain* **1975**, *11*, 177.)

(a) $[Co_4Ni_2(CO)_{14}]^{2-}$
  $4 \times 9e + 2 \times 10e + 14 \times 2e + 2e - 6 \times 12e = 14e = 7$ pr; *closo*

(b) $[Fe(CO)_3]B_4H_8 = [Fe(CO)_3](BH_4)(H)_4$
  $2e + 4 \times 2e + 4e = 14e = 7$ pr; *nido*

(c) $(CpCo)C_2B_7H_{11} = (CpCo)(CH)_2(BH)_7(H)_2$
  $2e + 2 \times 3e + 7 \times 2e + 2e = 24e = 12$ pr; *nido*

(d) $Os_5(CO)_{16}$
  $5 \times 8e + 16 \times 2e - 5 \times 12e = 6$ pr; *closo*

(e) $[(Et_3P)_2Pt(H)]B_9H_{10}S$

The electron-counting rules must be modified for complexes involving Pt^{II} which ordinarily forms 16-e rather than 18-e species. Hence, the number of framework electrons for $[(Et_3P)_2Pt(H)]^+$ is $v + x - 10 = 4$. For $B_9H_{10}S^-$, the number is 24. Hence, we have 14 e pairs and 11 vertices for an *arachno* structure.

(f) $Rh_6(CO)_{16}$
  $6 \times 9e + 16 \times 2e - 6 \times 12e = 14e = 7$ pr; *closo*

(g) $[Rh_9P(CO)_{21}]^{2-}$
  $9 \times 9e + 5e + 21 \times 2e + 2e - 9 \times 12e = 22e = 11$ pr; *nido*

17.14 Problem: Sketch the predicted geometry of the species in Problem 17.13.

17.14 Solution: The only prediction which can be made from Wade's rules is which atoms are at polyhedral vertices. No information is provided about how many carbonyls are bridging and how many terminal nor about the exact placement of bridging CO or H. The descriptions given here are those of actual structures and include details not predictable from Wade's rules.

(a) $[Co_4Ni_2(CO)_{14}]^{2-}$: octahedron; contains μ-CO (V. G. Albano, G. Ciani, and P. Chini, *J. Chem. Soc., Dalton Trans.* **1974**, 432).

(b) $[Fe(CO)_3]B_4H_8$: square pyramidal structure of B_5H_9 with $Fe(CO)_3$ replacing the apical BH (N. N. Greenwood, *et al.*, *Chem. Comm.* **1974**, 718).

(c) $(CpCo)C_2B_7H_{11}$: not known; would predict replacement by CpCo of one BH group in the $C_2B_8H_{12}$ framework which would be isoelectronic and isostructural with $B_{10}H_{14}$.

(d) $Os_5(CO)_{16}$: Trigonal bipyramid of four $Os(CO)_3$ groups having an $Os(CO)_4$ group in the plane which distorts the TBP framework (B. E. Reichert and G. M. Sheldrick, *Acta Cryst.*

1977, *B33*, 173).

(e) $[(Et_3P)_2Pt(H)]B_9H_{10}S$

The *arachno* structure for eleven atoms results from excision of two vertices of a 13-vertex polyhedron shown here. The structure observed results from excision of vertices numbered 1 and 4 with Pt and S at the 2 and 6 positions, respectively, (A. R. Kane, L. J. Guggenberger, and E. L. Muetterties, *J. Am. Chem. Soc.* **1970**, *92*, 2571.)

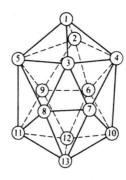

(f) $Rh_6(CO)_{16}$: octahedron of $Rh(CO)_2$ groups with remaining four CO's triply bridging alternate octahedral faces (E. R. Corey, L. F. Dahl, and W. Beck, *J. Am. Chem. Soc.* **1963**, *85*, 1202).

(g) $[Rh_9P(CO)_{21}]^{2-}$: Archimedian antiprism of Rh(CO)'s with one apex capped by Rh(CO). Three sets of four μ-CO's bridge apical and upper plane Rh's, upper and basal plane Rh's and basal plane Rh's, respectively. P is at center of Rh_9 framework (J. L. Vidal, W. E. Walker, R. L. Pruett, and R. C. Schoening, *Inorg. Chem.* **1979**, *18*, 129).

17.15 Problem: Consider $B_4H_4^{n-}$ in a tetrahedral geometry. Derive the MO energy-level diagram and predict the even values of $0 \leq n \leq 6$ for which this geometry will be stable.

17.15 Solution: The irreducible representation generated by the manifold of the 20 B valence orbitals under T_d symmetry is (See Box on pp. 828-830 in DMA, 3rd ed.):

	E	$8C_3$	$3C_2$	$6S_4$	$6\sigma_d$
$\Gamma_{x,y,z}$	3	0	–1	–1	1
Γ_s	1	1	1	1	1
$\Gamma_{val.\ orbs.}$	4	1	0	0	2
# unshifted atoms	4	1	0	0	2
Γ_{total}	16	1	0	0	4

This reduces to $2A_1 + E + T_1 + 3T_2$. Subtracting out the irreducible representations for the *exo sp* hybrids which are $A_1 + T_2$, the framework molecular orbitals span the irreducible representations $A_1 + E + A_2 + T_1 + 2T_2$. The order of energies of these MO's corresponds to the order for a single central atom as follows.

	s	p_x, p_y, p_z	$d_{x^2-y^2}, d_{z^2}$	d_{xy}, d_{xz}, d_{yz}	f_x^3, f_y^3, f_z^3
	↓	↓	↓	↓	↓
T_d	a_1	t_2	e	t_2	t_1

Hence, the energy-level diagram is as follows:

$$
\begin{array}{ll}
t_1{}^* & \text{———} \\
(f_x{}^3, f_y{}^3, f_z{}^3) & \\
t_2{}^*(2) & \text{———} \\
(d_{xy}, d_{xz}, d_{yz}) & \\
e & \text{——} \\
(d_{x^2-y^2}, d_{z^2}) & \\
t_2(1) & \text{———} \\
(p_x, p_y, p_z) & \\
a_1 \ (s) & \text{—}
\end{array}
$$

$E\uparrow$

Thus, the predictions are as follows:

n	No. of e pairs	Geometry
0	4	T_d; filled through $t_2(1)$; isoelectronic with B_4Cl_4 which has this geometry.
−2	5	two electrons in e MO; subject to Jahn-Teller distortion.
−4	6	T_d; filled through the e orbitals; B_4H_8 is the *nido* relative of $B_5H_5^{2-}$; but not yet isolated. Isoelectronic with P_4 and tetrahedrane.
−6	7	B_4H_{10} undergoes Jahn-Teller distortion. *Arachno* structure observed.

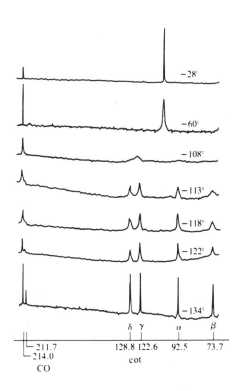

^{13}C NMR spectrum of (cot)Fe(CO)$_3$ at several temperatures. (Reprinted with permission from F. A. Cotton and D. L. Hunter, *J. Am. Chem. Soc.* **1976**, *98*, 1413. Copyright 1976, American Chemical Society.)

17.16 Problem: The figure above shows the ^{13}C NMR spectrum of (cot)Fe(CO)$_3$ at several temperatures. The signals at 214 and 212 ppm are attributed to CO, and the others are attributed to cot. Explain the appearance of the spectrum at –134°C, and its change with T. (See F. A. Cotton and D. L. Hunter, *J. Am. Chem. Soc.* **1976**, *98*, 1413.)

17.16 Solution: Cyclooctatetraene acts as a 4-*e* donor in order that Fe conform to the EAN rule. At –134° the static structure is frozen out. Two different kinds of terminal CO's exist in a 2:1 ratio on Fe. This leads to four non-equivalent pairs of C on the cot ligand. Both CO and cot signals begin to collapse at the same time indicating that the same process achieves equivalence for both CO and cot carbons. Presumably, this is the migration of the Fe(CO)$_3$ group around the cot ring. At –108°C, the speed of the migration process is of the same order of NMR time scale and the signal is broadened almost into the baseline. By –28°C, the migration process is much faster and a single averaged environment is seen for CO and for cot C's.

17.17 Problem: The figure on the next page shows the structure of [Cr$_3$S(CO)$_{12}$]$^{2-}$. Its ^{13}C NMR spectrum at several temperatures is also shown.
(a) Give the Polyhedral Skeletal Electron-pair Theory (PSEPT) electron count for the cluster and show how it is consistent with the observed structure.
(b) What is the source of the inequivalence of the two types of terminal CO's.
(c) Explain the changes in the NMR spectrum as the sample is warmed. (See D. J. Darensbourg and D. J. Zalewski, *Organometallics* **1984**, *3*, 1598.)

17.17 Solution: (a) Each Cr(CO)$_4$ unit contributes 2*e* (see Table below), μ_3-S contributes 4*e* and 2*e* come from the charge. The total is 12*e* = 6 pairs = *n* + 2 skeletal bonding pairs. Thus, we expect (and observe) the *nido* structure derived from the TBP, the tetrahedron.
(b) The carbonyls labeled *b* are on the opposite side of the Cr$_3$ plane from S while the *c* carbonyls are on the same side as S.
(c) At –100°C the available thermal energy is small enough to freeze out the three different types of CO environments. As the temperature is raised the signals due to *a* and *c* protons begin to broaden and disappear into the baseline indicating exchange of these carbonyls presumably by opening of CO bridges and conversion of terminal CO to μ-CO on re-closing. By –40° the two types of CO's are exchanging fast enough to give an averaged signal around 237 ppm. By –20° these carbonyls are beginning to exchange with the *b* carbonyls presumably via bridge opening and rotation on each Cr. Finally at +50° all CO environments are averaged by rapid exchange giving only one signal.

Number of skeletal bonding electrons (= $v + x - 12$) contributed by some transition-metal units[a]

Metal group	M	Cluster unit			
		M(CO)$_2$	M(η-Cp)	M(CO)$_3$	M(CO)$_4$
6(VIA)	Cr, Mo, W	–2	–1	0	2
7(VIIA)	Mn, Tc, Re	–1	0	1	3
8(VIII)	Fe, Ru, Os	0	1	2	4
9(VIII)	Co, Rh, Ir	1	2	3	5
10(VIII)	Ni, Pd, Pt	2	3	4	6

From K. Wade, *Adv. Inorg. Radiochem.* **1975**, *18*, 1.

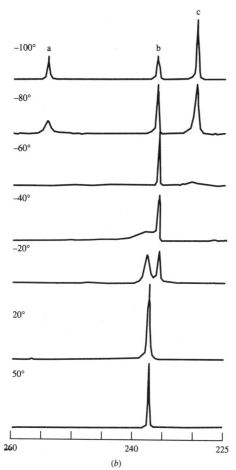
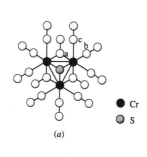

$[Cr_4(\mu_3\text{-}S)(CO)_{12}]^{2-}$. (a) Structure. (b) Variable-temperature ^{13}C NMR spectrum. (Reprinted with permission from D. J. Darensbourg and D. J. Zalewski, *Organometallics* **1984**, *3*, 1598. Copyright 1984, American Chemical Society.)

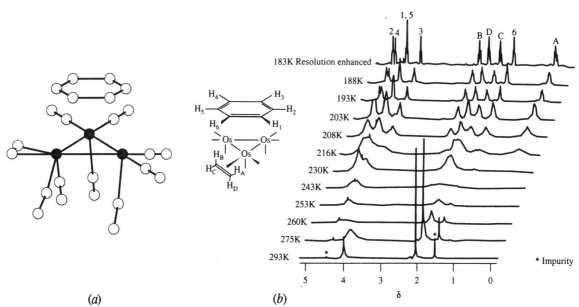

$[(\mu_3{:}\eta^2,\eta^2,\eta^2\text{-}C_6H_6)(\eta^2\text{-}C_2H_4)Os_3(CO)_9]$. (a) Static structure. (b) Variable-temperature ^1H NMR spectrum. (Reprinted with permission from M. A. Gallop, B. F. G. Johnson, J. Keeller, J. Lewis, S. J. Hayes, and C. M. Dobson, *J. Am. Chem. Soc.* **1992**, *114*, 2510. Copyright 1992, American Chemical Society.

17.18 Problem: The figure above shows the structure of [(μ$_3$:η^2,η^2,η^2-C$_6$H$_6$)(η^2-CH$_2$CH$_2$)Os$_3$(CO)$_8$]. Its ^{1}H NMR spectrum at several temperatures is shown along with a labeling scheme for the protons.
(a) Give the PSEPT electron count for the cluster and show how it is consistent with the observed structure.
(b) Account for the number of different kinds of protons observed at 188K.
(c) As the temperature is raised all the ring protons coalesce to a single NMR signal around 4 ppm. What kind of molecular motion would account for this?
(d) As the temperature is raised all the olefin protons coalesce first to two signals (not well resolved) at around 230K and finally to a single signal at 290K. What kind of molecular motion would account for this?
(e) If the temperature were further raised beyond 290K, would you expect the two signals to coalesce? Why or why not? (See M. A. Gallop, B. F. G. Johnson, J. Keeler, J. Lewis, S.J. Heyes and C. M. Dobson, *J. Am. Chem. Soc.* **1992**, *114*, 2510.)

PSEPT electron counts for some clusters with π-acid ligands

n (number of vertices)	Geometry	Electron count	Example
closo		(14n + 2)	
5	Trigonal bipyramid (TBP)	72	Os$_5$(CO)$_{16}$
6	Octahedron	86	Rh$_6$(CO)$_{16}$
7	Pentagonal bipyramid	100	—
8	Dodecahedron	114	—
9	Tricapped trigonal prism	128	—
10	Bicapped square antiprism	142	[Rh$_{10}$S(CO)$_{22}$]$^{2-}$
12	Icosahedron	170	[Rh$_{12}$Sb(CO)$_{27}$]$^{3-}$
nido		(14n + 4)	
4	Tetrahedron (TBP minus axial vertex)	60	Ir$_4$(CO)$_{12}$ [Fe$_4$(CO)$_{13}$]$^{2-}$
5	Square pyramid	74	Fe$_5$C(CO)$_{15}$
6	Pentagonal pyramid	88	—
8	Bicapped trigonal prism	116	—
9	Capped square antiprism	130	[Rh$_9$P(CO)$_{21}$]$^{2-}$
arachno		(14n + 6)	
3	Triangle (TBP minus both axial vertices)	48	Os$_3$(CO)$_{12}$
4	Butterfly (octahedron minus *cis* vertices)	62	[Fe$_4$C(CO)$_{13}$] [Fe$_4$C(CO)$_{12}$]$^{2-}$
6	Trigonal prism	90	[Rh$_6$C(CO)$_{15}$]$^{2-}$

17.18 Solution: (a) The benzene ring can be regarded as donating 2e to each Os; the three ligands provide 6e and Os-Os bonds 2e; 8 non-bonding e's give an electron-precise count of 18e/Os. Alternatively, the total e count of 6e from benzene (9 x 2e) from two-electron donor ligands and (3 x 8e) from Os's gives the 48e expected for a triangular cluster (See the table above).

(b) The molecule lacks a plane of symmetry. Hence, all the aromatic protons are inequivalent. The proton labeled 6 resonates far upfield of the other aromatic protons because of its proximity to the magnetic field provided by the π-system of the olefin. All the olefin protons are inequivalent because of the lack of any plane of symmetry. In particular, one can distinguish two "sets" of olefinic protons: A and B are closer to the benzene ring while C and D are farther away.

(c) The simplest motion leading to equivalence of the ring protons is rotation of the benzene ligand like a helicopter blade at the top of the molecule. The analysis given in the reference shows that this occurs via a straightforward 1,2 ring hopping.

(d) A propeller-like rotation around the olefin-Os bond would exchange A with C and B with D. If this were the only exchange process occurring, two olefin proton signals would be expected. In order to achieve a single signal some process must also interchange the "up" protons A and B with the "down" protons C and D. One such possibility would be a turnstile rotation which exchanges the olefin between the two equatorial sites on Os. In order to avoid steric problems with the benzene ring in the transition state, a possible lower-energy path would involve rotation of the olefin first to the axial site and then to the other equatorial site (anticlockwise in the Figure). In addition to the above intramolecular process, the data given here do not rule out intermolecular exchange with free olefin in solution. (However, see the reference.)

(e) Any process leading to the coalescence of the signals remaining at 290K would have to interchange aromatic and olefinic protons. One might imagine that this could proceed via oxidative addition and reductive elimination of C-H bonds. However, this would surely be a very high energy process since orienting the aromatic ring to give a Ph-Os^{II}-H group would require ring slippage which would leave other Os atoms electron-deficient when they could no longer coordinate to the benzene π-system. Hence, collapse to a single ^{1}H signal on warming would not be expected.

17.19 Problem: The compound $Co_3Rh(CO)_{12}$ is a tetrahedral cluster.

(a) At −85°C, its ^{13}C NMR spectrum displays seven signals in intensity ratio 1:2:2:2:3:1:1. The second and last two signals are coupled to ^{103}Rh ($I = 1/2$). What is the structure of the species "frozen out" at this temperature?

(b) On warming to +10°C, two signals appear--a single line of relative intensity 2 showing coupling to ^{103}Rh and one of relative intensity 10 which is somewhat broadened. At +30°C, only a single broad resonance is visible. Account for these observations. (See B. F. G. Johnson, J. Lewis and T. W. Matheson, *J. Chem. Soc., Chem. Commun.* **1974**, 441.)

17.19 Solution: (a) The molecule is isoelectronic with $Co_4(CO)_{12}$. At −85°C, a static structure is seen. Three non-equivalent CO's are coupled to ^{103}Rh. The signal of intensity two corresponds to two μ_2-CO's and the other two non-equivalent terminal CO's. There is a signal of relative intensity 3 which must correspond to CO's on the unique Co. Hence, the following static structure accounts for the spectrum. The other signals of intensity 2 would correspond to c and c' while the remaining one is due to b'.

(b) As the temperature is raised to −30°C a process of CO interchange among Co atoms interconverts a, b,b', c, and c'. The Rh-based CO's do not participate in this interchange and appear as a single signal. However, d and d' are rendered symmetry equivalent by equivalence of the 3 Co's. At +30° all carbonyl ligands are involved in the interchange.

290

17.20 Problem: The figure below shows the structure of $Cp^*Ru(acac)(\eta^2\text{-}CF_2CF_2)$ and its ^{19}F NMR spectrum at various temperatures.

(a) Account for the appearance of the spectrum at –50°C.

(b) As the temperature increases, the signals coalesce to give *two* signals at 100°C. What does this indicate about the process which interchanges F's? (See O. J. Curnow, R. P. Hughes and A. L. Rheingold, *J. Am. Chem. Soc.* **1992**, *114*, 3153.

17.20 Solution: (a) In the static structure there are two sets of F's: A and A' and B and B' which are "exo" and "endo" with respect to the Cp^* ring. The environmental differences between these two sets are more significant than the difference between the F's of each set which are a result of the orientation of Cp^*. Thus, two groups of signals are seen: one for A, A' (centered at $\delta = -123$) and one for B, B' (centered at $\delta = -134$). Since F (like H) has $I = 1/2$, each signal is split into a doublet by coupling to the *cis* F; $J_{AB} = J_{A'B'} = 140$ Hz. Each doublet is split into another doublet by coupling to *trans* F: $J_{AB'} = J_{A'B} = -52$ Hz. (Small couplings due to geminal F's are unresolved in the Figure.) (Note that the identification of A' as the "endo" F, etc. would not change the above argument.)

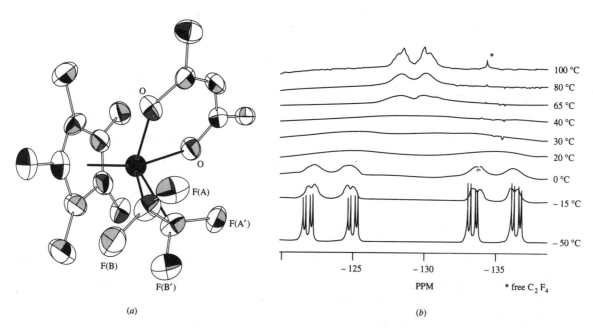

$Cp^*Ru(acac)(\eta^2\text{-}C_2F_4)$. (*a*) Static structure. (*b*) Variable-temperature ^{19}F NMR. (Reprinted with permission from O. J. Curnow, R. P. Hughes, and A. L. Rheingold, *J. Am. Chem. Soc.* **1992**, *114*, 3153.

(b) Since two signals remain at high T, the four different static environments must undergo pairwise exchange. A simple and attractive possibility would be a propeller motion around the Ru-C_2F_4 axis which would interchange A with B' and B with A', but not A with B or A' with B'. The signal at –134 ppm is due to free C_2F_4 in the solution. If F environments were interconverted by exchange of free and bound olefin, all F positions would be rendered equivalent leading to a single signal.

17.21 Problem: The figure on the next page shows the structures of some metal cluster compounds. Rationalize these observed structures by giving the PSEPT electron count.

Electron counts for some capped π-acid clusters

Geometry	Electron count	Example
closo **parent cluster**	$(14n + 2 + 12m)$	
Capped trigonal bipyramid	$72 + 12 = 84$	$Os_6(CO)_{18}$
Monocapped octahedron	$86 + 12 = 98$	$Os_7(CO)_{21}$
Bicapped octahedron	$86 + 24 = 110$	$[Os_8(CO)_{22}]^{2-}$
Tetracapped octahedron	$86 + 48 = 134$	$[Os_{10}C(CO)_{24}]^{2-}$
nido **parent cluster**	$(14n + 4 + 12m)$	
Bicapped tetrahedron = capped trigonal bipyramid	$60 + 24 = 84$	$Os_6(CO)_{18}$
Monocapped square pyramid	$74 + 12 = 86$	$H_2Os_6(CO)_{18}$

17.21 Solution: (See the table with Problem 17.17 and that above.)

(a) $[Os_8(CO)_{22}]^{2-}$ is a bicapped octahedron. It is expected to have 110 e's. The actual count is $(8 \times 8) + (22 \times 2) + 2 = 110$.

(b) $[Re_8C(CO)_{24}]^{2-}$ is also a bicapped octahedron, but this time opposite faces are capped. The expected electron count is 110. The observed count is $(8 \times 7) + 4 + (24 \times 2) + 2 = 110$.

(c) $H_2Os_6(CO)_{18}$ is a monocapped square pyramid and should have 86 electrons The actual count is $(2 \times 1) + (6 \times 8) + (18 \times 2) = 86$.

(d) $H_2Os_7(CO)_{21}$ is a square pyramid capped on one face and bridged on one edge. A capped square pyramid was shown above to require 86 electrons and the edge-bridging $Os(CO)_4$ group contributes 14 for a required total of 100. The actual count is $(2 \times 1) + (7 \times 8) + (21 \times 2) = 100$.

(e) $PtOs_5(PPh_3)(CO)_{15}(\mu_4\text{-S})$ has both transition-metal and main-group (S) vertices. The structure is an Os_5S octahedron face-capped by a $(PPh_3)(CO)Pt$ fragment. The octahedral electron count should be $(n - 1)(14) + 2 + 2 = 74$ and the capping group should contribute 12 more for a total electron count of 86. The actual count is (8×5) [Os] + 10 [Pt] + (15×2) [CO] + 2 [PPh_3] + 4 [μ-S] = 86.

(f) $[Rh_7(CO)_{16}I]^{2-}$ is a capped octahedron with μ_2-I. The expected electron count (Table above) is 100. The actual count is $(7 \times 9) + (16 \times 2) + 3 + 2 = 100$.

(g) $[Re_6C(CO)_{19}]^{2-}$ is an octahedron with interstitial C and expected electron count of 86. The actual count is $(6 \times 7) + 4 + (19 \times 2) + 2 = 86$.

(h) $[Rh_6C(CO)_{15}]^{2-}$ is a trigonal prism with interstitial C; this is the *arachno* structure derived from the nine-vertex *closo* tricapped trigonal prism and is expected to have 90 electrons. The actual count is $(6 \times 9) + 4 + (15 \times 2) + 2 = 90$.

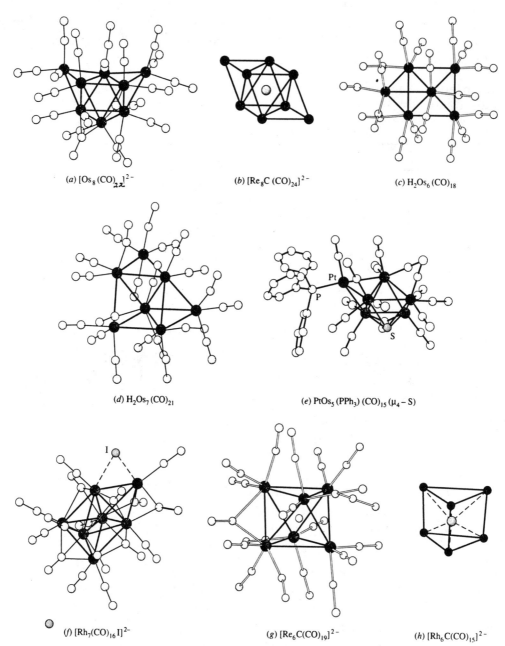

Structures for Problem 17.21. (Structure a is from P. F. Jackson, B. F. G. Johnson, J. Lewis, and P. R. Raithby, *J. Chem. Soc., Chem. Commun.* **1980**, 60. Structure c is from M. McPartlin, C. R. Eady, B. F. G. Johnson, and J. Lewis, *J. Chem. Soc. Chem. Commun.* **1976**, 883. Structure d is from B. F. G. Johnson, J. Lewis, M. McPartlin, J. Morris, G. L. Powell, P. R. Raithby, and M. D. Vargas, *J. Chem. Soc. Chem. Commun.* **1986**, 429. Structure e is reprinted with permission from R. D. Adams. J. E. Babin, R. Mathab, and S. Wang, *Inorg. Chem.* **1986**, 25, 1623. Copyright 1986, American Chemical Society. Structure f is from V. G. Albano, G. Ciani, S. Martinengo, P. Chini, and G. Giordano, *J. Organomet. Chem.* **1975**, 88, 381. Structure g is from J. Beringhelli, G. D'Alfonso, H. Molinari, and A. Sirone, *J. Chem. Soc. Dalton Trans.* **1992**, 689.)

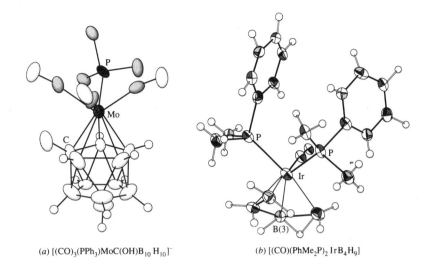

(a) [(CO)$_3$(PPh$_3$)MoC(OH)B$_{10}$H$_{10}$]$^-$

(b) [(CO)(PhMe$_2$P)$_2$IrB$_4$H$_9$]

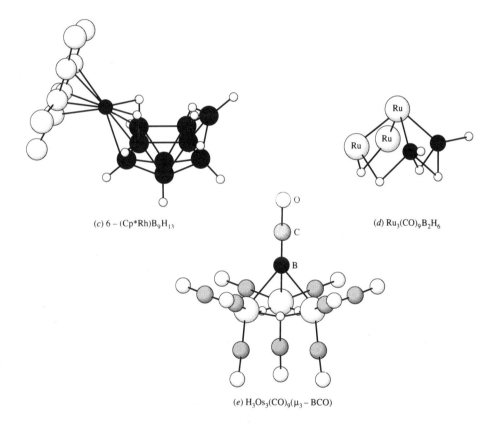

(c) 6-(Cp*Rh)B$_9$H$_{13}$

(d) Ru$_3$(CO)$_9$B$_2$H$_6$

(e) H$_3$Os$_3$(CO)$_9$(μ_3-BCO)

Structures for Problem 17.22. (Structure a is from X. L. R. Fontaine, N. N. Greenwood, J. D. Kennedy, P. I. MacKinnon, and I. Macpherson, *J. Chem. Soc. Dalton Trans.* **1987**, 2385. Structure b is reprinted with permission from S. K. Boocock, M. A. Toft, K. E. Inkrott, L.-Y. Hsu, J. C. Huffman, K. Folting, and S. G. Shore, *Inorg. Chem.* **1984**, *23*, 3084. Copyright 1984, American Chemical Society. Structure c is from X. L. R. Fontaine, H. Fowkes, N. N. Greenwood, J. D. Kennedy, and M. Thornton-Pett, *J. Chem. Soc. Dalton, Trans.* **1986**, 547. Structure d is from A. K. Chipperfield, C. E. Housecroft, and D. M. Matthews, *J. Organomet. Chem.* **1990**, *384*, C38. Reprinted from C. E. Housecroft, *Adv. Organomet. Chem.* **1991**, *33*, 1. Structure e is from D. P. Workman, D.-Y. Jan, and S. G. Shore, *Inorg. Chem.* **1990**, *29*, 3518. Reprinted from C. E. Housecroft, *Adv. Organomet. Chem.* **1991**, *33*, 1.)

17.22 Problem: The figure on the previous page shows the structures of some metallaborane cluster compounds. Rationalize these observed structures by giving the PSEPT electron count. With which borane is each cluster isoelectronic?

17.22 Solution:

(a) $[(CO)_3(PPh_3)MoC(OH)B_{10}H_{10}]^-$ is an icosahedron. From the Table on p. 286 the $(CO)_3(PPh_3)Mo$ fragment contributes $2e$; $C(OH)$ (Table 17.3 in DMA, 3rd ed.) contributes $3e$ and 10 BH contribute $20e$. With $1e$ for the charge, this totals $26e = 13$ pairs as expected for a twelve- vertex *closo* structure isoelectronic with $B_{12}H_{12}^{2-}$.

(b) $[(CO)(PhMe_2P)_2IrB_4H_9]$ has an apical IrL_3 contributing $3e$, four BH's $8e$ and "extra" H's $5e$ for a total of $16e$ or 8 pairs. This is the *arachno* structure isoelectronic with B_5H_{11}.

(c) Cp^*Rh contributes 2 skeletal bonding electrons, nine BH's $18e$ and "extra" H's $4e$ for a total of $24e = 12$ pairs $= n + 2$ pairs. This is the *nido* structure isoelectronic with $B_{10}H_{14}$.

(d) $Ru_3(CO)_9B_2H_6$ is a square pyramid with equatorial B's. Each $Ru(CO)_3$ contributes $2e$ (Table 17.7), two BH's $4e$ and there are 4 "extra" H's for a total of $14e = n + 2$ pairs. This is the *nido* structure isoelectronic with B_5H_9.

(e) In $H_3Os_3(CO)_9B(CO)$, BCO contributes $3e$ (See Table 17.13 in DMA, 3rd ed.); each $Os(CO)_3$ contributes $2e$ and there are 3 H's for a total of $12e = 4 + 2$ skeletal bonding pairs. This is the *nido* structure derived from the *closo* TBP—i.e. a tetrahedron.

17.23 Problem:

(a) Give the electron count and sketch the geometry for 1,2- $B_9C_2H_{11}^{2-}$ and 1,7-$B_9C_2H_{11}^{2-}$.
(b) Both of these ligands form a number of compounds with transition metals. Give the electron count and predict the structure for $CpCo(1,2-C_2B_9H_{11})$.
(c) MO calculations indicate that six electrons are available on the open faces of the dicarbollide ligands for donation to transition metals making them equivalent to Cp. Show that the following obey the EAN rule for the metal:
$(\eta^4-Ph_4C_4)Pd(1,2-C_2B_9H_{11})$, $[(1,2-C_2B_9H_{11})Re(CO)_3]^-$, and
$[(1,2-C_2B_9H_{11})Mo(CO)_3W(CO)_5]^{2-}$

17.23 Solution: (a) Both have 13 pairs of framework electrons and exhibit a *nido* structure obtained by removing one vertex of an icosahedron. The 1,2- and 1,7- numbering reflects the preparation of these anions starting with 1,2- and 1,7-$C_2B_{10}H_{12}$ rather than the usual rules of nomenclature.

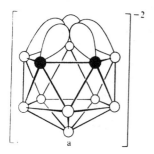
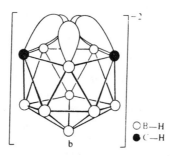

Structures of $[B_9C_2H_{11}]^{2-}$

○ B—H
● C—H

(b) $Cp^- + Co^{III} + [1,2-C_2B_9H_{11}]^{2-}$
   $6e + 6e + 6e = 18e$ (around Co)

If we view the compound as a cluster, then the e count is:

$CpCo^{2+} + 2(CH) + 9(BH) +$ charge
$0e + 2 \times 3e + 9 \times 2e + 2e = 26e$ (skeletal count)

$CpCo^{2+}$ contributes 0 e since Co is Co^{III} here. We have $p + 1$ pairs where $p = 12$ and a *closo* structure (an icosahedron) is predicted by Wade's rules (and observed).

(c) $(\eta^4-Ph_4C_4)Pd(1,2-C_2B_9H_{11})$

$Pd^{IV} + \eta^4-Ph_4C_4^{2-} + 1,2-C_2B_9H_{11}^{2-}$
$6e + 6e + 6e = 18e$

$[(1,2-C_2B_9H_{11})Re(CO)_3]^-$

$Re^I + 3\,CO + 1,2-C_2B_9H_{11}^{2-}$
$6e + 3 \times 2e + 6e = 18e$

$[(1,2-C_2B_9H_{11})Mo(CO)_3W(CO)_5]^{2-}$

$Mo^0 + 1,2-C_2B_9H_{11}^{2-} + 3\,CO$
$6e + 6e + 3 \times 2e = 18e$

The $[(1,2-_2B_9H_{11})Mo(CO)_3]^{2-}$ group is equivalent to $[CpMo(CO)_3]^-$ and can be considered to be a $2e$-donor toward $W(CO)_5$, giving it $18e$.

17.24 Problem: What do the following experimental results suggest about the substitution mechanisms in the clusters?

(a)
Reaction	Medium	$\Delta V^{\ddagger}(cm^3 mol^{-1})$
$[HRu_3(CO)_{11}]^- + PPh_3$	thf	+21.2
$[Ru_3(CO)_{11}(CO_2Me)]^- + P(OMe)_3$	90/10 thf/MeOH	+16
$[Ru_3(CO)_{11}(CO_2Me)]^- + P(OMe)_3$	10/90 thf/MeOH	+2.5

(See D. J. Taube, R. van Eldik and P. C. Ford, *Organometallics* **1987**, *6*, 125.)

(b) For the reaction
$M_4(CO)_9[HC(PPh_2)_3] + L \rightarrow M_4(CO)_8L[HC(PPh_2)_3] + CO$

M = Co, Rh, Ir, rate = $(k_1 + k_2[L])[M_4(CO)_9[HC(PPh_2)_3]]$ for M = Co, Rh, but the k_2 term dominates for Ir. Activation parameters are as follows:

M	L	$\Delta H_1^\ddagger$ (kJ/mol)	$\Delta S_1^\ddagger$ (J/mol K)	$\Delta H_2^\ddagger$ (kJ/mol)	$\Delta S_2^\ddagger$ (J/mol K)
Co	^{13}CO	101	–2.2		
	P(n-Bu)$_3$	105	31	43.1	–144
Rh	PPh$_3$	110	96	21.3	–121
	P(n-Bu)$_3$			25.5	–82.8
Ir	P(OMe)$_3$			59.4	–170
	P(n-Bu)$_3$			47.2	–192

(See J. R. Kennedy, P. Selz, A. L. Rheingold, W. C. Trogler and F. Basolo, *J. Am. Chem. Soc.* **1991**, *111*, 3615.)

17.24 Solution: (a) Since CO, PPh$_3$ and P(OMe)$_3$ are all neutral, they are not likely to be tightly solvated. A slightly positive $\Delta V^\ddagger$ would be consistent with an **a** mechanism for CO displacement in [HRu$_3$(CO)$_{11}$]$^-$ since the transition state would still be ionic and solvated; however, its larger size would decrease solvent structure. However, such a large postive $\Delta V^\ddagger$ is more consistent with a **d** mechanism--probably even **D**. The second reaction has a large positive $\Delta V^\ddagger$ suggesting again a **d** (and probably **D**) mechanism. The very small $\Delta V^\ddagger$ for the third reaction is surprising. There is no reason to imagine that the mechanism should change drastically with the different solvent system. One possibility is that alcohol is somehow more effective in solvating the transition state. This could be so if H-bonding involving the alcohol and the ester O were important.

(b) Similar $\Delta H_1^\ddagger$ values and (mostly) large positive $\Delta S_1^\ddagger$ values point to a **D** mechanism for the k_1 path. The smaller $\Delta H_2^\ddagger$ values and very large negative $\Delta S_2^\ddagger$ values point to an **a** mechanism for the k_2 path. In all likelihood this mechanism is also **A** since M-M bond breaking in the clusters could lead to intermediates of formula M$_4$(CO)$_9$[HC(PPh$_2$)$_3$]L which obey the 18-e rule. The fact that this pathway dominates only for large Ir is interpretable on steric grounds.

17.25 Problem: Using the approach developed by Mingos for cage and ring compounds (Section 17.11, DMA, 3rd ed.), predict plausible structures for the following.
(a) S$_8$ (c) P$_4$
(b) P$_4$(C$_6$H$_{11}$)$_4$ (d) [Fe(NO)$_2$]$_2$(SEt)$_2$

17.25 Solution: (a) S$_8$: Each S atom contributes four skeletal electrons ($v + x - 2$, see Problem 17.5) for a total of 32 e = 16 pairs. Starting with a cube of S atoms bonded via S–S bonds gives 12 edge bonds requiring 12 e pairs. The four "extra" e pairs can occupy four S–S antibonding orbitals breaking opposite pairs of bonds on the upper and lower cube faces leading to the S$_8$ ring structure.

(b) $P_4(C_6H_{11})_4$: Total number of framework e pairs is eight. Arranging the four $P(C_6H_{11})$ groups at the vertices of a tetrahedron, we have sufficient e pairs to break two of the six P–P bonds to afford a butterfly structure.

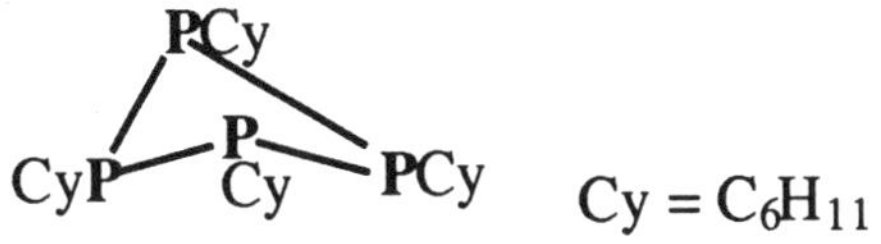

$$\text{CyP}-\overset{\overset{\displaystyle PCy}{|}}{\underset{\underset{\displaystyle Cy}{|}}{P}}-PCy \qquad Cy = C_6H_{11}$$

(c) P_4: Number of framework e is $4 \times 3e = 12e$ = 6 pairs, precisely enough to fill all six bonding orbitals for four P's arranged at the vertices of a tetrahedron.

(d) $[Fe(NO)_2]_2(SEt)_2$: Each $Fe(NO)_2$ group contributes $2e$; each SEt group contributes $5e$ ($v + x - 2$) for a total of 14 framework electrons. Arranging the groups at the corners of a tetrahedron, we have the six pairs needed to form single bonds along all tetrahedral edges plus one to occupy an antibonding orbital breaking one bond giving a butterfly structure.

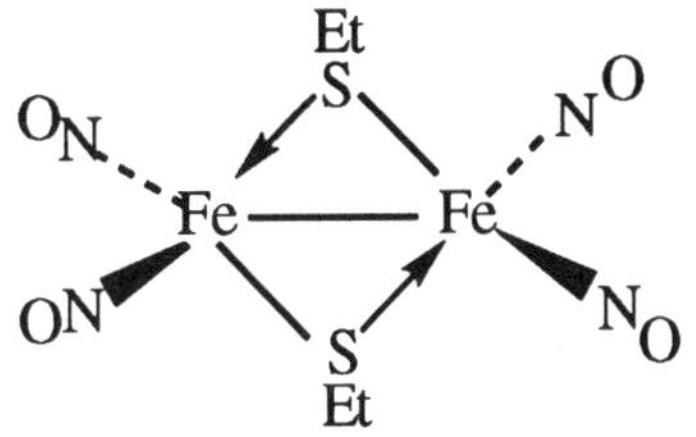

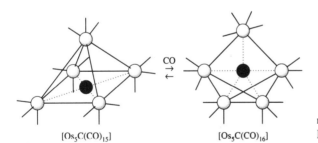

17.26 Problem: The figure shows the structures of $Os_5C(CO)_{15}$ and its carbonylation product $Os_5C(CO)_{16}$. Rationalize these structures on the basis of their PSEPT electron counts. (See B. F. G. Johnson, J. Lewis, J. N. Nicholls, W. J. H. Nelson, J. Puga, P. R. Raithby, M. J. Rosales, and M. D. Vargas, *J. Chem. Soc., Dalton Trans.* **1983**, 2447.)

The reversible carbonylation of $[Os_5C(CO)_{15}]$. (From J. N. Nicholls, *Polyhedron*, **1984**, *3*, 1307.)

17.26 Solution: $Os_5C(CO)_{15}$ is a 74-e *nido* cluster. Adding an additional CO gives a 76-e cluster having one $Os(CO)_4$ group. Its structure can be rationalized as a capped butterfly which is predicted to require $62 + 14 = 76e$.

17.27 Problem: Some metal carbonyl clusters are essentially close-packed units of the metal stabilized by coordination of CO. In the clusters $[Rh_{13}H_3(CO)_{24}]^{2-}$ and $[Rh_{13}H_2(CO)_{24}]^{3-}$ the Rh atoms are in a hexagonal close-packed arrangement. Describe the expected arrangement of atoms within the very symmetrical Rh_{13} cluster.

17.27 Solution: This is the simplest *hcp* unit giving the full C.N. 12 of one atom. The Rh at the center has a hexagonal arrangement of six Rh in one layer with triangular arrangements of three Rh above and below this layer. The triangular groups are eclipsed giving the *ABA* or *hcp* arrangement. The unit has D_{3h} symmetry.

17.28 Problem: Diborane will react with $NaBH_4$ in diglyme (a polyether) to form NaB_2H_7 in solution. No reaction occurs without the solvent present. Propose a role for the solvent.

17.28 Solution: The polyether complexes the metal ion, thus reducing the ion-ion interaction in $NaBH_4$. There is little loss in ion-pair energy when the larger $B_2H_7^-$ ion forms from the BH_4^- ion. In contrast, with solid $NaBH_4$ the loss of lattice energy on forming the $B_2H_7^-$ ion is greater than the energy released in forming the (isolated) ion, so the reaction does not occur.

VI. SELECTED TOPICS
18
Bioinorganic Chemistry

18.1 Problem: For each of the following elements, identify one significant role in biological processes: Fe, Mn, Cu, Zn, I, Mg, Co, Ca, and K.

18.1 Solution: Fe—hemoglobin for O_2 transport. Fe is present in Fe–S proteins and cytochromes, both are important in oxidation-reduction processes.
Mn—redox process producing O_2 in photosynthesis
Cu—hemocyanin—O_2 carrier in invertebrates
    There are many Cu enzymes, for example cytochrome oxidase.
Zn—active center in carboxypeptidase and many other enzymes
I—thyroxine (secreted by the thyroid gland)
Mg—chlorophyll—photosynthesis
    Mg^{2+} is a major cation within cells
Co—Vitamin B_{12}
Ca—bones
    Ca^{2+} plays an important role in the transmission of nerve impulses
K^+—major cation within cells

18.2 Problem: (a) Why is iron used as the edta complex rather than a simple salt for supplying iron for plants in basic soil? (b) Why are edta and other chelating ligands used for the removal of toxic and radioactive metal compounds?

18.2 Solution: (a) $Fe(OH)_3$ is very insoluble at high pH so the soluble $[Fe(edta)]^-$ is used to make Fe available to plants.
(b) Chelating ligands such as edta coordinate or sequester metal ions forming soluble complexes for their removal.

18.3 Problem: High-spin iron(II) is too large for the opening of the porphyrin ring, but low-spin iron(II) can be accommodated in the opening. Why does high–spin iron(II) have a larger radius?

18.3 Solution: Low-spin octahedral Fe^{II} has the t_{2g}^6 configuration with the e_g orbitals, directed along the x, y, and z axes, empty. High-spin octahedral Fe^{II} has the $t_{2g}^4 e_g^2$ configuration. The metal e_g electrons are antibonding, or, in other terms, the electrons in these orbitals along the axes provide screening of the positive metal center from the ligands, giving a larger radius.

18.4 Problem: What prevents simple iron porphyrins from functioning as O_2 carriers? What features of successful models of Fe–porphyrin O_2 carriers have made it possible to avoid this problem?

18.4 Solution: Simple Fe porphyrins tend to be oxidized irreversibly through dimerization. Successful models have substituents (picket fence or canopy) which prevent dimerization.

18.5 Problem: How is iron stored and transported in mammals? What is the oxidation state of iron for storage and for transfer?

18.5 Solution: Fe is stored as ferritin and transported as transferrin. Fe^{III} is stored, and transported, but it seems to be transferred as Fe^{II}. (See DMA, 3rd ed., p. 901)

18.6 Problem: Identify two chemical types of siderophores.

18.6 Solution: The siderophores are very stable iron complexes used by lower organisms for Fe transport. Fe is coordinated to O in hydroxamate and catechol type complexes:

Iron(III) hydroxamate complex

Iron(III) catechol complex

18.7 Problem: The formation constant of Fe^{III} enterobactin is about 10^{52}. Calculate the concentration of Fe^{3+} in equilibrium with $10^{-4}\ M$ Fe^{III}enterobactin and $10^{-4}\ M$ ent^{6-}. This corresponds to how many liters per Fe^{3+} ion? The volume of the hydrosphere (all bodies of water, snow, and ice) is *ca.* 1.37×10^{21} L. (K. N. Raymond *et al.*, *J. Am. Chem. Soc.* **1979**, *101*, 6097)

18.7 Solution: $Fe^{3+} + enterobactin^{6-} \rightleftharpoons Fe^{III}(enterobactin)^{3-}$

$$K = 10^{52} = \frac{[Fe^{III}(ent)^{3-}]}{[Fe^{3+}][ent^{6-}]} = \frac{10^{-4}}{[Fe^{3+}](10^{-4})}$$

$[Fe^{3+}] = 10^{-52}$ mol/L or 6×10^{-29} ions/L or *ca.* 10^{28} L per ion! This is for the ligand anion. Enterobactin is a very weak acid and exists as the neutral H_6ent at pH 7.

18.8 Problem: What is the cytochrome chain? What are the advantages of such a complex system?

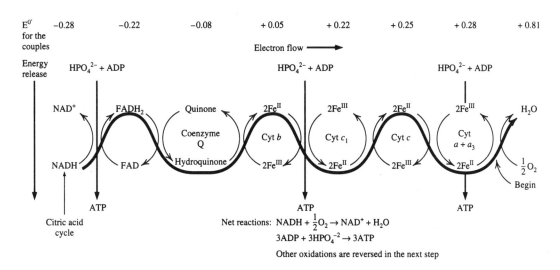

The respiratory chain, showing electron transport and oxidative phosphorylation.
The $E^{0'}$ values for couples are shown above the figure.

18.8 Solution: The cytochromes are a group of electron-transferring proteins that act in sequence to transfer electrons to O_2. The prosthetic group in cytochromes is heme, which undergoes reversible Fe^{II}–Fe^{III} oxidation. The cytochromes in a sequence differ from one another in electrode potentials by about 0.2 volt or less, with a total potential difference of about 1 volt. Differences in potentials result from changes in porphyrin substituents or axial ligands.

The many steps in the chain serve to break down the large amount of energy involved in the reduction of O_2 into smaller units that can be stored as ATP and keep the reactants well separated. High biological specificity is achieved by the chain. Photosynthesis uses a similar cytochrome chain.

18.9 Problem: What are the prosthetic groups of cytochromes and hemoglobin?

18.9 Solution: Heme is the prosthetic group of cytochromes and of hemoglobin.

18.10 Problem: What are the two important systems for biological electron-transfer processes?

18.10 Solution: Cytochromes and Fe–S non-heme proteins are among the most important systems for biological electron transfer.

18.11 Problem: What chemical properties of Fe and Cu make them suitable for redox processes in biological systems?

18.11 Solution: Fe and Cu have two easily accessible oxidation states, making them suitable for redox processes in biological systems.

18.12 Problem: The conversion of carbonic acid to $CO_2 + H_2O$ is a spontaneous process; why is carbonic anhydrase needed?

18.12 Solution: Uncatalyzed dehydration of H_2CO_3 is too slow for respiration. Carbonic anhydrase (a Zn enzyme) accelerates the process greatly.

18.13 Problem: Give an example of the substitution of Co^{II} for Zn^{II} in an enzyme to provide a "spectral probe" for study of the enzyme.

18.13 Solution: Co^{II} replaces Zn^{II} in carboxypeptidase with an *increase* in activity. The Co^{II} absorbs in the visible region so that spectral studies provide direct information about the active site geometry.

18.14 Problem: The direct reduction products of O_2, H_2O_2 (or HO_2^-) and O_2^-, are toxic. How are these products handled in biological systems?

18.14 Solution: H_2O_2 is decomposed by peroxidase or catalase and O_2^- is decomposed by superoxidase. Oxygen is toxic to organisms lacking these enzymes.

18.15 Problem: Wilson's disease causes the accumulation of Cu in the body. How can symptoms be relieved?

18.15 Solution: Wilson's disease is hereditary, resulting in a deficiency of ceruloplasmin. Cu accumulates in the liver, brain, and kidneys. A strong chelating agent, such as edta or penicillamine, is used to form a stable complex to remove the accumulated Cu.

18.16 Problem: Give an example of each of two types of reactions brought about vitamin B_{12}.

18.16 Solution: B_{12} can undergo one-, two-, and three-electron transfers:

$$B_{12}\text{--}Co\text{--}R \longrightarrow \begin{cases} B_{12}\text{--}Co^I + R^+ \\ B_{12}\text{--}Co^{II} + {}^\bullet R \\ B_{12}\text{--}Co^{III} + :R^- \end{cases}$$

One carbon transfer — The introduction or transfer of one carbon unit.

$$\underset{\text{homocysteine}}{HO_2CCHCH_2CH_2SH \atop |\atop NH_2} \xrightarrow[\text{+ enzyme}]{\text{coenzyme B}_{12}} \underset{\text{methionine}}{HO_2CCHCH_2CH_2SCH_3 \atop |\atop NH_2}$$

$$\underset{\text{glycine}}{H_2N\text{--}CH_2CO_2H} \xrightarrow[\text{+ enzyme}]{\text{coenzyme B}_{12}} \underset{\text{serine}}{H_2N\text{--}CHCO_2H \atop |\atop CH_2OH}$$

Isomerization —moving a substituent along a carbon chain.

$$\underset{\text{glutamic acid}}{\begin{array}{c}CO_2H\\|\\HC-NH_2\\|\\H_2C-CH_2-CO_2H\end{array}} \xrightarrow[\text{+ enzyme}]{\text{coenzyme B}_{12}} \underset{\text{methyl aspartic acid}}{\begin{array}{c}CO_2H\\|\\HC-NH_2\\|\\H_3C-CH-CO_2H\end{array}}$$

18.17 Problem: What metals are at the active centers of nitrogenase? Name some reduction processes other than that of N_2 that are accomplished by nitrogenase.

18.17 Solution: Nitrogenase is an Fe–Mo enzyme. It reduces N_2, C_2H_2, N_3^-, and N_2O. Nitrogenase activity of a preparation of the enzyme is usually monitored by following the reduction of acetylene, since this can be followed easily.

18.18 Problem: What electron transport systems are used in photosynthesis?

18.18 Solution: Cytochromes and Fe–S proteins are used as electron-transport systems in photosynthesis.

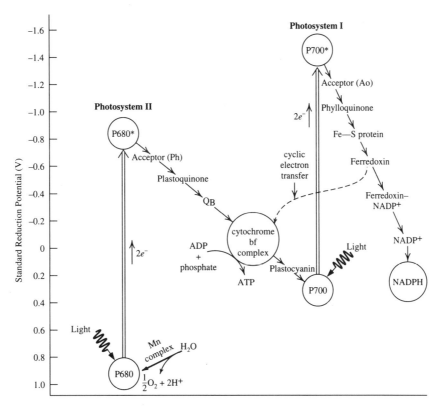

The "Z scheme" for the pathway of electron transfer from H_2O to $NADP^+$ in the two photosystems for green plants. Acceptor Ph is pheophytin, Q_B is a second quinone in Photosystem II, and Acceptor A_0 is an electron acceptor chlorophyll. The cytochrome bf complex contains a cytochrome chain and a Fe-S protein. (Adapted with permission from A. L. Lehninger, D. L. Nelson, and M. M. Cox, *Principles of Biochemistry*, 2nd ed., Worth, New York, 1993, p. 582.)

18.19 Problem: Magnesium has important roles as a cellular cation and in enzymes. What are likely reasons for the toxicity of Be?

18.19 Solution: Be^{2+} with smaller radius than Mg^{2+} is probably too small for cellular transport processes required for Mg^{2+} and Ca^{2+}. Be^{2+}, with higher charge density, is likely to displace Mg^{2+} at the active sites of enzymes. A delicate balance of lability/stability is required for metal ions at active sites for rapid turnovers in catalytic processes.

18.20 Problem: The enzyme alkaline phosphatase contains Zn^{2+} and catalyzes the hydrolysis of orthophosphate monoesters. In order to elucidate the coordination geometry about the metal ion, Zn^{2+} can be substituted by Co^{2+}. The $d \to d$ spectrum of the resulting complex contains the following peaks:

Wavelength (nm)	ε
640	260
605(shoulder)	220
555	708
510	335

What can you say about the coordination of Co^{2+} from these data?

18.20 Solution: Obvious candidates for consideration would be tetrahedral and octahedral geometries. Low-spin tetrahedral complexes are rare. Low-spin d^7 octahedral complexes are unlikely—especially because the O-containing groups found in biological systems are not high in the spectrochemical series. Consulting the Orgel diagram in Problem 10.5, we see that three bands are possible for either tetrahedral or octahedral high-spin d^7. Any of these bands might not be observed because they are too high or too low in energy. In particular, the first band of some tetrahedral Co^{2+} complexes is sometimes located in the infrared. The number of bands makes it likely that there are two distinct coordination environments. The rather large intensities indicate tetrahedral or coordination of lower symmetry than octahedral. (The actual situation is somewhat more complex. A more complete treatment, as well as a good example of the kinds of arguments usually made from such data can be found in M. L. Applebury and J. E. Coleman, *J. Biol. Chem.* **1969**, *244*, 709).

18.21 Problem: In birds air flows in one direction through the lungs, air sacks and hollow bones during inhalation and exhalation. Why should this offer an advantage at high altitude over our "batch" breathing process?

18.21 Solution: At high altitude the partial pressure of O_2 is low and a continuous process provides more air flow for more efficient utilization of O_2. (See K. Schmidt–Nielsen, "How Birds Breathe", *Sci. Am.,* December, **1971**, 73.)

18.22 Problem: Why might we expect some elements for life at low concentration to be toxic at higher concentrations?

18.22 Solution: At high concentration, in addition to serving its essential role, an element can complete with and displace other essential elements from their roles. (See D. E. Carter and Q. Fernando, "Chemical Toxicology", *J. Chem. Educ.* **1979**, *56*, 490.)

305

18.23 Problem: Radiopharmaceuticals containing gamma-ray-emitting nuclides such as ^{99m}Tc, ^{201}Tl, ^{43}K, etc., have been used to locate areas of bone cancer, to produce heart images showing areas of myocardial infarct (heart attack), and to aid in the location of specific tumor growth. To be used in the above fashion, the radiopharmaceutical must, of course, concentrate in the region of interest. Explain why ^{99m}Tc is a promising nuclide for such use for reasons other than its radioactivity. How might a bone imaging agent differ chemically from a heart imaging agent?

18.23 Solution: The metal ion should be complexed with ligands that are lipophilic (fat loving) or ossophilic (bone loving) so that the complex would seek out fatty tissue or bone tissue. Pyrophosphate or diphosphates form ossophilic complexes. Amino acids such as glycine, cystine, etc., give lipophilic complexes. Tc shows a wide variety of stable oxidation states, and thus permits a wider variety of complexes to be prepared than the other metals mentioned.